Risk, Governance and Society

Volume 22

The book series "Risk, Governance, and Society" is devoted to interdisciplinary and cross-cutting studies addressing comprehensive risk analysis and risk management strategies. The focus is on integration: risk is viewed from scientific, economic, social, and cultural perspectives. The series addresses all types of risks – technological developments, natural hazards, chemical and biological threats, human activities at work and at home, as well as social risks such as terrorism. Although wide in scope and diversity, the series emphasizes the common elements of risks. The series highlights the interaction between natural and social sciences, governmental regulation, and societal actors from industry and civil society. It encourages the building of bridges between the disciplines and provides a home for both empirical research and theoretical analyses that can contribute practical advice for risk governance in plural societies and is able to advance our knowledge in the conceptualization and analysis of risks.

Natalie Rauscher • Welf Werner
Editors

Natural Disasters in the United States

Making Sense of Risks and Vulnerability

Springer

Editors
Natalie Rauscher
Würzburg University
Würzburg, Germany

Welf Werner
Heidelberg Center for American Studies
Heidelberg University
Heidelberg, Germany

ISSN 2512-3076 ISSN 2512-3084 (electronic)
Risk, Governance and Society
ISBN 978-3-031-96435-0 ISBN 978-3-031-96436-7 (eBook)
https://doi.org/10.1007/978-3-031-96436-7

This Springer imprint is published by the registered company Springer Nature Switzerland AG
The registered company address is: Gewerbestrasse 11, 6330 Cham, Switzerland

Contents

Contributors

Jacob Birken University of Cologne, Cologne, Germany

Lawrence Culver Utah State University, Logan, UT, USA

Lea Espinoza Garrido John-F.-Kennedy Institute, Free University Berlin, Berlin, Germany

Jan Hansen Humboldt-University Berlin, Germany

Andrew C. Isenberg University of Kansas, Lawrence, KS, USA

Alan Lessoff Illinois State University, Normal, IL, USA

Mikael Lönnborg Södertörn University, Huddinge, Sweden

Uwe Lübken Ludwig-Maximilian-Universität, Munich, Germany

Edward Manger Heidelberg Center for American Studies, Heidelberg University, Heidelberg, Germany

Robin Pearson University of Hull, Kingston upon Hull, UK

Natalie Rauscher Würzburg University, Würzburg, Germany

James M. Turner Wellesley College, Wellesley, MA, USA

Welf Werner Heidelberg Center for American Studies, Heidelberg University, Heidelberg, Germany

Katharina Wood Technical University Dortmund, Dortmund, Germany

Introduction

Natalie Rauscher and Welf Werner

Abstract The United States faces recurring and often catastrophic natural disasters, including hurricanes, floods, wildfires, tornadoes, and earthquakes. Despite the exceptional economic strength and technological capabilities, the U.S. has developed throughout history, the country remains highly vulnerable to these events, with disaster impacts persisting over time. The contributions to this book explore the historical factors contributing to the ongoing struggle with disaster resilience in the United States, emphasizing the social, political, and economic structures that shape vulnerability. Through an interdisciplinary approach, the volume examines governance failures, cultural perceptions of disasters, and the limitations of mitigation efforts, past and present. The chapters highlight how short-term economic priorities, media narratives, and individualistic attitudes have hindered long-term disaster preparedness and response. Divided into three sections, the book investigates institutional governance, cultural representations of disasters, and case studies of past events to understand the country's disaster landscape comprehensively. Ultimately, this work helps to understand the historical roots of the limited role that the U.S. has played in international climate negotiations so far and underscores the urgent need for proactive risk management, policy reform, and a cultural shift toward resilience in the face of escalating climate risks.

N. Rauscher (✉)
Würzburg University, Würzburg, Germany
e-mail: Natalie.rauscher@uni-wuerzburg.de

W. Werner
Heidelberg Center for American Studies, Heidelberg University, Heidelberg, Germany
e-mail: wwerner@hca.uni-heidelberg.de

N. Rauscher, W. Werner (eds.), *Natural Disasters in the United States*, Risk, Governance and Society 22, https://doi.org/10.1007/978-3-031-96436-7_1

Introduction

The United States faces extreme weather and other large-scale natural events regularly. Hurricanes and tropical storms, followed by storm surges and flooding, tornadoes, winter storms, droughts, forest fires, or earthquakes, have all been common occurrences throughout its history. Yet, unlike other countries that have made significant strides to reduce casualties and losses due to these events (Ritchie 2024), the United States remains highly vulnerable to them. Although the United States reduced fatality rates due to disaster in absolute terms from the early twentieth century (from a low overall level due to the sparse population across a large terrain), its rates have remained rather stagnant compared to other countries in the world, such as those on the Asian continent, which have lowered these rates considerably. On top of that, the number of people affected or injured by disaster in the United States has actually risen over the last decades (Ritchie and Roser 2014; Rauscher and Werner 2025). Even though the United States with exceptional economic and technological resources as well as a range of largely functioning institutions at its disposal, could have been expected to be the world's leading nation in improving resilience, this is not the case. The United States displays an ongoing strong hazard potential when it comes to naturally occurring events. These hazards habitually turn into massive disasters with enormous effects on society, influenced by underlying political, social, and economic structures (Quarantelli 1996; National-Research-Council 2006, 14; Tierney 2019, 5–6).

This volume aims to illuminate the history and current impact of natural disasters in the United States and what role risk[1] and vulnerability play in this context. Whereas disaster resilience is defined as a concept that refers to the ability of social units to absorb disaster shocks, cope with their impact, and adapt or improve in the aftermath of disasters, (social) vulnerability in disaster studies seeks to study the "potential for experiencing adverse impacts from disasters and having poorer outcomes after the occurrence of disasters" (Tierney 2019, 7). While massive differences in people's vulnerability remain inside and between societal groups,[2]

[1] Risk is a term that is studied in many different fields, including in the context of natural hazards, technological threats, working conditions, health, crime, or pollution (Renn 2008, 12). Risk revolves around the scope of negative effects and what the possibility (uncertainty) of an occurrence is, but the conceptualization of the term includes thinking about how to aggregate negative outcomes of occurrences in order to compare, set priorities, and communicate risk (Renn 2008, 12). The term risk refers to the possibility of losses but also the potential for positive outcomes (e.g., risking money on the stock market with the potential of massive gains). The concept of risk is accompanied by uncertainty—losses from risks may happen but we can only estimate their likelihood, an intersection where the work of risk management and risk analysis are located (Tierney 2019, 7).

[2] In recent decades, studies of the concept of social vulnerability have emphasized the underlying social factors that increase the risk for humans during disasters, e.g., through researchers like Susan Cutter (2016), Kathleen Tierney (2019) or Alice Fothergill (Fothergill and Peek 2004). These researchers have shown that there are specific groups in society (women, poor people, racial/ethnic minorities, the elderly, people with disabilities, children) that are more vulnerable in disaster situations than others.

American society as a whole faces disastrous events on a regular basis—and remains extremely vulnerable to them over time. In this book, vulnerability is approached from different angles, referring to the underlying factors and conditions that make US society ill-equipped to cope with disasters. Social and also cultural conditions rather than just risk exposure are taken into account (Bergstrand et al. 2015, 392). Thus, this volume sheds light on the issue of high vulnerability to disasters in the United States, taking into account both historical developments in confronting substantial risks and the necessity of proactive risk management as well as political, societal, and cultural aspects of the US's vulnerability malaise. Looking at the developments throughout the country's history, it can be argued that the United States has largely been unable to effectively "adapt or improve"—and thus increase resilience—in the aftermath of disasters.

Scope of This Volume

As the vulnerability to disasters is entangled in a great variety of different and often opposing societal, cultural, and socioeconomic developments, the contributions to this volume trace them in three spheres: Governance in political institutions (Part I), representation and perception of natural disasters in American culture and public discourse (Part II), and experience with the impact of and the response to prominent disasters and natural hazards (Part III).

While local and state governments in the United States were often overwhelmed by natural disasters, the federal government did not have a lucky hand in handling these emergencies or strengthening resilience in the medium and long run, as the case studies in the first part of this volume indicate. Three contributions address prominent episodes of unfortunate federal interventions and motions, and a fourth showcases the limited options for intervention on the community level.

Forging a bridge from the colonial era to the postwar Sunbelt migration in the twentieth century, Lawrence Culver shows how the environment and exposure to natural disasters have always been a distant second concern in federal policies compared to the omnipresent and strong commercial interests of gaining, securing, and developing land and territory. Strong foundations for a philosophy of short-sighted economic exploitation regardless of environmental consequences and exposure to natural hazards were laid in the nineteenth century, when the federal government was the owner of vast public lands, when exploring, surveying, and selling lands was a primary preoccupation of the new government and when, before the Civil War, the sale of western lands was the main source of revenue. Federal policies were designed to accommodate national ambitions to profit from land and to ignore climate and other environmental risks when they interfered with this goal.

It was not until the middle of the twentieth century that major legislation with a nationwide reach was enacted in Washington to counteract the ever-rising casualties and costs from natural disasters. Although the establishment of the National Flood Insurance Program (NFIP) was a significant step forward in disaster mitigation for the United States, passed during historically unique conditions under the Johnson Administration, the system remains flawed. For more than half a century after the introduction of the NFIP, the federal government's foremost role has remained that of emergency aid giver of last resort, disincentivizing lower governments to take responsibility for disaster preparedness and prevention. Uwe Lübken traces the history of the NFIP vis à vis programs elsewhere, and points to structural deficits, problematic incentive structures, inadequate funding, and the strong presence of partial interests in the formulation and resistance to the modifications of the NFIP. The chapter also addresses the limits of insuring effectively against natural hazards in a time of accelerating climate change.

While the 1960s and 1970s brought heightened awareness of environmental concerns, several landmark environmental laws and a fragile compromise on limiting the production of domestic fossil fuels from federal lands, Andrew C. Isenberg and James M. Turner show that the traumatic experience of the oil crises of the 1970s stalled these developments in the decades to come and led both political parties to accept excessive risks of climate change to minimize risks to economic growth. Once the 1970s crises ended the short-lived environmental consensus among political actors, the issue of unstable access to foreign energy supplies lingered on and so did the overriding imperative of economic concerns. While the Reagan and Trump administrations stand out as examples of a new consensus of the last 50 years that put energy security way ahead of environmental challenges, Democrats, while being more devoted to environmental concerns, have embarked on an ambivalent path, developing climate initiatives while simultaneously backing fracking of natural gas and expanding offshore oil drilling.

Jan Hansen discusses the challenges of proactive management of natural disasters on the community level looking into early endeavors in Los Angeles. While the city is better known as a symbol for the relentless and breathtaking economic ascent of the American West, the greater LA region has been grappling with significant wildfires, earthquakes, periodic flooding, and at the same time, severe droughts. Early flood control efforts along the Los Angeles River "from below" were undertaken by citizens, business owners, and especially railroad companies, representing a dominant self-help attitude. Yet this element of the "frontier myth" stood in the way of a comprehensive and effective answer to the flood danger. Predating federal intervention in the 1930s, these early piecemeal initiatives did not reach their goal of better preparing the region against flooding. Only large-scale flood control measures initiated and financed by the federal government in 1935 effectively tamed the river, making it at once almost unrecognizable as a natural watercourse. The federal government's answer to the flood risk of the Los Angeles River was as efficient in terms of flood control as it was radical in other ways, narrowing the river to a 51-mile concrete riverbed. An alternative design, the Olmsted-Bartholomew Plan, that had been introduced in 1930, was not implemented. Along the lines of more integrated urban planning, it had recommended preserving several floodplains and prohibiting construction within them to create an urban web of greenbelts that would have made the city more livable.

While the lack of suitable and timely government management of natural catastrophes already sheds some light on the specific perceptions of natural disasters in the American context, in the second part of this volume, four contributions take a closer look at these representations.

Reports on the San Francisco earthquake of 1906 by local media oscillate between narratives of shattering devastation and excessively optimistic expressions of hope, as Jacob Birken explains. Downplaying the devastating earthquake on the one hand and exhibiting prominent and overly optimistic perceptions of growth and urban renewal on the other, these reports likely hampered further disaster mitigation and prevention in a region that displays a significant lack of disaster preparedness until today. Additionally, the chapter shows how despite severe challenges with national disaster mitigation and response, narratives of American heroism and innovation framed the event in contrast to other countries where the alleged "backward" ways led to devastating catastrophic effects and an inadequate disaster response. Much like today, it seems that the United States has often found it difficult to look inward critically and learn from disasters.

Perceptions of fortune and misfortune, their determinants, and the choices people can make to cope with the vicissitudes of life are in many societies strongly determined by theodicy, which attempts to vindicate God in the face of the existence of evil. Confronted by momentous hurricanes in the twentieth and twenty-first centuries, Christian writing and thinking, always a dominant factor in American society, culture, and politics, reflect major movements and trends in American Christianity at any given time, as Edward Manger points out. He argues that Protestant responses to natural disasters have moved through various phases from stressing God's agency to service and to the experience of suffering. He concludes that each response, by confronting or ignoring the causes of natural disasters, contributed in its own way to a cultural acceptance of vulnerability as they all share a focus on the aftermath of events rather than the options for minimizing exposure in the future.

Turning to the recent past, the *New York Times* bestselling book series "I Survived," which came out in the 2010s, mirrors dominant sentiments in emblematic American fictional writing on natural catastrophes with the motto "when disaster strikes, heroes are made." As Lea Espinoza Garrido shows, the focus in this strand of fictional writing is once more on postdisaster recovery instead of disaster risk reduction. Similar to public risk management measures that often have discriminatory racial and socioeconomic undertones, skipping over neighborhoods of minorities and the poor, the storytelling of this book series furthers environmental education on the one hand, while it on the other perpetuates the erasure of Indigenous peoples and the stereotypical depictions of heroism in the form of White bravery.

Finally, a discourse analysis of the coverage of natural disasters by major US news outlets covering the late 2010s mirrors once more familiar patterns of prominent American perceptions surrounding natural disasters. Journalistic reporting by the *New York Times*, *Fox News Network*, and *ABC News* reviewed by Natalie Rauscher, hardly ever reflects on underlying causes of high vulnerability or the long history of policy failure, as in the case of the NFIP. Except for a few background pieces, media narratives surrounding hurricanes that often lead to significant destruction, casualties, and flooding have a short attention span, focusing on the human element of suffering and on more or less sufficient emergency help. For most Americans, there is thus little exposure to the fundamentals of proactive disaster risk reduction, which might ultimately contribute to the fact that as voters and taxpayers, they are unlikely to support the often expensive disaster adaptation and mitigation measures.

In the third part of this volume, three contributions focus on the impact of and responses to natural disasters.

With an individualistic view on risk and risk management, a distinctive appreciation of markets as compared to government intervention, American expectations for adequate disaster risk management often hinged on the performance of insurance and reinsurance markets. While the task of reinsurance is to help insurers deal with cumulative events, these specialized companies repeatedly faced problems in coming to terms with the enormous magnitude of natural disasters in the United States (Werner 2016). As a consequence, they retreated from American markets or found themselves incapable of offering policies for certain risks to begin with. As Robin Pearson and Mikael Lönnborg show, it was already the San Francisco earthquake and fire of 1906 that brought to light the limits of and special challenges for mostly Western European (re-)insurers in dealing with major US events adequately. As a result of the 1906 devastations, citizens and businesspeople of San Francisco were left with a conglomerate of different and often conflicting company reactions, with some companies paying out claims while others refusing to do so, referring to various legal explanations, interpretations of the 1906 course of events, and expectations of the public of correct and moral conduct. Some companies stayed while others quit the Californian market in the face of existential losses and difficulties to enforce adequate premium rate hikes in the future. All in all, the 1906 experience was a foretaste of the risk management options that the insurance industry had to offer in the twentieth and twenty-first centuries to mitigate the consequences of natural hazards—offerings that in many cases remained remarkably opaque and inadequate to citizens and businesspeople alike.

Moving to developments on the Gulf Coast in the postwar era, it was the engineering of an industrial "petroleumscape" that rendered this region remarkably vulnerable to storm surges and flooding in the last 70 years. As Alan Lessoff exemplifies once more, it is society that creates the conditions in which natural disasters wreak exceptional damage. While man-made waterways along the Texan Gulf Coast offered abundant space for freight yards, pipelines, storage tanks, refineries, loading docks, petrochemical plants, equipment fabricators, and other support industries, severe weather events in this area have shown that the concentration of Texas' industrial "petroleumscape" on the Gulf Coast has created conditions for environmental disaster. In other words, while oil and gas made the modern Gulf Coast a thriving commercial area, they also threatened it—not least through their continuous contributions to global warming.

Finally, scattered ambitious projects such as the "Kendeda" building on the Georgia Tech campus, which opened in Atlanta in 2019, give a taste of American capabilities for inspired innovation. Green and resilient residential construction could be put to good use elsewhere in the United States if societal awareness and political dynamics created the right incentives. Katharina Wood shows in which ways the Kendeda building contributes to the knowledge about climate-sensitive building practices and how building standards can reduce disaster risks. At the same time, she cautions that single "eco-cathedrals" will not suffice as a mitigation strategy, but that a more widespread application of green building standards is necessary to achieve climate goals and create resilient communities.

Adverse Consequences

All in all, the case studies in this volume provide ample examples of deeply rooted deficiencies in the adaptive capacities of the United States surrounding the management of natural disasters. Although the United States has had unsurpassed economic and technological resources at its disposal and should have been able to make use of them through its federal, state, and local governments, when it came to risk management and building resilience capacity against natural disasters, the United States has remained a laggard. A casual glance at the historical development of population densities in the United States shows already that large parts of the American population have more or less systematically moved to or settled in the country's most exposed regions: Coastal areas in the Southeast and the West (Census-Bureau 2012). As the contributions of this volume show, instead of addressing structural deficiencies and approaching sustainable solutions, specific political constellations and dynamics, deeply rooted values of individualism as well as short-sighted "consumption" of disasters through mass media or cultural production often dominate. The contested notion of a remarkable abundance of (economic) opportunities in a vast and empty country blessed with seemingly unlimited resources, which was characteristic of the Early Republic, seems to have survived on a remarkably grand scale. Yet, the adherence to the existence of this myth, which often still dominates public and political discourse in the United States today, clashes with the realities of a narrowing time horizon to correct the directions of trodden paths and to devote more attention to the largest resilience project of the planet: Responding to climate change in the face of imminent natural hazards of global and apocalyptic dimensions.

Although Washington remains an indispensable player in international climate negotiations, its contributions have been wholly inadequate for decades. The case studies of this volume help to understand the frustrating multilateral negotiations by illuminating the historical baggage that the United States brings to the table. The contributions help to explain the historical roots of the United States as a nation that has been one of the most resistant actors when it comes to acknowledging and managing the substantial risks that natural disasters pose. The causes that hinder the United States from taking appropriate responsibility in international climate negotiations are thus twofold: On the one hand, there is a deeply rooted (political) culture of failed risk mitigation efforts at home that lives on. On the other hand, and inextricably connected to the history of domestic failure, stands the fact that the United States after more than 200 years of a negligent approach to environmental challenges, displays per capita emissions of greenhouse gases today that puts the country in the most significant position for their reduction. US governments have to weigh between international demands for a significant reduction of American greenhouse gases and expectations of voters at home, quite a few of whom are not willing to throw their deeply rooted ignorance vis-à-vis environmental challenges overboard and to shoulder the cost of changing course. Thus, the country continues to leave its role in effectively reducing emissions—and proactively managing natural disasters—unfulfilled. While some improvements have been made under the Biden administration, experts see these advances as inadequate overall (Vaidyanathan 2022). Additionally, there is no doubt at the time of writing that the Second Trump Administration will aggressively reverse these efforts, which in March of 2025 already made significant strides toward overturning some of the successes in fighting climate change and decarbonizing the US economy. President Trump once again enacted a withdrawal from the Paris Climate Accord as one of the first official acts after entering the White House in early 2025. He has also vowed to increase oil and gas production in the United States and take back many climate regulations (McGrath 2025). Additionally, Donald Trump with the help of the newly founded Department of Government Efficiency (DOGE) has called into question the integrity of the federal bureaucracy on many levels, including federal disaster response. He indicated to critically assess the work of agencies such as FEMA and to push more responsibility to the state level in disaster situations. Although under normal circumstances, a president would not have the power to solely eliminate an agency such as FEMA without congressional approval, the administration can certainly freeze many of its operations and hinder any work that is done by the agency (Barrón-López, Popat and Barnhart 2025; Rozdilsky 2025; White House 2025). The Trump administration, under EPA Administrator Lee Zeldin, is currently starting to roll back dozens of crucial environmental regulations, including measures on air and water quality, emissions limits for coal plants, and vehicle standards. Zeldin remarked that these actions are aimed at "stabbing climate activists in the heart," a statement reflecting the administration's destructive stance on climate policy (Friedman and Tabuchi 2025). All these developments certainly do not bode well for the United States as a country so vulnerable to disasters and environmental challenges on many levels. As we discuss in this book on several occasions, the lack of effective institutions and federal activity in the field of disaster mitigation has already been a significant weakness in effectively addressing disasters and climate challenges long before Trump became president for a second time.

Our Volume in the Field of Disaster Research

Of course, research on natural disasters and environmental degradation in the United States is much broader in its reach than the chapters in this volume show. The focus of the contributions is on factors contributing to high vulnerability to disasters and a lack of building resilience capacity over time. While the interdisciplinary endeavor of this book seeks to illuminate various perspectives on the study of risk, disasters, and vulnerability in the United States, we see our contributions in the tradition of the different research angles taken in disaster studies over the last decades. Disaster research as a discipline began in the 1920s and 1930s, yet was first systematically funded by the US military in the 1940s, faced with the potential of nuclear war in the aftermath of WWII and the beginning of the Cold War (Tierney 2019, 15). At this time, researchers became conscious that the United States faced enormous challenges due to natural hazards and large-scale disasters that needed addressing. As we see in the contributions of our volume, political actors often ignored the potential for disasters and hazards in the nineteenth century in favor of economic expansion (Chapter "Disaster Made Manifest: The Federal Role in US Natural Disasters"), whereas the early twentieth century brought about some mitigation efforts on the local level, for instance, in Los Angeles. These efforts improved flood management but simultaneously brought about enormous changes to the natural environment and consciously favored White parts of the population over racial minorities (Chapter "Letting Nature Do Its Thing: Early Flood Control and Environmental Disaster Along the Los Angeles River"). The early twentieth century also brought about one of the most significant disasters lingering on the US consciousness until today, the 1906 San Francisco earthquake and fire. This event features heavily in historical accounts of disasters. In this volume, contributions are concerned with its influence on the American insurance and international reinsurance market (Chapter "Multinational Insurers and Catastrophic Loss: Responses to the San Francisco Disaster of 1906"), as well as its impact as a mediatized international event with the capacity to rewrite disasters into a national narrative of progress and innovation (Chapter "Hopeless Optimism. Framing Early Twentieth-Century Earthquakes as News and History in the Making"). Thus, disaster events are often used—also in hindsight—to perpetuate historical American myths. Today, one can see a continuation of this in contemporary literary production which tries, on the one hand, to advance environmental consciousness, but on the other hand continues to emphasize narratives of white heroism and the erasure of Indigenous roles in disaster situations (Chapter "Through the Eyes of the White, Innocent Child: Whiteness, Vulnerability, and (Environmental) Crisis in Lauren Tarshis's *I Survived* Series").

As disasters and national security concerns became part of the political agenda of the 1940s and after, events such as chemical accidents or natural events such as tornadoes were classified as (potential) emergencies where the government was called upon to help. This was the time of the first larger-scale endeavors to directly study these events and how they impacted the affected population on the ground, for example, through researchers such as Enrico (Henry) Quarantelli at the Disaster Research Center (DRC) founded at Ohio State University (1963), which later moved to the University of Delaware (Tierney 2019, 16–17). Defense considerations were important for the American government when it came to disaster mitigation but soon extreme weather and other natural events were deemed significant (and costly) enough—and maybe even more likely—than imminent nuclear attack from outside. Therefore, studies of things like floodplains were conducted to address the rising costs in the aftermath of events like hurricanes or flooding around the United States (Knowles and Kunreuther 2014). Here, geographer Gilbert White was crucial in conducting the first studies in this context. Even at this time, however, White and his colleagues emphasized that human decisions and actions determined the impact of naturally occurring events, not just their physical force (Tierney 2019, 21–25). As we see in the chapter by Uwe Lübken in this volume (Chapter "Nature by Numbers: The National Flood Insurance Program"), White later advised the US government in forming the NFIP, where only parts of his concerns about possible moral hazard in the program's structure made it into the final version of the program (National-Research-Council 2006, 26; Knowles and Kunreuther 2014). Although it was an enormous achievement of the time to move from ad hoc disaster aid to a more systematic approach in the form of the NFIP, the program remains inefficient until today. Reform efforts have stalled in Congress on a regular basis, not least aided by the lack of exposure to structural challenges and necessary reform of disaster mitigation in public discussion, for example, in newspaper and TV reporting, as one of our contributions shows (Chapter "Reporting Catastrophes: Mainstream Media and Vulnerability to Disasters in the United States").

While the 1960s and 1970s had been marked by a certain consensus in disaster research as well as a short-lived and delicate awareness of environmental concerns in national policies under the Nixon Administration, the 1980s did not only bring about significant critique inside the field of disaster research but also a turnaround in national policies. After the oil crises of the 1970s, energy independence and economic considerations dwarfed the concern for the environment—a constellation that has continued more or less until today, despite some efforts to address the most glaring environmental challenges to some degree (Chapter "All of the Above: Energy, Environmental Politics, and US Climate Change Policy since the 1970s Oil Crises"). The consequences of efforts to increase energy independence can also be seen in specific regions such as along the Gulf Coast of Texas, where the oil and gas industry grew tremendously, despite continuous dangers posed by the region's exposure to storm surges and floods (Chapter "Oil Ports and Hurricanes Along the Texas Gulf Coast").

In disaster research, the 1980s was a time that deemed the field too technocratic and an emblem of the managerial state, while weak in actual application of research results (Tierney 2019, 26). At this time, key contributions to the field were made by shifting the focus to the concept of (social) vulnerability to disasters, taking into account that different groups of people in society face different kinds of risks and impacts of extreme events. The emphasis at this time was on class and the developing world, whereas efforts in later years also included focuses on gender and race, as well as a focus on social vulnerability in the US context. Vulnerability of different groups, it was argued, was the result of a combination of root causes such as lack of political power or economic resources and the result of social, political, and economic inequalities. Thus, the study of disaster events became part of the study of the social fabric itself, rather than the study of isolated occurrences (Tierney 2019, 28–29). In this volume, we emphasize that it is not only political or economic circumstances that foster vulnerability but that cultural factors such as faith (Chapter "Hurricanes and Theodicy in Twentieth and Twenty-First Century American Protestantism"), literary production (Chapter "Through the Eyes of the White, Innocent Child: Whiteness, Vulnerability, and (Environmental) Crisis in Lauren Tarshis's *I Survived* Series"), or media discourses (Chapter "Reporting Catastrophes: Mainstream Media and Vulnerability to Disasters in the United States") also play a significant role as these factors can frame people's perception of events and their response to them, and thus have a lasting effect on decision-making by democratically elected political representatives. Many more avenues to the study of disaster impacts exist, for example, in the form of the study of the improvisational nature of disaster responses (Kreps 1985, 1989) or the emergence of response and rescue networks (Drabek et al. 1981)—both strands of research that became highly relevant once more in the aftermath of the 9/11 attacks (Tierney 2019, 34–35).

In the 1990s and early 2000s, contributions to the field of disaster research came from the study of land-use planning (Burby 1998), hazard insurance (Kunreuther and Roth 1998), geographic dimensions of disaster (Cutter 2001), disaster preparedness (Perry, Lindell, and Tierney 2001) as well as from studying the intersection of vulnerability factors such as race, gender, and poverty (Fothergill 1996; Fothergill et al. 1999; Fothergill and Peek 2004).

The twenty-first century brought new inflection points for disaster researchers as the world was confronted with disasters of massive proportions including the 9/11 terrorist attack, the 2004 earthquake and tsunami in the Indian Ocean, or Hurricane Katrina in 2005. For the United States, Hurricane Katrina and its devastating aftermath brought to light the notions studied in disaster research for decades, namely that disasters are not external events only but come about through "endogenous political and economic forces" (Tierney 2019, 36). Additional hurricane seasons of 2012 and 2017 in the United States highlighted issues of vulnerability further, especially also in light of efforts to address climate change (Tierney 2019, 37). Our volume reflects on some contemporary efforts to bring about a more sustainable future and to take into account hazards and disasters in the warming climate of the twenty-first century. Efforts in green and resilient building are noteworthy and offer the potential for risk mitigation and sustainability. However, these efforts still lack the scale to be anything more than a solitary model for the future (Chapter ""Becoming Green": Resilient and Green Building as Risk Mitigation in Atlanta, Georgia").

As we can see, interdisciplinary study of disasters has been conducted along many avenues. Research is done in a wide array of fields, very intensively in sociology and geography. But fields like anthropology, psychology, economics, political science, or urban planning are also active in illuminating different perspectives when it comes to risk and vulnerability to disasters. As this volume shows, apart from the social sciences, efforts to study US disasters and environmental hazards from an interdisciplinary perspective are similarly done in the humanities, for example, in the field of American Studies. Apart from this volume, some specific projects combine both cultural, literary, and historical analysis of disasters, for example, at the Rachel Carson Center for Environment and Society including publications such as Christof Mauch's and Sylvia Mayer's *American Environments: Climate, Cultures, Catastrophes* (2012) or Sylvia Mayer's and Alexa Weik von Mossner's edited volume *The Anticipation of Catastrophe: Environmental Risk in North American Literature and Culture* (2014). A lot more research is done in the field of Environmental Humanities in American Studies, for example, lately in the doctoral program *Rethinking Environment* (2024) at the University of Augsburg and Munich, at *the Käte Hamburger Centre for Apocalyptic and Post-Apocalyptic Studies* at Heidelberg University, and of course by many individual researchers all across the field of American Studies, including the researchers featured in this volume.

In the twenty-first century, it seems that disasters of massive proportions are everywhere, especially also in the US context. At the same time, the events seem to continue to blindside the public and political stakeholders despite scientific knowledge being available. In the face of these realities, we need to acknowledge what many disaster researchers both in the social sciences and in the humanities have investigated for a long time: Disasters and their impacts display a many-faceted interplay with social dynamics and processes. From such a perspective, social, political, economic, and cultural factors all play a role in rendering an event a disaster (Tierney 2014, 5). In line with this notion, this volume contributes to the research in this field by combining perspectives on risk and vulnerability to disasters in the United States from an interdisciplinary perspective including contributions from history, literary studies, religious history, art history, risk management, political science, geography, and urban planning. All these contributions will further illuminate the US's vulnerability malaise and its continuing challenge to create a more resilient society.

References

Barrón-López, Laura, Shrai Popat, and Eliot Barnhart. 2025. *Trump questions need for FEMA and says states should 'take care of disasters.'* PBS News Hour. Accessed Feb 13, 2025. https://www.pbs.org/newshour/show/trump-questions-need-for-fema-and-says-states-should-take-care-of-disasters.

Bergstrand, Kelly, Brian Mayer, Babette Brumback, and Yi Zhang. 2015. Assessing the Relationship Between Social Vulnerability and Community Resilience to Hazards. *Social Indicators Research* 122 (2): 391–409.

Burby, Raymond J. 1998. *Cooperating with Nature: Confronting Natural Hazards with Land-Use Planning for Sustainable Communities*. Washington, DC: Joseph Henry Press.

Census Bureau. 2012. *Coastline County Population US Census Bureau.*

Cutter, Susan L. 2001. *American Hazardscapes: The Regionalization of Hazards and Disasters*. Washington, DC: Joseph Henry Press.
Cutter, Susan L. 2016. The landscape of disaster resilience indicators in the USA. *Natural Hazards* 80:741–758.
Drabek, T. E., H. L. Tamminga, T. S. Kilijanek, and C. R. Adams. 1981. *Managing Multiorganizational Emergency Responses: Emergent Search and Rescue Networks in Natural Disaster and Remote Area Settings*. Boulder: Natural Hazards Center, University of Colorado.
Fothergill, Alice. 1996. Gender, Risk and Disaster. *International Journal of Mass Emergencies and Disaster* 14 (1): 33–56.
Fothergill, Alice, and Lori Peek. 2004. Poverty and Disasters in the United States: A Review of Recent Sociological Findings. *Natural Hazards* 32:89–110.
Fothergill, Alice, Enrique Maestas, and Joanne Darlington. 1999. Race, Ethnicity and Disasters in the United States: A Review of the Literature. *Disasters* 23 (2): 156–173.
Friedman, Lisa, and Hiroko Tabuchi. 2025. *E.P.A. Targets Dozens of Environmental Rules as It Reframes Its Purpose*. The New York Times. Accessed March 17, 2025. https://www.nytimes.com/2025/03/12/climate/epa-zeldin-rollbacks-pollution.html.
Knowles, Scott Gabriel, and Howard C. Kunreuther. 2014. Troubled Waters: The National Flood Insurance Program in Historical Perspective. *The Journal of Policy History* 26 (3): 327–353.
Kreps, Gary. 1985. Disaster and the Social Order. *Sociologial Theory* 3 (1): 49–64.
Kreps, Gary. 1989. *Social Structure and Disaster*. Newark: University of Delaware Press.
Kunreuther, Howard C., and Richard J. Roth Sr. 1998. *Paying the Price: The Status and Role of Insurance Against Natural Disasters in the United States*. Washington, DC: Joseph Henry Press.
Mauch, Christof, and Sylvia Mayer, eds. 2012. *American Environments: Climate – Cultures – Catastrophe, American Studies – A Monograph Series*. Heidelberg: Winter.
Mayer, Sylvia, and Alexa Weik von Mossner, eds. 2014. *The Anticipation of Catastrophe: Environmental Risk in North American Literature and Culture, American Studies – A Monograph Series*. Heidelberg: Winter.
McGrath, Matt. 2025. *Trump Vows to Leave Paris Climate Agreement and 'Drill, Baby, Drill.'* BBC News. Accessed Feb 13, 2025. https://www.bbc.com/news/articles/c20px1e05w0o.
National-Research-Council. 2006. *Facing Hazards and Disasters: Understanding Human Dimensions*. Washington, DC: The National Academies Press.
Perry, Ronald W., Michael K. Lindell, and Kathleen Tierney. 2001. *Facing the Unexpected: Disaster Preparedness and Response in the United States*. Washington, DC: Joseph Henry Press.
Quarantelli, Enrico L. 1996. *The Future is not the Past Repeated: Projecting Disasters of the 21st Century from Present Trends*. Preliminary Paper #229. Disaster Research Center, University of Delaware, Newark.
Rauscher, Natalie, and Welf Werner. 2025. "Hurricanes and Floods: The Case of the United States and the National Flood Insurance Program." In Teresa da Silva Lopes, Paul Duiguid, Robert Fredona (Eds.) *Climate Change and Business: Historical Perspectives*.
Renn, Ortwin. 2008. *Risk Governance: Coping with Uncertainty in a Complex World*. London: Earthscan.
Ritchie, Hannah. 2024. *The World Has Become More Resilienct to Disasters, But Investment is Needed to Save More Lives*. Our World in Data. Accessed June 14, 2024. https://ourworldindata.org/the-world-has-become-more-resilient-to-disasters-but-investment-is-needed-to-save-more-lives.
Ritchie, Hannah, and Max Roser. 2014. *Our World in Data: Natural Disasters*. Accessed Feb 13, 2025. https://ourworldindata.org/natural-disasters.
Rozdilsky, Jack L. 2025. *Trump's Plan to Eliminate FEMA is a Very Bad Idea*. The Conversation. Accessed Feb 13, 2025. https://theconversation.com/trumps-plan-to-eliminate-fema-is-a-very-bad-idea-248293.
Tierney, Kathleen. 2014. *The Social Roots of Risk*. Stanford: Stanford University Press.
Tierney, Kathleen. 2019. *Disasters: A Sociological Approach*. Cambridge: Polity Press.

UmweltDenken. 2024. *Doctoral Program UmweltDenken/Rethinking Environment*. Accessed Sept 9, 2024. https://rethinking-environment-idk.de/.

Vaidyanathan, Gayathri. 2022. Biden Signs Historic Climate Bill as Scientists Applaud. *Scientific American Accessed Sept* 12:2024. https://www.scientificamerican.com/article/biden-signs-historic-climate-bill-as-scientists-applaud/.

Werner, Welf. 2016. Natural Catastrophes and their Effects on Reinsurance. In *Managing Risk in Reinsurance: From City Fires to Global Warming*, ed. Niels Viggo Haueter and Geoffrey Jones. Oxford: Oxford University Press.

White House. 2025. *Council to Assess the Federal Emergency Management Agency*. Executive Order, January 24. Accessed Feb 13, 2025. https://www.whitehouse.gov/presidential-actions/2025/01/council-to-assess-the-federal-emergency-management-agency/.

Governance

Disaster Made Manifest: The Federal Role in US Natural Disasters

Lawrence Culver

Abstract Natural catastrophes occur worldwide, yet the United States suffers a striking number and variety of natural disasters. These environmental phenomena intersect with human society, human environments, and human choices. In the United States, a society and political system occupying a disaster-prone continent repeatedly made choices exacerbating natural disasters. This chapter illuminates the federal role in worsening natural disasters. The US experiences unusually high vulnerability to natural catastrophes because at key moments the federal government repeatedly made choices that exposed Americans to natural hazards and catastrophes, a pattern that predates the nation's founding. National leaders and institutions were preoccupied with gaining more territory and profiting from the development of new land. This chapter thus focuses on two overriding concerns of the US federal state: land and climate. It then examines key examples of environmental risk and natural disaster, primarily aridity, wildfire, severe weather events including hurricanes and tornadoes, and the policies and infrastructure the government implemented to manage, mitigate, or sometimes even conceal these hazards. Government science institutions and knowledge grew, but state-generated science was sometimes stifled for the sake of expansion and profit. Chronologically, federal attitudes toward natural disasters and environmental risk can be divided into two parts. From its creation through the nineteenth century, the federal government often minimized potential environmental risks. By the early twentieth century, the federal government, along with local governments and business interests, shifted to actively managing, and sometimes minimizing, actual disasters and hazards, and subsidizing risk. Greatly simplified, the natural catastrophe and disaster risk history of the United States can be divided into two phases: encouraging bad choices and grappling with the consequences of bad choices.

L. Culver (✉)
Utah State University, Logan, UT, USA
e-mail: lawrence.culver@usu.edu

N. Rauscher, W. Werner (eds.), *Natural Disasters in the United States*, Risk, Governance and Society 22, https://doi.org/10.1007/978-3-031-96436-7_2

Introduction

Natural catastrophes are a global phenomenon. Yet the United States suffers a striking number and variety of natural disasters. Some of these are a byproduct of climate and geography, from the unusual frequency of tornadoes in North America to the earthquake faults underlying some of the nation's largest urban areas. Natural disasters are environmental phenomena, but ones that intersect with human society, human environments, and human choices. In the United States, I will argue, a society and political system occupying a disaster-prone continent has repeatedly made choices that exacerbated natural disasters and their consequences.

The United States, from its federal to state and local governments, is hardly alone in allowing the market to lead to poor disaster planning, or to leave its poorest citizens at risk. Other governments also make dubious planning choices, such as allowing development in floodplains or fire-prone areas. Nor does it have a monopoly on poorly planned or constructed infrastructure, such as dams and levees, or haphazard enforcement of seismic building codes. Sadly, twenty-first-century news coverage provides ample examples worldwide of poor planning, inequality revealed or worsened by disaster, or poorly built infrastructure. One can also point to the role of governments' indirect or direct involvement in fossil fuel extraction, from tax breaks and incentives favoring energy companies to partially or wholly owned state energy companies, making governments complicit in the climate crisis. From a longer historical perspective, the United States is also not unique in being a product of settler colonialism, a nation where many arrived from elsewhere, unaware of natural hazards, largely oblivious to Indigenous peoples' knowledge of those hazards, and often actively dismissive of knowledge gained from long-term Indigenous tenure in these landscapes.

What then are the main reasons for high risks and vulnerability to natural catastrophes in the United States? If its historical experience is not unique, why does it remain particularly vulnerable to these disasters? Drawing on environmental history and especially climate history, I argue that the US experience of natural disasters is distinctive in one respect. At key moments, the federal government of the United States has, again and again, made choices that exposed Americans to natural hazards and catastrophes, a pattern that dates to the nation's founding, long before more recent examples of dubious governmental action or inaction we might draw from the twentieth and twenty-first centuries. In this chapter, I will focus on two preoccupations of the US federal state, land and climate, and a few key examples of environmental risk and natural disaster, primarily aridity and drought, wildfire, severe weather events including hurricanes and tornadoes, and the policies and infrastructure the government implemented to manage, erase, and sometimes even conceal these hazards. Understanding how federal policies were shaped to fit national ambitions to secure and profit from land, and to elide or ignore climate and other environmental risks whenever it interfered with this goal, will help illuminate the federal role in worsening and even precipitating natural disasters.

Here, I discern a distinct difference between earlier and more recent natural disasters and environmental risk history in the United States. Since 1945, and especially since the 1970s, the federal government has subsidized environmental hazards and risks. Further, it has divorced risk from geographic location or proximity to natural hazards. As historian Ted Steinberg argues in *Acts of God: The Unnatural History of Natural Disaster in America*, environmental risk has been "de-placed." Federal flood insurance means that taxpayers far removed from flood zones subsidize development in dubious places. Likewise, until relatively recently, private insurers and the federal regulation of the insurance market distributed risk broadly, with less concern about locales with higher risk. That has now changed with remarkable speed, with insurers abandoning whole states such as hurricane-prone Florida or wildfire-prone California or charging property owners rates so exorbitant that they can no longer afford coverage. Ironically, Florida, one of the nation's most politically conservative states, seemingly allergic to government regulation, takes pride in its free market and free-range capitalism, yet is increasingly dependent on a state-subsidized property insurance system, one that may prove woefully inadequate for all the hurricanes and storm surges to come.

Certainly, there are multiple reasons for this change in regulatory approaches to risk, and the federal role of natural disasters. One is the rise of a much larger administrative and regulatory state across the twentieth century, as well as the rise of large corporations, and their ever-larger role in national political life. We might suppose that more scientific data, and a better scientific understanding of disasters, would have also resulted in changes in planning for disasters and risk. Such hopes are sometimes misplaced. Actual science sometimes has little to do with how the US plans for natural hazards and disasters, and that is a constant across US history.

Instead, the change I perceive between earlier and more recent phases of disaster and environmental risk history in the United States has more basic causes. In the nineteenth century, and indeed into the early twentieth century, the federal government was preoccupied with understanding new places: new territories annexed, purchased, or otherwise added to the republic. As we shall see, the federal government had a vested interest in surveying and preparing this land for sale, thus making it as appealing as possible for prospective settlers. Even into the early twentieth century, as the federal government shifted from surveying and describing new lands to actively altering them, particularly through the hydraulic engineering projects undertaken by the Newlands Reclamation Act of 1902 and later expanded under the Bureau of Reclamation, federal policies and initiatives were still aimed at the future, promoting areas for settlement, encouraging investment, or promising federal projects that would ensure that regions with more challenging climates or topography could still be transformed into profitable and settleable lands.

Yet as the nation's population grew, territories became states, and small settlements became cities, climate and environmental risks ceased to be abstractions to be analyzed, described, or minimized and became facts to be dealt with and experienced. It was one thing to promote development in drought-prone Arizona. It is another thing entirely to grapple with the reality of metropolitan regions, with millions of residents, dependent upon federal irrigation schemes conceived more than a century earlier. Across the nineteenth century, the federal government of the United States was actively engaged in minimizing potential environmental *risks*. As that century drew to a close and the twentieth century advanced, the federal government, along with local governments and business interests, shifted to actively managing—and sometimes minimizing—actual *disasters*. Greatly simplified, the natural disaster and environmental hazard risk history of the United States can be divided into two phases: encouraging bad choices and grappling with the consequences of bad choices.

The Prehistory of Bad Federal Choices: Attitudes Toward Land and Climate Before the United States

Wishful climatic thinking predated 1776, or the federal government created under the new Constitutional system of 1789. The climate of North America represented both promise and problem for European settlers and their descendants. Their evolving conceptualization of climate was rooted in prior European climatic cultural knowledge and expectations, the immediate need for personal, family, or community survival, and the longer-term desire for success. This success was conceptualized as both individual and national. Settlers hoped to secure a plot of good land, a reliable income, and stability for their family and descendants, described in frontier parlance as achieving a "competency." Many were also deeply vested in the success of the United States as a national project. What that project would become, however, was a core point of contention between divergent regions, economic interests, and political ideologies in a contentious republic where climate served as a proxy for many other interests and ambitions. Even after the Civil War and the consolidation of federal authority over a continental nation, climate remained a concern. A key contradiction was that even as climate science advanced and the scientific apparatus of the state grew exponentially, denial and fiction endured and even prospered. Fact could exist alongside fabulation, a core contradiction that would linger.

The first English settlements in North America proved disastrous. Roanoke was lost, its fate unknown. In the years after its founding in 1607, Jamestown suffered an eighty percent fatality rate, with evidence of cannibalism. Only later would tobacco and slavery provide an economic foundation for the colony. Company investor and settler expectations, inflated by Spanish imperial success, predicted the presence of large Native cities, like those in Mesoamerica. They further expected that gold would be plentiful and that an easy passage to the Pacific lay nearby. They also made wildly incorrect predictions about the climate. These expectations, however, were less fantastical. Instead, they were rooted in conceptualizations of climate dating to the classical world. For the ancient Greeks and Romans, climate was synonymous with geography. Climatic belts girdled the globe, from torrid regions at the equator to frigid ones in the Arctic. Expectations of a Mediterranean climate where orange trees would bloom were not irrational for a colony due west of

Gibraltar, nor could island residents of a marine climate be expected to understand a continental climate, with much greater seasonal extremes in temperature. In response to a climate that did not match expectations, colonists could respond with either curiosity or wishful thinking. They would engage in both.

In ways that might seem surprising to modern readers, colonial Americans, in fact, thought that climates could change and that the climate of eastern North America was changing. They even thought that human actions played a role in this change, though they were not imagining anthropogenic climate change as we understand it in the twenty-first century. They thought, in essence, that climate could be changed through good works, that European-style agriculture and industriousness would improve the land and the climate. They even imagined that this new and improved climate would favor lighter European complexions over darker Indigenous or African ones. Their cultural, religious, and racial prejudices were inextricably connected to their perceptions of climate and landscape. Harvard professor, minister, and polymath Samuel Williams, for example, wrote and corresponded about these climatic ideas, part of the climate change discourse evident before and after the Revolution.[1] And though it remains a contested issue, a sad irony of this era is that if the climate in North America or elsewhere did indeed change, it did so not because of European actions but because of the sudden loss of Indigenous populations through disease and the disruption of their long traditions of landscape management. It is even possible, as Koch, et. al and Charles Mann have argued, that the rapid reforesting of eastern North America, caused by the abrupt dissapearance of Indigenous people and their fire practices, may have resulted in a sharp increase in carbon dioxide absorption by trees from the atmosphere and potentially intensified the Little Ice Age.[2]

Of the political leaders and thinkers Americans would later revere as "founding fathers," Benjamin Franklin's interest in the natural world is best known for his investigations of lightning, electricity, and tornadoes. Thomas Jefferson also kept weather records at Monticello, particularly recordings of temperature.[3] So did James Madison at his plantation, Montpelier, where he recorded temperature, wind direction, and facts related to planting, harvesting, and gardening, and sightings of the

[1] Samuel Williams, "Change of Climate in North America and Europe," undated [1770s?]. Papers of Samuel Williams, HUM 8, Box: 1, Folder: 10. Harvard University Archives. https://id.lib.harvard.edu/ead/c/hua05010c00010/catalog. Accessed September 23, 2023.

[2] Alexander Koch, Chris Brierley, Mark M. Maslin, Simon L. Lewis, "Earth System Impacts of the European Arrival and Great Dying in the Americas after 1492," *Quaternary Science Reviews*, Volume 207, 2019, Pages 13–36. Charles C. Mann elaborates on this possibility in *1493: Uncovering the New World Columbus* Created (New York: Vintage, 2011), 38–43.

[3] The Thomas Jefferson Papers project at Princeton University has helpfully made these records available, both providing a great deal of weather-related data, but also reminding us of the limitations of this early weather recordkeeping. Measurements of something like temperature might be irregular, either in frequency or by varying the location where temperature records were taken: https://jefferson-weather-records.org/

northern lights.[4] Yet many of this generation were interested in the natural world in at least one respect; they were avid buyers of and speculators in the land. It is sometimes difficult to separate national and personal ambitions in land acquisition among this cohort. Some, like George Washington, were trained land surveyors with experience reading and analyzing landscapes. Cartographic literacy was also a key marker of social refinement and engaged citizenship, and wealthy landowners both commissioned and collected maps. This literacy would continue and flourish—one example was the popularity and impact of Alexander von Humboldt's explorations and publications in the nineteenth-century United States, as well as the accounts of British explorers, from James Cook in the Pacific to David Livingstone in Africa.

Whether the owners of vast slave plantations or humble pioneers, American colonists and settlers attempted to determine climate by "reading" landscape. By looking at vegetation, soils, surface water, and other factors, they tried to determine if land would be useful for agriculture. In this regard, the more it resembled good agricultural land in Europe, the more promising it seemed. Eastern prairies in future states such as Illinois, for example, which lacked trees but were humid with good soil, were initially viewed as wastes, since treeless land was presumably infertile.

The History of Bad Federal Choices

After independence, while the nation was briefly governed by the Articles of Confederation, Alexander Hamilton made a shrewd proposal that became the Northwest Ordinance of 1787, enacted two years before the United States adopted the Constitution. Under it, the thirteen original states would give up western land claims in all the land ceded by the British between the Appalachians and the Mississippi, land once promised to their Native American allies. In exchange, the new national government would assume all state government debts. From this point forward, all new western lands absorbed in the United States would become federal land. The Northwest Ordinance delineated a system by which this land would be surveyed and sold. In terms of physical geography, this was one of the most lasting landmarks of the Enlightenment. A Cartesian grid spread across the US portion of North America. The square mile survey portions are still visible from aircraft, whether one flies across the agricultural fields of the Great Plains or the vast gridded subdivisions of Sunbelt cities like Phoenix.

After Thomas Jefferson's gargantuan Louisiana Purchase in 1803, interest in the nation's new territorial holdings reached a fever pitch. For much of the rest of the nineteenth century, exploring, surveying, and selling western lands would be a primary preoccupation of the new national government of the United States. (Here, one is reminded of French political scientist and sociologist Emile Boutmy, who

---

[4] American Philosophical Society, "Historical Meteorological Records at the APS," https://www.amphilsoc.org/museum/exhibitions/historic-meteorological-records-aps

observed in this era that the United States was less a nation than a vast real estate enterprise.) Indeed, before the Civil War, the sale of these lands was the primary source of revenue for the federal government. The government would fund a series of surveys of western lands. These surveys gathered valuable scientific and geographical data, but that was not their only purpose. Survey findings, bound in impressive volumes filled with maps, diagrams, analyses of soil and water, and attractive illustrations, were on the one hand compendiums of cartographic knowledge, and on the other hand glossy guides to new American real estate. In fact, the federal government spent more on these lavish publications than on the survey expeditions themselves.

While Americans were most interested in determining the climate in the territory of their growing republic, they were also interested in global climate events. One example was 1816, the year without a summer. A massive volcanic eruption, Mount Tambora in Indonesia, meant that warm weather never arrived in North America, and crops failed. In Europe, this led to food riots in a region still recovering from the Napoleonic wars. In the United States, it would spur movement out of New England, which had a cold climate in normal years, and all the best agricultural land had long since been claimed. The exodus from New England would lead many settlers to New York state, the Midwest, and points beyond. The year without a summer would also lead to a great deal of transatlantic correspondence as Americans and Europeans alike tried to understand the strange climatic events of that year, which also included strangely colored skies caused by volcanic ash, later recorded in the paintings of J.M.W. Turner, and a cold rainy summer that gave birth to one of the first and most famous works of science fiction, Mary Shelley's *Frankenstein, or the Modern Prometheus* (1818).

For many Americans, however, the "Year without a Summer" was a climatic hiccup in an inevitable march toward national greatness. By the 1820s and 1830s, the nation embraced the ideology of Manifest Destiny, the idea that White Protestant Americans were racially, politically, religiously, and economically superior to all other peoples, and their advance across the continent as "the great nation of futurity" was foreordained.[5] The racism and thinly veiled genocidal impulses apparent in this ideology have been capably analyzed by many historians. What is less fully understood is the degree to which Manifest Destiny encompassed the natural world too. Anglo-Americans did not merely expect the other inhabitants of North America to cooperate with their ambitions: They expected the same of the continent itself.

Yet it soon became apparent that the continent and its climates might not accommodate American ambitions. The Lewis and Clark expedition found no easy route to the Pacific Ocean, and John Jacob Astor's scheme, supported by Jefferson, for fur trading and empire building in the Pacific Northwest foundered. Washington Irving, the popular author of "Rip Van Winkle" and "The Legend of Sleepy Hollow," wrote an engaging account of the Astoria Expedition, and while Americans enjoyed its

[5] John L. O'Sullivan, "The Great Nation of Futurity," *The United States Magazine and Democratic Review* VI (November 1839): 426–430.

tales of adventure, they were less enamored with a term Irving coined for much of the West traversed by Astor's expedition on its way to the Pacific coast. He stated that much of the interior of North America was a "Great American Desert," a wasteland destined to always be inhabited by nomads.[6] The Zebulon Pike expedition to the Front Range of the Rockies and South into Nuevo Mexico and New Spain would likewise report that much of the Southern Plains and near Southwest were barren and apply the "Great American Desert" to that region as well.

Decades before the Civil War, a regional split was already apparent in the United States, and it centered on interpretations of climate. This would rapidly come to influence federal policy, both for future survey expeditions and for ambitions for territorial expansion, whether through purchase or conquest. Many Northerners embraced the idea of the Great American Desert, a wasteland that would halt the spread of slavery. Meanwhile in the South, plantation owners were relentlessly intent on expanding their slave and cotton empire to the Pacific Ocean. Their climatic fixation originated in part from the fickle nature of their cash crop—cotton. Cotton is highly climate sensitive, requiring large and regular amounts of rain and an average temperature no lower than 25 °C/77 °F, day or night. As a result, cotton could only be grown in the Deep South, from the Carolinas through Georgia and Alabama, the Mississippi Delta, and later in East Texas. Even in this reliably warm and humid region, only short-staple cotton could survive. This cotton, with shorter fibers, was less profitable and more labor-intensive than the long-staple cotton, which could be grown in more tropical climates. Eli Whitney's cotton gin of 1793, usually only remembered as a labor-saving device for how it separated cotton seeds and fiber, was in fact a technological climatic adaptation, making possible the profitable processing of short-staple cotton. That profitability also spurred a massive expansion of slavery, both in the numbers of enslaved people and the acreage of land under slave cultivation.

Southern slaveowners flatly refused to accept accounts of an arid, inhospitable West. Instead of the Great American Desert, they imagined a vast plantation garden. This vision would lead them to Mexican Texas, where there was indeed a good climate for cotton cultivation. The government of newly independent Mexico offered a vast land grant in East Texas to Stephen Austin, who led an influx of Anglo-Americans into the area, ignoring Mexican law, which banned slavery. That in turn led to an Anglo slaveowner revolt in Texas, eventually its annexation into the United States, and finally to the US–Mexico War and the loss of more than half of Mexico's territory.

In this moment of triumph, Southern slaveowners would tolerate no bad news. A new government survey was created to delineate the new boundary between the United States and a diminished Mexico. When the leader of the US Boundary Commission, John Russell Bartlett, a New Yorker, reported that much of this land was barren, worthless desert and that the new Southwest had not been worth a war, Southern members of the survey party rebelled. They complained to their members

[6]Washington Irving, *Astoria* (New York: George P. Putnam, 1849), 216–217.

of Congress, and Bartlett was removed from command, and replaced with William Emery, a Southerner, who did not write federal survey climate fiction, but instead claimed that even though the Southwest was more arid than the Southeast, with slavery it could definitely prosper.[7]

The irony of all this is that by gobbling up the Mexican North, the United States undid itself. The question of whether all that new land would be slave or free helped precipitate the Civil War and its catastrophic losses. Yet climatic inventions and illusions did not end with national union. With Southerners no longer represented in Congress, the North pursued its own Western climatic ambitions, using a federal government no longer dominated by the desires and designs of Southern slaveowners. These included deciding the route for the new transcontinental railroad, creating land grant universities funded by the sale of western lands (and the dispossession of Native Americans), and the Homestead Act of 1864. This act was intended to facilitate rapid settlement of Western lands. For a filing fee of $18, settlers could claim 160 acres of land as long as they improved it by building a home and attempting to cultivate it. Kansas, Colorado, the Dakotas, and Montana would see the largest number of filed claims. Many of these individuals and families had previously homesteaded at points further east, in Ohio, Illinois, or Iowa. The High Plains, however, were different. These grasslands were much more arid and experienced much greater extremes of heat and cold, as well as cyclical extended droughts.

John Wesley Powell, a Civil War veteran, explorer of the Grand Canyon and Colorado River, and first head of the United States Geological Survey, warned that eastern-style homesteading would not succeed in this new, harsher West. His reports combined Victorian travelogues and descriptions of scenery with a clear-eyed and decidedly unromantic view of the limited potential of arid western lands for settlement and agriculture. The culmination of his writings was his *Report on the Lands of the Arid Region of the United States* (1879). In it, Powell argued that the arid West called for a new government approach and a new kind of homesteading. Unlike in the East, homesteading land allocations would have to be much larger, and the government would have to plan large-scale irrigation projects: "To a great extent, the redemption of all these lands will require extensive and comprehensive plans … individual farmers, being poor men, cannot undertake the task."[8] Powell thought that the arid West, properly settled through careful planning, could be home to a fairly small population of farmers and ranchers overseeing rangelands and irrigated farms. He urged the federal government to adopt the successful irrigation agriculture that made possible the Church of Jesus Christ of Latter-day Saints (Mormon) settlement near the Great Salt Lake in 1847, and the dry farming practiced by Pueblo

[7] Amy S. Greenberg, "Domesticating the Border: Manifest Destiny and the 'Comforts of Life' in the U.S.-Mexico Boundary Commission and Gadsden Purchase, 1848–1854." In *Land of Necessity: Consumer Culture in the United States–Mexico Borderlands*, edited by Alexis McCrossen, 83–112 (Durham NC: Duke University Press, 2009).

[8] John Wesley Powell, *Report on the Lands of the Arid Region of the United States, with a More Detailed Account of the Lands of Utah*, second ed. (Washington, DC: Government Printing Office, 1879), viii.

Indians in northern New Mexico and the valley of the Rio Grande. Furthermore, he advocated the vast expansion of homesteading tracts, up to 640 acres, and advised settlers to pursue raising livestock for grazing rather than farming crops. In challenging years, livestock grazing could be supplemented with purchased feed, and unlike planted crops, cattle and other livestock could be moved if conditions warranted it. Powell also believed that the regional irrigation culture in Utah would have to be scaled up to a vast federal water management system, building dams, reservoirs, canals, and irrigation projects. In essence, Powell was arguing for a federal hydrological reengineering of the West, one that could triumph over aridity.

Congress and many homesteaders, however, ignored Powell. His plans would take time and cost money. Congress—not for the last time—when confronted with science it did not like, found science it did like. Railroad companies, which had been paid for railroad construction with government land, likewise had no interest in long-term and expensive investments in land they wanted to sell rapidly. Instead, national economic ambitions led politicians, settlers, and railroad executives to seize upon a disastrous climatic myth: the claim that rain would follow the plow. In this climate fantasy, plowing the earth would release moisture into the air. This would then fall as rain, which plowed earth more easily absorbed. This would achieve a permanent transformation of the climate. Though new, this theory harkened back to old ideas from the era of colonial settlement about how landscapes and climates could be transformed through labor. Of course, it was utter climatic fantasy, but appealing fantasy nevertheless, leavened with a very small amount of actual scientific fact. Some scientific experiments at the new land grant universities, focused on agriculture as part of their institutional mission, did in fact demonstrate that plowed earth retained more moisture from rainfall than unplowed land—though agricultural scientists made no claims of climate change based on this finding. Regional "boosters" and land sellers would ignore any such scientific scruples. The popularity of "rain follows the plow" also coincided with two decades of greater-than-average precipitation on the western plains. It seemed the climate myth was proving true.

The myth even spread worldwide, spurring agricultural settlement in the fatally dry Australian interior. Climatic thinking in the United States spread across the Anglophone world and beyond. One of the more fascinating facets of the emergence of climate science in the United States and its preoccupation with the Great Plains and American West was a similar development of state-sponsored science in the Russian Empire, as that expansionist state encountered the Asiatic Steppe and arid Central Asia. As historian David Moon has documented, despite profound differences in economic and governmental regimes, climate scientists in the United States and Russia, and later the Soviet Union, continued a long correspondence and collaboration as two nations with "built-in" territorial empires tried to determine how best to economically and politically assimilate lands alien to European eyes.[9]

[9] David Moon, *The Plough that Broke the Steppes: Agriculture and Environment on Russia's Grasslands, 1700–1914* (New York: Oxford University Press, 2013).

Drought and blizzards in the 1880s would wreak havoc on vast speculative cattle enterprises on the western plains, where millions of cattle died. (Millions of sheep would likewise die in Australia, where the same climate fantasy had taken root.) But even those calamities would not keep settlers at bay. From the 1890s to the end of World War I, new homesteaders were lured by growing crop commodity prices, new farming technology and machinery, and cheap land.

They could also claim to have science on their side. Agricultural scientists at Western land grant universities promoted dry farming, which used deep plowing and early planting to try to coax a crop out of the soil before the heat and drought of summer descended. Huge numbers of European immigrants, with no experience with the climate of North America, were lured by tracts making wild claims like one, excerpted and reprinted in German, asserting that the soils of the valley of the Platte River in Nebraska were more fertile than that of the Rhine or Nile.[10] Settlers poured in. The United States Bureau of the Census famously declared the frontier closed in 1890, a statistical factoid much remarked upon by historian Frederick Jackson Turner in his 1893 "The Significance of the Frontier in American History" essay, but more Americans would homestead between 1890 and 1920 than in all previous decades. New farm machinery and "dry farming"—relying on deep plowing and ground moisture to coax out an early crop in arid regions—made these marginal lands more promising. A huge influx of European immigrants, utterly unfamiliar with the climate of the western plains, also took up farming in this marginal region. This led to yet another speculative boom on the High Plains, one that would crash first in the agricultural depression that followed World War I, and then, by the early 1930s, in one of the most infamous environmental disasters in US history, the Dust Bowl.

New National Forests and Fire Policies

In the same decades, an effort to conserve timber, forests, and watersheds led to obsessive efforts to prevent wildland fires in the new National Forest System and the United States Forest Service created in 1905. For conservationists, fire was dangerous and an unconscionable waste. We might imagine the new Forest Service was preoccupied with fighting fire in rural or mostly uninhabited regions, but in reality, the flashpoints for fire control and fire policy early in the agency's history were most pressing in the most inhabited places. In a premonition of things to come, wildfires, lapping at the edges of expensive Southern California suburbs or mountain resort communities in the twenty-first century, actually threatened wealthy suburbanites early in the history of the US Forest Service. Fire policy evolved in no small part in the San Gabriel Mountains, northeast of Los Angeles, in affluent suburbs including

[10] Charles Dana Wilber, The Great Valleys and Praries of Nebraska and theNorthwest (Omaha, NE: Daily Republican Print, 1881), 34; 37.

Pasadena and Altadena, where homeowners, most recently relocated from the Midwest, abutted highly flammable forested mountain landscapes.

For that matter, early fire policy explicitly concerned urban areas. Though little remembered today—primarily because it occurred the same day as the more infamous Chicago fire of 1871—one of the greatest losses of life of any natural disaster in the United States was the huge Peshtigo Fire, which burned through timbered lands where logging debris, slash-and-burn agriculture, and railroad construction crews using fire to clear routes conspired to create a giant tinderbox after a hot, dry summer. The fire grew uncontrollably, ultimately burning 1.5 million acres in Wisconsin and Michigan and completely obliterating the small city of Peshtigo, claiming 800 lives there and at least 1200 lives in all.[11]

On October 20, 1918, another terrifying fire struck the Midwest, after yet another hot and dry summer. Apparently ignited by a passing train, the fire roared through land that had been logged, devouring small communities and ultimately entering the city of Duluth. At least 1000 people were killed.[12] Events this large and deadly could not be minimized or ignored. They gave impetus to ever more aggressive fire control and fire suppression policies.

But for western forests, which evolved to survive in a fire ecology and tended by Indigenous residents who set thinning fires, the United States Forest Service fire policy proved disastrous. In forestry, as in agricultural land policy, government policy shaped by economic concerns rather than ecological or climatic reality proved calamitous, growing even more dire as the twenty-first century progresses and wildland fires grow more fearsome. Conservationists, imagining that they were acting in the name of the greatest good for the greatest number and rejecting wanton waste of natural resources, might well be appalled by what they wrought—fires that now threaten forests, cities, and suburbia, worsening an "urban-wildland interface" made flammable. While the number of fires and threatened homes has grown, in some ways, what is most striking is the continuities of these fire disasters. Even before the creation of the US Forest Service, they threatened existing communities and were products of an inhabited landscape already being turned into property and commodities, rather than potential events in landscapes still being surveyed and comprehended.

More Climate Science and More Climate and Weather Denial

In one of the deepest contradictions of climatic thinking in the United States, the disastrous adoption of climate fantasy occurred at the same time the natural sciences began to apprehend the realities of climate, and the federal government of the United

[11] Kim Estep, "The Peshtigo Fire," *Green Bay Press-Gazette*, reprinted by the National Weather Service; https://www.weather.gov/grb/peshtigofire

[12] The Moose Lake and Cloquet Fires of October 1918," National Weather Service; https://www.weather.gov/dlh/MooseLake_Cloquet_1918Fires

States rapidly developed the agencies and apparatus of state-sponsored science. Already, in 1856, American Eunice Newton Foote determined that atmospheres with different compositions of gases retained sunlight-derived heat to different degrees. Those with more carbon dioxide got hotter. This research, presented at a science conference sponsored by the Smithsonian by a male delegate because she was not allowed to address the conference herself, laid out the basic physics of atmospheric climate change. Foote did not understand how this process could become anthropomorphic through human industrial omissions, and later scientists ignored her findings—or used them without attribution.[13]

By this point, the federal government was already interested in climate and weather. In 1849, the Smithsonian Institution provided weather-measuring instruments to telegraph companies and organized a weather-reporting network. One hundred and fifty volunteers were providing weather information to the Smithsonian regularly by the end of that year, and by 1860, five hundred stations furnished daily weather reports via telegraph. Interrupted by the Civil War, by 1869 a new telegraph service, headquartered in Cincinnati, began collecting weather data and producing weather reports. In 1870, Congress passed a resolution requiring the Secretary of War "to provide for taking meteorological observations at the military stations in the interior of the continent … and for giving notice on the northern lakes and on the seacoast, by magnetic telegraph and marine signals, of the approach and force of storms." President Ulysses S. Grant signed it into law, and though it was not yet a full-fledged federal agency, a new national weather service had been born within the US Army Signal Service's Division of Telegrams and Reports for the Benefit of Commerce. (Though unsurprising, it is interesting to note how explicitly economic in orientation this new weather-data–collecting entity would be.) In 1890, it became a civilian agency when its responsibilities were shifted from the military to the newly created US Weather Bureau in the Department of Agriculture. By 1898, the Weather Bureau, like the nation, would expand geographically beyond its continental confines, when President William McKinley ordered the Weather Bureau to create a hurricane warning network in the Caribbean.[14]

An avid imperialist, McKinley was acting to protect American interests in the Caribbean, such as sugar plantations in Cuba, as well as the coastal United States, but his interest in hurricane prediction proved premonitory for one of the greatest natural disasters in US history. On September 8, 1900, what would later be classified as a category four hurricane hit Galveston on the Texas coast. The town had rebuffed suggestions to build a seawall, and city government, already hobbled by corruption, proved utterly and catastrophically ineffective. The total death toll was estimated at 6000, including many townspeople who were swept out to sea by the

[13] Eunice Foote, "Circumstances Affecting the Heat of the Sun's Rays," *The American Journal of Science and Arts* Second Servies, 22 (1856), 382–83; Anne K. Armstrong, Marianne E. Krasny, and Jonathon P. Schuldt, *Communicating Climate Change: A Guide for Educators* (Ithaca: Cornell University Press, 2018), 10–11.

[14] "History of the National Weather Service," National Weather Service, National Oceanic and Atmospheric Administration, https://www.weather.gov/timeline

storm surge. After the water receded, dead bodies were piled on barges pushed out into the ocean. This proved spectacularly ineffective, and soon remaining residents were afflicted with the horror of dead bodies washing back ashore, threatening the survivors with disease. Galveston was a stark reminder that while hurricanes were natural disasters, their damage was exacerbated by human choices and actions—poor planning and the corruption of local government. Galveston would serve as a rallying cry in the push for more effective and less corrupt urban governments in the United States during the Progressive Era. It did not, however, lead to major changes in the federal disaster response.[15]

Over a few decades, the federal government had organized and systematized the collecting and reporting of instrument-based weather data and began generating weather forecasts. It had also witnessed ample proof for why this data, and the weather it could predict, was so valuable. Yet even this new scientific bureau was hobbled by a kind of climate denial or erasure. Tornadoes occur with greater frequency in North America than anywhere else. In the late nineteenth century, they were poorly understood, difficult to predict, and utterly terrifying. As a result, in 1883, the Telegraph Signal Service (later the Weather Bureau) was banned from forecasting them, blaming damage on them, or even mentioning them by name. For a weather phenomenon that was terrifyingly destructive but poorly understood, predictions were dubious at best. Then, as now, repeated weather predictions warning of danger that proved incorrect increased the likelihood that accurate ones would be disregarded. More grimly, perhaps government officials thought that there was no point in causing mass panic with erratic tornado forecasts, and most citizens, especially rural ones, would still have no warning until they saw a funnel or heard the terrifying roar.

This ban would remain in place until 1938, and forecasters would continue to avoid the term through the 1940s. This finally began to change in 1948 when a tornado severely damaged an Air Force base near Oklahoma City. The Air Force demanded more reliable forecasts of tornadoes and received them. These even included predicted tornado watch areas, familiar to any American television viewer since the 1950s. Finally, the federal government was willing to forecast tornadoes, but only after climate science and the innovation of radar made forecasting possible.[16]

Until then, however, scientific fact could not compete with weather denialism, or, if one wishes to be more charitable, a desire to manage public fears and panic through selective and self-censoring forecasting. Just as the federal government and railroad companies colluded to assuage the concerns of prospective settlers, forecasters were ordered to manage fear of terrifying natural disasters. The fact that one of the most famous children's novels ever written in the United States, L. Frank Baum's *The Wonderful Wizard of Oz* (1900) would make a tornado its central plot

---

[15] Jessie Kratz, "Herald of the Storms: Isaac Cline," National Archives "Pieces of History," https://prologue.blogs.archives.gov/2015/11/05/herald-of-the-storms-isaac-cline-2/; Erik Larson, *Isaac's Storm: A Man, a Time, and the Deadliest Hurricane in History* (New York: Random House, 2000).

[16] Steinberg, *Acts of God*, 152–53.

device by delivering Dorothy Gale to Oz was proof that whatever the government might say or not say, tornadoes still loomed large in the national imagination.

Reclamation: Looking Backward and Forward from Roosevelt Dam

By the later nineteenth century, the federal government had begun to accumulate a systematized infrastructure for collecting weather data. By the early twentieth century, it would begin to construct actual infrastructure to cope with the realities of the arid West. Finally, the federal government had begun to adopt some of the ideas of John Wesley Powell. The Newlands Reclamation Act of 1902 funded the construction of dams and irrigation projects in 16 arid and semiarid states: Arizona, California, Colorado, Idaho, Kansas, Montana, Nebraska, Nevada, New Mexico, North Dakota, Oklahoma, Oregon, South Dakota, Utah, Washington, and Wyoming. In 1906, Texas was added to the states covered by the act. The first project undertaken was a dam on the Rio Salado, the Salt River, in southern Arizona, which facilitated the growth of metropolitan Phoenix. This dam, soon to be named Roosevelt Dam in honor of President Theodore Roosevelt, is a fascinating place to pause and look backward and forward in time.

Historians tend to draw a sharp divide between federal initiatives of the nineteenth century, intended to encourage farming and settlement, and federal initiatives of the twentieth century, often intent on facilitating urbanization, suburban growth, and homeownership. Yet this divide is illusory, and the transition from a government intent on creating a nation of farmers to a government intent on creating a nation of homeowners is more of a gradual transition—and even more of a continuum—than a change from one ideology to another. For that matter, though reclamation was an ideology focused on modern technology and engineering, it was filled with far older ideas, even still including the earliest ways colonial American settlers had perceived land and climate and imagined their beneficial transformation through labor. Through ingenuity, industry, and enterprise, even the most hostile lands could be transformed and "reclaimed" to fulfill their highest purpose: abundant and prosperous homelands for Anglo-Americans.

Reclamation was an idea imbued with far more than just aspirations of irrigated agriculture. Reclamation was intended to bring new and "better" inhabitants and more profitable and healthful crops. With the completion of the dam, the Salt River Valley was rebranded as the Valley of the Sun, and it played a key role in the development of Phoenix. The Newlands Reclamation Act was conceived to foster farming and farmers, but was still profoundly influential in a later era, when government policy had shifted to create a nation of homeowners. Irrigated cotton fields might seem very different from the FHA-financed suburban housing tracts that replaced them by the 1950s and after, but they were linked by the waters held by the same dam and by the complex ideas that undergirded reclamation.

While a growing chorus supported private and public irrigation schemes, there were significant divisions within the irrigation movement, or what would come to be called "reclamation." Two individuals personified this divide: John Wesley Powell and William Ellsworth Smythe. For more than two decades, Powell had warned of the dangers of unplanned settlement and argued for exactly the kind of large-scale dam and irrigation project construction and infrastructure the Reclamation Act would facilitate. Even so, he did not imagine the act as utterly transformative or facilitating a vast urban future. Instead, Powell thought that the arid West, properly settled through careful planning, could be home to a fairly small population of farmers and ranchers overseeing rangelands and irrigated farms.

This modest, slow, and carefully planned future did not appeal to some Americans, who thought that irrigation could be transformative—it could utterly remake the West into a place as populous and agriculturally productive as the East. Here, even centuries later, there remained that undercurrent of Colonial Era thinking, imagining blessings bestowed through good works: that American industry could transform the landscape and allow it to ascend to its highest form, land suitable for farming and homes, inhabited by White Americans.

A vociferous champion of this view was journalist William E. Smythe. Smythe agreed with Powell on the importance of irrigation. However, he saw irrigation as a *deus ex machina*, a wondrous solution that could transform the desert into a garden, make farms productive, and create cities as large as any in the East. His magnum opus—printed as a book and in serial or excerpted form in numerous regional and national magazines, was *The Conquest of Arid America* (1900). In it, he essentially resuscitated Manifest Destiny for a new century. Instead of divine providence, irrigation technology was now the force that would lead Americans to their continent-conquering destiny. For Smythe, the Mormon settlement in Utah had created one of the "real utopias of the arid West."[17] He likewise praised the growth of Southern California, which had transformed from a rural backwater in the 1870s to a booming urban and agricultural region in the early twentieth century. He toured the country extolling the virtues of irrigation, telling fervent believers that irrigation had freed them from climate. He even claimed that aridity was a virtue, because it forced individualistic Americans to work together. Better yet, aridity "compels the use of irrigation. And irrigation is a miracle!"[18] Originally conceived as a way to combat the realities of climate, irrigation, according to Smythe, could now transcend it and render it impotent to stop American progress and profits.

Irrigation would create wealth, encourage civilization and science, and replace isolated farms with irrigated cities and farmlands interspersed together, merging city and country into an irrigated utopia. He would later champion the "Little Landers" movement, which aimed to place American families on irrigated suburban home lots where they would both reside and farm for subsistence or additional

[17] William E. Smythe, *The Conquest of Arid America*, New and Revised Ed. (New York: Macmillan Company 1907), 49.

[18] Smythe, *Conquest of Arid America*, 40.

income. That would fizzle, but in broad outlines, the new landscape he envisioned, in its sprawling decentralized form, also looked very much like future Sunbelt suburbia. Readers eagerly bought what Smythe was promising. So did the nation's political class.

Pressure mounted for the federal government to undertake storage and irrigation projects. Congress had already invested in America's infrastructure, and Westerners wanted the federal government also to invest in irrigation projects. The irrigation movement demonstrated its strength when pro-irrigation planks found their way into both Democratic and Republican platforms in 1900. Congress passed the Reclamation Act on June 17, 1902. It created the United States Reclamation Service within the US Geological Survey. The Act required that water users repay construction costs for projects they benefited from, with residents of arid regions ultimately paying for local reclamation.

Not surprisingly, we remember the term "reclamation" because it is in the act's name. The most vocal sponsoring congressman's name, Francis Newlands, "New Lands," also seems felicitous. Yet another term, "homemaking" was also a key argument for supporters of reclamation and appeared repeatedly in their rhetoric. Irrigation's supporters believed reclamation programs would encourage Western settlement, making homes for Americans on family farms. This was also a rhetorical shift from "homesteading," a term that of course also appeared repeatedly in the language of the act. And in the longer term, the majority of those homeowning families would be suburbanites rather than homesteading farmers.

Smythe crowed that the Reclamation Act "is perhaps the only measure in the history of American legislation enjoying a popularity so absolute and unquestionable that the only possible controversy between the two parties was as to which was entitled to the greater credit for bringing it to pass."[19] President Roosevelt even said that it was the single most important piece of legislation he signed in his entire presidency. The first project completed under the new law, on the Salt River in Arizona, was named for Roosevelt. Anglo settlers renamed the Salt River Valley the Valley of the Sun, and metropolitan Phoenix was born. By 2010, Phoenix—in the middle of one of the driest deserts on earth—had surpassed Philadelphia to become the fifth-largest city in the United States. Powell likely would have found this unsustainable and foolhardy, but in some ways, his vision of the West had lost, and Smythe's had won.

It is worth noting that Senator Newlands, a Democrat from Nevada, had his own history of minimizing natural hazards and environmental risks. He was a large real estate investor in San Francisco, and after the catastrophic 1906 quake, he was one of the loudest voices in the political and business establishment, claiming that it had been fire, rather than quake, that had demolished large parts of the city. He even went so far as to claim that only three percent of the damage had been caused by the earthquake, and everything else had been caused by fire. (His source for this dubious statistic, if there was one, is unknown.) Earthquakes were mysterious acts of

[19] Smythe, *Conquest of Arid America*, 287.

God, unpredictable and poorly understood. Urban fires, on the other hand, could be combatted with better-equipped fire departments, a bigger network of fire hydrants, and improved building codes. Under a fire regime, disaster could be planned for and prevented, and property could be preserved, its wealth continuing to accumulate through the magic alchemy of American real estate values. If earthquakes were a major hazard, however, continued investment in San Francisco was a dubious choice. Newlands' claims might be suspect, but his prioritization of profit over environmental realities or an accurate assessment of environmental hazards was entirely consistent with long-standing attitudes in the United States.[20]

Supporters of the act could point to Roosevelt Dam as an important early success. From 1902 to 1907, the Reclamation Service began dozens of projects in Western states. Yet many projects encountered problems: soils unsuitable for irrigation; land speculation that leapfrogged and disrupted development plans; irrigators and landowners who could not meet repayment plans due to unexpected costs; settlers who were inexperienced in irrigation farming; and even waterlogging of irrigable lands that then required expensive drainage projects. Good works in the form of reclamation were supposed to bring blessings, better lands, and prosperity, but these ideals could not always triumph over the realities of Western climate and landscape.

The difficulties encountered in the early years resulted in changes to the act. In 1923, the agency was renamed the "Bureau of Reclamation." In 1928, Congress authorized the Boulder Canyon (Hoover Dam) Project, and large appropriations began, for the first time, to flow to Reclamation from the general funds of the United States. Now, reclamation would depend not on repayments from irrigators, but rather on the economic and political power of the federal government of the United States. The huge New Deal dams on the Colorado, the Columbia, and elsewhere, the behemoth landmarks of the Bureau of Reclamation at its apogee, were all byproducts of this new federally funded era of reclamation, projects built on a scale far beyond the ability of individual farmers to ever repay. Though they were decidedly place-based, located where rivers could be dammed to create huge reservoirs to feed the potable water, irrigation, and electricity needs of the modern West, they nevertheless "de-placed" aridity. Now, all American taxpayers would pay for reclamation, not just western farmers.[21]

So what about that first project, the Roosevelt Dam on the Salt River? In terms of facilitating agriculture, economic development, and urban growth, it was certainly a success. Yet the dam and its consequences also reflected relations of power in Arizona and the Southwest. Indigenous people would be forced out by the reservoir created by the dam. While we remember the calamity of Allotment and the mass transfer of tribally held reservation lands to individual property owners—and that so

---

[20] Steinberg, *Acts of God*, 32–33.

[21] William D. Rowley, *The Bureau of Reclamation: Origins and Growth to 1945*, Volume 1 (Washington, D.C.: Government Printing Office, 2006) provides an institutional history of the Bureau. Marc Reisner's *Cadillac Desert* provides a far less flattering but incisive history of the agency and water in the West.

much reservation land fell into Anglo hands as a result—we cannot forget just how much reservation land and other landscapes that Indigenous people had used for subsistence, and places with material and cultural meaning, were lost to dams, reservoirs, irrigation projects, and settler dreams and schemes.

Here, at the very beginning, there were clear winners and losers. The same was true at the scale of cities. Predominantly Anglo Phoenix secured federal funding for its dam, while predominantly Mexican Tucson did not, even though Tucson was a far older and much more established community. Phoenix possessed a larger water supply and its predominantly Anglo population more easily gained favor with the federal government. The Roosevelt Dam was the first of multiple federal dam and irrigation projects, amid many other examples of federal largesse, that would permit the growth of Phoenix. With water, federal aid, and relentless promotion, Phoenix surpassed Tucson in population by 1920.

The community and its boosters (not unlike William Smythe) also possessed an irrepressible showboating flair worthy of the most shameless Western boosters. Like Los Angeles, Phoenix used water to create the fiction of Midwestern greenery, and even tropical verdure, in an arid environment. Tucson had embraced conservation and an awareness of scarcity early on. Phoenix instead emphasized unlimited growth and the assumption that technology and federal money would always provide water. While in hindsight, Tucson's approach was obviously more environmentally sustainable, it was perhaps not so obvious in the heady days of federal reclamation and wonders such as Hoover Dam. American will and technological know-how, it seemed, could conquer aridity and the desert.

The Phoenix-Arizona Club, created in 1923, sold the city in national magazines such as *Better Homes and Gardens* as "Delightful Phoenix, the Garden Spot of the Southwest." Local boosterism enlisted homeowners as well. Phoenix homeowners were urged to plant grass, roses, and other Eastern or European ornamental landscaping. New modern resorts, like the Arizona Biltmore Hotel, served as nuclei for surrounding tourist, residential, and commercial developments.

Phoenix would grow from a group of settlements and winter resorts to a booming regional metropolis. It also continued its tradition of seizing on federal largess whenever possible. In the early twentieth century, this facilitated the development of water and irrigation. By the 1930s, 1940s, and 1950s, federal money would instead be used to construct houses. When President Franklin Roosevelt unveiled the original National Housing Act in 1934, which would later be greatly expanded as the FHA loan program, many bankers and builders reacted with caution or hostility.

In Phoenix, however, the reception was warmer. Del Webb, the owner of a large construction company and a housing developer who would later create the "active adult" retiree community of Sun City, foresaw the potential of these federal loans to aid homebuyers, banks and builders, and metropolitan Phoenix. Webb's firm eagerly and indiscriminately snagged government contracts, particularly during World War II. One of the largest was a contract to construct the Poston Internment Camp for Japanese Americans, which at full capacity was the third largest "city" in Arizona. For Webb, and for Phoenix, the FHA program was yet another federal windfall. "Homemaking" was a central goal of the Reclamation Act. Ranch houses on the suburban frontier might be a world away from rural farmsteads, but they were very much a product of the original intent of the Reclamation Act. The nature of home-making could change, from homesteading farmers to new families in suburban starter homes, but both are reminders of the cultural continuities and environmental wishful thinking that underlay successive pieces of federal legislation and major federal initiatives over time. Here, it seemed that big science and big engineering sponsored by a powerful national government had finally triumphed over the harsh climate of the American West and erased the Great American Desert once and for all. Now, farms, cities, and suburbs could prosper.

Migrating Toward Hazard and Disaster in the Sun Belt

Combined with the US victory in World War II, cementing the United States as a superpower for the remainder of the twentieth century and unleashing the Atomic Age, Americans might have felt that in some sense Manifest Destiny had indeed come true. They had become the "great nation of futurity" on a scale antebellum Americans could scarcely have imagined. The irony of this is that in the postwar era, a population that spent generations thinking about, arguing about, and worrying about climate stopped thinking about it entirely. With colossal dams and the advent of air-conditioning, climate was now simply a benign commodity or amenity, the balmy summers, mild winters, and recreational assets of the booming suburbs of the Sunbelt from Florida to California.

Yet the Sunbelt migration, far larger than the homesteading migration had ever been, delivered huge numbers of American homeowners to places of dire climatic peril. In the nineteenth century, the federal government of the United States tried to create successful farmers. Now, it wanted to create middle-class suburban home-owners. With federal FHA and VA loans, they bought new homes in hurricane-prone Florida and Texas, the drought-prone Southwest, and fire-prone California. This was in its own way a product of a newer kind of pernicious climate fantasy, one that imagined technology had rendered climate impotent.

In these same decades, the American oil industry would actively combat and suppress research in climate science and the growing awareness among scientists of climate change. Richard Nixon would be briefed on the dangers of climate change, but the Republican Party would become handmaidens of climate denial. Oil corporations would use the same tactics tobacco companies had used to obscure the science linking tobacco smoking with cancer and other health problems. The possible future profits of tobacco led the Virginia Company to embrace slavery starting in 1619. Three centuries later, potential future oil and coal profits would lead American energy companies to pursue an agenda that may well prove catastrophic for global civilization. The federal government has proven as complicit in promoting the goals and profits of oil companies as it once did with land speculators and plantation owners. From Jamestown to Exxon, climate reality often succumbed to climate fantasy if there was profit at stake, and even the scientific and physical infrastructures created by the federal government to assess and combat environmental risks and natural disasters would be shaped and sometimes constrained by false ideas and the pursuit of profit. This hard fact underlay government policy over and over, across time and geography. Understanding this history helps us better understand the problematic relationship between the US federal government and US natural disasters.

References

American Philosophical Society. *Historical Meteorological Records at the APS*. Accessed September 23, 2023. https://www.amphilsoc.org/museum/exhibitions/historic-meteorological-records-aps.

"Change of Climate in North America and Europe." [1770?]. Papers of Samuel Williams, HUM 8, Box: 1, Folder: 10. Harvard University Archives. Accessed September 23, 2023. https://id.lib.harvard.edu/ead/c/hua05010c00010/catalog.

Duluth, Minnesota Weather Forecast Office. The Moose Lake and Cloquet Fires of October 1918. National Weather Service. Accessed September 23, 2023. https://www.weather.gov/dlh/MooseLake_Cloquet_1918Fires.

Estep, Kim. The Peshtigo Fire. *Green Bay Press-Gazette*, n.d. Reprinted by National Weather Service. Accessed September 23, 2023. https://www.weather.gov/grb/peshtigofire.

Foote, Eunice. 1856. Circumstances Affecting the Heat of the Sun's Rays. *The American Journal of Science and Arts. Second Series*. 22.

Greenberg, Amy S. 2009. Domesticating the Border: Manifest Destiny and the 'Comforts of Life' in the U.S.-Mexico Boundary Commission and Gadsden Purchase, 1848–1854. In *Land of Necessity: Consumer Culture in the United States–Mexico Borderlands*, ed. Alexis McCrossen, 83–112. Durham, NC: Duke University Press.

Irving, Washington. 1849. *Astoria*. New York: George P. Putnam.

"The Jefferson Weather and Climate Records." *The Papers of Thomas Jefferson*. Princeton University. Accessed September 23, 2023. https://jefferson-weather-records.org/.

Koch, Alexander, Chris Brierley, Mark M. Maslin, and Simon L. Lewis. 2019. Earth System Impacts of the European Arrival and Great Dying in the Americas after 1492. *Quaternary Science Reviews* 207.

Kratz, Jessie. Herald of the Storms: Isaac Cline. *National Archives "Pieces of History"* (blog). Accessed September 23, 2023. https://prologue.blogs.archives.gov/2015/11/05/herald-of-the-storms-isaac-cline-2/.

Larson, Erik. 2000. *Isaac's Storm: A Man, a Time, and the Deadliest Hurricane in History*. New York: Random House.

Mann, Charles C. 2011. *1493: Uncovering the New World Columbus Created*. New York: Vintage.

Moon, David. 2013. *The Plough that Broke the Steppes: Agriculture and Environment on Russia's Grasslands, 1700–1914*. New York: Oxford University Press.

National Weather Service National Headquarters. *History of the National Weather Service*. National Weather Service, National Oceanic and Atmospheric Administration. Accessed September 23, 2023. https://www.weather.gov/timeline.

Powell, John Wesley. 1879. *Report on the Lands of the Arid Region of the United States, with a More Detailed Account of the Lands of Utah*. 2nd ed. Washington, DC: Government Printing Office.

O'Sullivan, John L. The Great Nation of Futurity. *The United States Magazine and Democratic Review* VI (November 1839): 426–430.

Rowley, William D. 2006. *The Bureau of Reclamation: Origins and Growth to 1945*. Vol. 1. Washington, D.C.: Government Printing Office.
Smythe, William E. 1907. *The Conquest of Arid America*. Rev. Ed. New York: Macmillan Company.
Steinberg, Ted. 2006. *Acts of God: The Unnatural History of Natural Disaster in America*. 2nd ed. New York: Oxford University Press.

Nature by Numbers: The National Flood Insurance Program

Uwe Lübken

Abstract In response to ever-growing material damages caused by floods over the last two centuries, societies have sought techniques and methods to prevent, minimize, or at least mitigate the physical and social havoc wreaked by such events. It was not until the middle of the nineteenth century, however, that insurance was added to the toolbox of flood management practices. In the United States, despite the large number of physical flood protection devices such as levees, floodwalls, etc., the amount of flood damage continued to increase in the late nineteenth and twentieth centuries. As a response to these developments, the National Flood Insurance Program (NFIP) was created by Congress in 1968. This article sheds light on the conflicted history of the NFIP and especially the challenges resulting from structural deficits, funding issues, the definition of hazard and, more recently, climate change.

When the four protagonists in John Boorman's Hollywood classic Deliverance (1972) set off on a weekend trip to the mountains of Georgia to take their canoes down the (fictional) Cahulawassee River, it was supposed to be the perfect break from the dull daily routine of the office. For the Atlanta businessmen, this is their last chance to conquer the wild, unregulated, and rapids-strewn river, as work to build a dam is already in full swing, and in a few weeks, a huge reservoir would bury the entire area. Arriving at the river, Lewis Medlock (Burt Reynolds), the daredevil and outdoorsy leader of the quartet, expresses doubts about the fitness of Bobby (Ned Beatty). Ed (Jon Voight) assures him, however, that Bobby is one of the best in his line of work, the insurance business, whereupon Lewis laconically remarks: "Insurance? I've never been insured in my life. I don't believe in insurance. There's no risk."

According to Lewis, then, those who insure themselves live low-risk lives, and one of the last ways to escape this "insurance mentality" is to take on the challenges of untamed nature. Boorman's constructed opposition of rational, bureaucratic risk management and natural processes that run counter to this control apparatus is

U. Lübken (✉)
Ludwig-Maximilian-Universität, Munich, Germany
e-mail: Uwe.Luebken@carsoncenter.lmu.de

N. Rauscher, W. Werner (eds.), *Natural Disasters in the United States*, Risk, Governance and Society 22, https://doi.org/10.1007/978-3-031-96436-7_3

paradigmatic of common notions of the relationship between nature and society in modernity. But if one takes a closer look, it becomes apparent that the actuarial encounters with the dynamics of rivers and oceans are as great an adventure for the insurance industry as the canoe trip was for the four men from Georgia.[1]

The Taming of Chance

Insurance is an innovative instrument for managing risk and uncertainty (Bonß 2013). The prospect of financial compensation after the occurrence of a damaging event generates planning security and frees up capital and energies that would otherwise have had to be spent on individual and collective precautionary measures and coping capacities. For a long time, however, the insurance principle was anything but self-evident. For centuries, the idea of insuring oneself against the economic consequences of a random stroke of fate by paying the price, the premium, was considered sinful. This was especially true of life insurance, for to put a monetary value on the body of a free human being was contrary to both Roman law and Christian morality (Arps 1965, 16; Rotman 1979; Bernstein 1996).

As religious interpretive authority subsided, however, the future was no longer predestined but fundamentally open, and thus the demand for "technologies of risk" increased (Ewald 1991, 198). By merging individual risks, insurance companies were able to statistically manage and thus financially control more and more uncertainties of life—from sudden death to job loss and car accidents.[2] In modern society, insurance is a key element in the "taming of contingency" (Makropoulos 1990; Hacking 2014); for some, the organizing principles of insurance even constitute a new form of social contract, an "insurance society" (Ewald 1989).

Insurance has also become an instrument of protection against the financial risks of natural hazards, albeit with different degrees of success. While the insurance industry has learned to deal very successfully with fire and hail damage and, with some reservations, can also calculate the risk of storms and earthquakes (Oberholzner 2006, Mohr 1963), floods continue to pose an immense and largely unmet challenge. However, it is precisely the exploration of these limits of insurability that makes the history of flood insurance interesting.

[1] Interestingly, no insurance company was willing to insure the extremely dangerous filming, so that the actors, who performed all the stunts themselves, were uninsured.

[2] Francois Ewald (1991) has reminded us that the main goal of insurance is not, as one might think, financial compensation for damaging events of the past, but much rather the management of uncertainty. Furthermore, it should be kept in mind that insurance itself is also a risk. When the hazard that a person is insured against, does not become an actual damage, premiums have been paid in vain (Luhmann 1996).

Flood Insurance: Origins and Challenges

For centuries, private charity and government assistance were the only lifeline for those affected by floods. It was not until the mid-nineteenth century that an attempt was made to add a new instrument to disaster management: flood insurance. The first European policy was issued in 1845 by the Azienda Assicuratrice in Trieste. The Austro-Hungarian company's venture did not last long, but others soon followed suit, especially after severe floods hit large parts of France in 1865 (Rommel 1950, 24f.).

These early flood insurance projects and companies were a response to changing damage patterns after natural catastrophes, especially river floods, in the second half of the nineteenth century. As Stéphane Castonguay and Matthew Evenden (2012: 1–2) have pointed out, industrialization and urbanization have subjected rivers "to a new era of modern engineering and rationalization; new modes of transportation and power generation affected river courses; and the water supply and sewage needs of urban centers altered the flow and ecology of urban rivers as well as distant sources." As a result, rivers have become infrastructural corridors along which industries, roads, bridges, telephone and telegraph lines, and residential quarters were concentrated. These developments changed not only the rivers and the floodplains but also the character of floods themselves. "Floods like these have happened before," the New York Times wrote about the Ohio River flood of 1884: "They have happened, to some extent, when the country was almost untouched by the hand of man. But, unquestionably, the great disasters of recent years are the result of the condition of the earth's surface as modified by human action" (New York Times 1884).

For a long time, the main answer to this challenge was structural flood protection. With the passage of the Flood Control Act in 1936, flood control in the United States was explicitly placed within the purview of the federal government, which had previously pursued it only sporadically. But despite the construction of hundreds of new levees, dams, protective walls, and a few detention basins, economic losses from flooding continued to rise steadily in many parts of the United States, and even faster than spending on flood control measures. With each new flood, it became clearer that the flood problem could not be solved by concrete and heaped-up earth alone (HUD 1966, 44, Luebken 2007).

The financial burden of the federal government was further increased with the enactment of the Disaster Relief Act in 1950. With this legislation, the federal government became systematically involved in help for disaster victims, rather than just on a case-by-case basis as before (Rozario 2007, 150–55, Platt 1999, 70). "Whenever a major disaster strikes anywhere in the United States today," the Department of Housing and Urban Development noted in 1966, "substantial public help and relief are available" (HUD 1966: 54). Also, this availability of relief tended to diminish the risk awareness of businesses and homeowners in floodplains. Furthermore, structural flood protection had a counterproductive effect in part by evoking a false sense of security and promoting development behind levees (Platt 1999, 70; Peralta and Scott 2020). Development in hazardous locations did not help either. By 2010, the number of Americans living in coastal shoreline counties had increased by 40% over the previous four decades (Knowles and Kunreuther 2014, 327).

The combination of ineffective yet costly flood protection and increasing aid payments generated massive pressure to act, especially since floods, as a HUD report marked in 1966, were "by a wide margin the most serious of the natural disaster hazards facing the people of the United States" (20 f.). Since the early 1950s, therefore, plans and ideas for the establishment of flood insurance had been circulating repeatedly in the United States (Langbein 1953, Grutzner 1955), influenced not least by the findings of a group of researchers led by the Chicago geographer Gilbert F. White, who called for a more comprehensive approach to solving the flood problem (Platt 1999, 70, White, 1945; Hinshaw, 2006). Insurance in particular seemed to be an ideal tool to provide relief to those affected by floods while promoting responsible use of floodplains (Langbein 1953, 330, Rozario 2007, 170).

However, any flood insurance scheme, whether private or government funded, had to deal with several substantial challenges that were well known for a long time. When, for example, agricultural circles in Prussia demanded protection by flood insurance after a series of devastating inundations in the late nineteenth century (Annalen des Deutschen Reiches 1899a, b), the Prussian Insurance Advisory Council did not rule out flood insurance in principle, but it considered it "not feasible at present" because it contradicted "two fundamental principles that are indispensable for insurance" (Annalen des gesamten Versicherungswesens 1899b): the lack of statistical material and the spatial concentration of bad risks. Insuring against natural catastrophes without extensive information about the specific hazard resembled a "leap into the dark," as the Association of German Private Fire Insurance Companies noted (Verband 1898). The verdict of the trade press was even more negative. One important journal even demanded that the government forbid private insurance companies "to engage in such a dangerous, completely unpredictable business, since it could only endanger the safety of those insured against fire and hail damage" (Annalen des gesamten Versicherungswesens 1899a; Deutsche Versicherungs-Zeitung 1900).

While the lack of knowledge about floods could be (and partially has been) overcome by collecting data and modeling future floods, the second problem was of a more substantial nature. Tornadoes, earthquakes, or hailstorms threaten entire regions and thus create a large hazard community without clearly defined boundaries in which no one can be sure of not being among the victims when the event occurs. In the case of floods, however, it is not uncommon for the degree of hazard to be expressed in feet. In Wareham, Massachusetts, for example, the distance from hazard zone A, where floods occurred on average every five years or more, to hazard zone D, which flooded on average only every 25 to 50 years, was less than 500 feet. In narrow river valleys, these distances could be even less. (HUD 1966, 28 f.)

This spatial concentration meant that only a fraction of all households in the United States were exposed to significant risk from flooding, and within this small group, the burden was distributed extremely unevenly. Less than 1% of these households bore half of all flood risks, while the total residual risk was distributed among

10% (HUD 1966, 28 f.). It also meant that flood insurance schemes literally attracted bad risks, i.e., those properties with a high likelihood of being flooded. Companies that offered flood insurance policies therefore needed a relatively large capital stock, since losses accumulated quickly (Gerdes 1963, Kalenberg 1998, 6–8). Even the absence of floods was problematic. Several "event-free" years caused interest in insurance coverage to drop dramatically, as the few companies that offered flood insurance policies in the United States soon found out (American Insurance Association 1966). In fire insurance, on the other hand, "thousands of fires occurring daily with fairly regular distribution over large areas" kept interest alive. "It was only the realization of the constant danger that encouraged the wide spread of insurance against fire damage," the private German fire insurers noted as early as the end of the nineteenth century (Verband 1898).[3] This dilemma also haunted the National Flood Insurance Program (NFIP), which was formally introduced in the United States in 1968. While a series of disasters in 1964 and 1965, especially Hurricane Betsy, fostered public and Congressional support for a flood insurance project, "the need for severe disasters to spur legislation has made it difficult to sustain policy implementation when nature is not creating havoc" (Knowles and Kunreuther 2014, 345).

Unsurprisingly, the history of flood insurance on a private-sector basis in the United States, as in Europe, is a history of failure. In Cairo, Illinois, in 1899, the business model of a joint-stock company founded 2 years earlier to insure against floods sank in the Mississippi floodwaters together with the company buildings (Rommel 1950, 25, Grutzner 1955). In the 1920s, several companies offered flood insurance; however, their activities were suddenly interrupted when severe floods hit first the Mississippi Valley and then New England in 1927 (Barry 1997; Hutchinson 2000).[4] Early experiments in Switzerland and Austria did not fare better. The Austrian company Providentia had begun operations immediately before World War I. During its four-year existence (1914–1917), the company sold 53 policies and collected about 100,000 crowns in premium payments. Policyholders reported a total of 20 claims, which resulted in payments of 190,000 crowns for the Vienna company. The Swiss National Insurance Company did not suffer as much from current losses as its Austrian counterpart, but the scale of the Helvetic enterprise remained small and unprofitable. After a short time, both companies abandoned their flood insurance activities (Münchner Rück 1927; Quante 1926).

By mid-century, the insurance industry in the United States and in other countries had repeatedly made it clear, with reference to these structural deficits, that in their view flood damage was not insurable and that no initiative could be expected from their side (American Insurance Association 1966; HUD 1966, 64). However,

---

[3] For the transformation of urban conflagrations from a catastrophic (in insurance terms) into a manageable risk in the nineteenth century, see Frost and Jones (1989); Bramwell (2023), Wermiel (2012).

[4] For the end of the flood insurance business of the Globe & Rutgers Fire Insurance Company as well as the Hartford Fire Insurance Company, see *New York Times* (1927a) and *New York Times* (1927b).

as the chain of severe flood disasters - both coastal and riverine - did not break, eyes turned more and more to the federal government.

The National Flood Insurance Program

Initial tentative attempts by the Truman administration to introduce a national insurance fund after record floods in Kansas and Missouri in the summer of 1951 failed (Langbein 1953, 328), but a series of other flood disasters lent urgency to the arguments for introducing flood insurance (Gerdes 1963). In 1955, it was hurricanes in the northeastern United States and flooding in California that brought the issue back on the table. After a comprehensive report was prepared by the staff of the United States Senate Banking and Currency Committee (1956), the Federal Flood Insurance Act was passed in August 1956, providing for an insurance system subsidized by the federal government, but to be carried out by private companies. Congressional funding failed to materialize, however, because of doubts about its feasibility. In 1957, the Federal Flood Indemnity Administration was dissolved (HUD 1966, 65, King 2005, 6 f.). The fact that a government insurance program was nevertheless established a good ten years later was primarily due to Hurricane Betsy, which left severe devastation in the Gulf region in September 1965. The Southeast Hurricane Disaster Relief Act, passed that same year, not only provided financial aid to flood victims but also authorized a feasibility study for a national insurance program. The two resulting comprehensive reports (HUD 1966; United States Task Force on Federal Flood Control Policy 1966) provided the basis for the National Flood Insurance Act, which passed Congress in 1968 (Knowles and Kunreuther 2014).

The basic concept of the NFIP has remained the same from its inception, although the specific design has been subject to considerable change over the decades. Cities and towns identified by the Federal Insurance Administration (FIA) as being at risk of flooding can participate in the program and offer their residents policies subsidized by the federal government in exchange for taking action in those areas to reduce flood hazards. In 1969, the FIA began cooperating with the National Flood Insurers Association (NFIA), an association of more than 100 private property insurers. NFIA administered the policies and settled claims, while the federal government provided financial support for the program and bore much of the risk. To determine whether a community was at risk of flooding, FIA prepared flood hazard boundary maps that roughly delineated the area at risk and arranged for further hydrologic and topographic studies to accurately calculate premiums. The result of this work, which could take several years, was a Flood Insurance Rate Map (FIRM) that assigned different hazard classes and thus different premiums to locations in the community (Power and Shows 1979, 62 f.).

In the early years, the response to the program was poor. When in 1972, "tropical storm" Agnes produced record floods and killed 128 people, there was great surprise that only a fraction of the flood victims were insured. In Pennsylvania, the state that bore the brunt of the storm, a mere 683 people and businesses were asking for compensation under the NFIP. While overall damages amounted to $3 billion, the program paid only $three million in claims. Lack of information about individual flood risks and premiums that were too high despite subsidies were cited as reasons for this shortfall (Platt 1999, 71, Power and Shows 1979, 64; Knowles and Kunreuther 2014, 337). To increase the number of insureds, the voluntary nature of the program was severely restricted in 1973. Proof of flood insurance became mandatory for homeowners who wanted to receive subsidized mortgage loans from the federal government. While only $5.5 billion in value was insured in December 1973, this number grew to $36 billion by November 1977 (Power and Shows 1979, 61). The number of policies issued rose from only 95,000 at the time of "Agnes" to 1.2 million in 1977, while the number of participating communities increased from 2200 to 15,000 during the same period (FEMA 2002).

Problems of the NFIP

When the NFIP was established, the plan was to have the program be self-funding, meaning that the return on premiums would be large enough to cover claims payments. If large catastrophes overwhelmed the NFIP's financial strength, it could borrow from the US Treasury. But this structure repeatedly ran into trouble, accumulating a net loss of $1.1 billion between 1986 and 1997 that was ultimately paid by US taxpayers. On the other hand, policyholders raised $6.4 billion themselves through their premium payments during the same period, so that, according to Platt (1999, 72), the NFIP was still more effective than government aid payments.

One reason for the NFIP's high payouts was "repetitive loss properties" (RLPs). Unlike the private insurance industry, which would cancel a policy if a policyholder had multiple claims in a short period of time, the NFIP continued to insure these bad risks. As a result, in 1998, more than a quarter of all claims to the NFIP came from RLPs, of which, in turn, 37% had received payouts three or more times. Of the payouts, as much as 40% went to these policyholders who had been affected by floods more than once. These included not only middle-class homeowners, but also—and this was of course noted with great interest in the press—quite a few establishments owned by Americans who were not exactly destitute. In Long Island, for example, a bar in a waterfront resort received two payments totaling $752,000 in 1995 and 1996—the largest amount ever paid out by the NFIP in Long Island at that time (Platt 1999, 73).

Participation in the NFIP has also been rather disappointing, despite the 4.5 million policies now in place. This is partly because there were few penalties for violations of the insurance requirement, so homeowners who failed to meet their premium payments dropped out of the program and lost their coverage but did not face penalties. Another reform law in 1994 made lenders more accountable, but "noncompliance" remained a weighty problem with the NFIP (King 2005). Moreover, lenders who were not regulated by the government were no longer subject to mandatory insurance than homeowners who paid for their property out of pocket. Voluntary participation, finally, was also not particularly attractive because the state had proven in the past during natural disasters that it would not abandon victims. Federal disaster relief spending increased from $five million in 1950 to $5 billion in 1999 despite the existence of the NFIP (Hood 2000).

When Hurricane Katrina hit the Gulf Coast in 2005, the inadequacies of the NFIP became fully apparent. In New Orleans, fewer than half of those eligible for insurance had taken advantage of the NFIP, which since 1979 has been run by the newly formed Federal Emergency Management Agency (FEMA). Yet despite these low numbers, the amount of insured damage as a result of the Category 3 hurricane was enough to rock the entire system (Washington Post 2005a, b). Ever since Katrina, the NFIP has been deep in the red. Its deficit increased from 1.5 to more than $20 billion after Katrina and to over $30 billion after Hurricane Sandy in 2012. Judith Kildow and Jason Scorse (2012) argued in the New York Times that the NFIP had become "a fiscal time bomb for the government." In 2017, the combined effect of hurricanes Harvey, Irma, and Maria widened the financial gap even further (Knowles and Kunreuther 2014, 328; Elliot 2021a, 165). "The dream of subsidized flood insurance driving a safety-minded land-use revolution was, if not dead, seriously injured," Scott Knowles and Howard Kunreuther have pointed out (340).

Congress responded to these challenges with several laws. Most importantly, in 2012, the Biggert-Waters Flood Insurance Reform Act (BW12) sought to make premiums better represent the actual flood risk of insured households and businesses. Also, it provided money to update flood insurance rate maps (FIRMs), phased out grandfathered rates, and tightened building requirements (Peralta and Scott 2020, 10–11; Knowls and Kunreuther 2014, 347–48). However, after it had become clear how hard some property owners would be affected by the new law, especially by the increased premiums, protest quickly formed that even included the coauthor of BW12, Democratic Representative Maxine Waters from California, who held in October 2013:

> I certainly did not intend for these types of outrageous premiums to occur for any homeowner. When I agreed to coauthor this legislation, our goal was to create a bipartisan solution to repair our National Flood Insurance Program. Neither Democrats nor Republicans envisioned it would reap the kind of harm and heartache that may result from this law going into effect. (quoted after Elliott 2021a, 156)

Insurance, Displacement, and Climate Change

Two years later, Congress passed the Menendez-Grimm Homeowner Insurance Affordability Act (MG14), which softened the most drastic provisions of BW12 (Abbott 2014, 13–14; Strother 2018, 467). With the new law, premiums did rise more slowly, but more and more insured people were nevertheless feeling the financial pressure of annually rising payments and some decided to sell their homes and move away. Selling a property that is burdened with mandatory flood insurance could be difficult though. Elisa Staton had purchased her small house in Norfolk,

Virginia's Larchmont-Edgewater neighborhood in 2005 for $320,000. The rates for her flood insurance, which she had to have as a recipient of a federally backed loan and with her property located in the 100-year floodplain, were modest. Since then, however, her annual premium has risen to $6000. When she decided to sell, a real estate agent told her that „she'd be lucky to sell the house for $180,000″ (Jarvis 2017). "It wasn't the risk of high waters that spooked buyers; it was the certainty of high premiums", the agent held.[5] Rising premiums are not only driving people out of the floodplains but they also influence people's decisions on where and when to retire. For Marti Beller Lazear, who bought a house near St. Petersburg, Florida, the projected increase of over $6000 annually marked a huge financial burden, even after the house had been paid down (Flavelle 2021).

Flood insurance is not the only actuarial challenge that may drive people away from the coastal shores. Property insurance is also increasingly difficult and costly to get in hazard-prone states such as Florida. The exposure to hurricane risk in particular has increased the cost of homeownership. In the Sunshine State, annual payments for property insurance in 2022 amounted to $4200—three times as high as the national average (Otte 2023). In the aftermath of the devastating Hurricane Andrew in 1992, Florida lawmakers had authorized the creation of several organizations that insured people against property damage who could not get insurance coverage elsewhere. Out of these entities grew the Citizens Property Insurance Corporation (Citizens) in 2002. Citizens is not only structurally similar to the NFIP in that it transfers individual risks into risks shouldered by the taxpayers of the entire state. It also had a similar history. In years without major property damages (which in Florida basically means years without hurricanes), the system worked reasonably well. But after seven insurance companies dropped out of the Florida market in 2022 and 2023, either because of insolvency or as a deliberate decision to minimize their risks after Hurricane Ian, Citizens was swamped with new members and had to increase its rates significantly (Otte 2023).

The displacement effects resulting from the interplay of rising insurance premiums and actual catastrophes are being intensified by climate change and rising sea levels. To remain in highly exposed coastal zones becomes more and more costly, so the number of people who can afford such a lifestyle will decrease. As a result, lower-risk neighborhoods will see an influx of people who will, in turn, displace poorer people (Elliot 2021a, 165). Somewhat ironically, flood insurance, the very device that is supposed to protect people against the consequences of hurricanes, river floods, etc., can significantly contribute to these processes, as Rebecca Elliot (2021a, 180) has pointed out. In addition to the displacement effects of rising premiums, building requirements, which are often an integral part of flood insurance schemes, are becoming stricter and add to the financial pressure on homeowners and businesses. Also, municipalities threatened by sea level rise and an increasing

[5] For Tampa real estate agent Jennifer Zales, rates were supposed to rise from $480 a year to $7417 (Flavelle 2021).

number of severe events find it more and more necessary to tax citizens for the construction of structural mitigation measures such as floodwalls.

An Exceptional Program

The history of the NFIP and especially its financial challenges is often written as the history of a well-intentioned and sound system that suffers from a combination of mismanagement, lack of (flood) knowledge, and political maneuvering. If only the homeowners and businesses in hazardous locations could be made to pay premiums that reflected their real risk and if only the mapping of hazard would be more accurate, the NFIP would work just fine. This assessment, however, ignores several structural problems that fundamentally inhibit the workings of the NFIP—the catastrophic and cumulative nature of the risk to be insured, the uneven distribution of hazards across a given territory, and the political dimension of the NFIP.

Floods, like terrorist attacks and unlike, say, car accidents, are "catastrophic" risks, i.e., the probability of occurrence is low while the extent of destruction is potentially very large (Grace and Klein 2007). Together with the accumulation of bad risks, this creates a residual uncertainty that cannot be managed by standard risk management strategies. If some form of insurance coverage is sought, this "residual risk" must be assumed by the state as the "ultimate risk manager" (Moss 2002), either by forcing people to become members of the risk community by way of a mandatory, all-hazard insurance scheme, as it has been practiced for example in Switzerland, Spain, and France for decades, or by creating and financing its own insurance program, as in the United States (Schwarze and Wagner 2009, 6–8; Lanz-Stauffer and Rommel 1936; Johner 1958; Fischer 2004).

Although mandatory all-hazard insurance is successfully practiced in Switzerland and other countries, it is tied to certain conditions that are rarely met elsewhere, most importantly a geography of hazard that is perceived as reasonably uniform in all regions of a given state. In other words, if one region feels it is unduly subsidizing another region's hazard management, compulsory insurance will be difficult to enforce. As early as 1903, the Neue Preussische Zeitung wrote with regard to the establishment of a government-funded flood insurance scheme: "In our opinion, [state flood insurance] should not appear to be advisable for political reasons, because the state, with regard to the equal rights of its citizens, may not favor the inhabitants of a single area of land, which is preferentially exposed to floods, by means of insurance to be carried out with state funds" (Neue Preussische Zeitung 1903). In the United States, the NFIP faced similar problems despite its mostly voluntary nature. "Somebody on a mountain in Colorado is helping the person in Virginia Beach live on the waterfront," as economist Larry Filer has pointed out (Jarvis 2017).

To the NFIP's credit, the goal of shifting risk from the state to those at risk has been at least partially successful in "normal" crisis years. However, extreme hydrological and meteorological events such as the 1993 Mississippi River flood regularly highlight the limits of resilience. Moreover, it has been shown time and again, most notably in the aftermath of Hurricane Katrina, that the very people who would need help the most are generally not insured.

Both within the US-American context and from an international perspective, the NFIP is an exceptional program. That the NFIP has come into existence at all is astonishing for two reasons. First of all, a system that transfers risk from individuals to the federal government and uses tax money to reimburse a very small part of the population for losses that were in most cases not entirely unpredictable seems to be at odds with fundamental American convictions of individual and public responsibilities. "There is a kind of irony to this outcome, given an observed American ideological preference, across policy domains, for market-based forms of social provision," Rebecca Elliott has noted (2021b, 118). Secondly, of all Western industrialized countries, the United States was alone in establishing a federally backed insurance system that focuses exclusively on floods. Most countries with similar flood problems opted for some kind of mandatory, all-hazard scheme or no program at all (Schwarze and Wagner 2009).

The National Flood Insurance Program has been created in historically unique conditions. Flood damages and relief payments seemed to be on an endless upward trend since the late nineteenth century. Neither structural flood control nor information campaigns about the hazards of living in a floodplain or the relocation of entire cities could solve this problem. Thus, homeowners, cities, states, and the federal government were yearning for innovative approaches to deal with this issue. In this situation, insurance seemed to be a smart way to transfer risk, at least partly to those living in a hazardous location. Furthermore, the creation of the NFIP benefitted from the overall social, cultural, and economic environment in the affluent postwar era when Congress and society at large were still willing to establish and maintain big government-funded programs run by the "sympathetic state" (Dauber 2013). It is probably fair to assume that a mere decade later, the NFIP would not have been created in the first place. With neoliberal mindsets and policies slowly but surely replacing Keynesian approaches since the late 1970s, the focus was more and more on deregulation, market solutions, and small government (even though the size of the federal government continued to grow consistently during this period) (Harvey 2005).

Yet, the question remains why the NFIP has not been swept away by the wave of neoliberal economic policies, especially after it had become clear that it would not enter a path of financial sustainability any time soon. One reason for the persistence of the program is certainly path dependence. Once the system was up and running and distributing benefits in the form of subsidized premiums, reforming or even terminating the NFIP meant increasing the financial burden for home and business owners. Unsurprisingly, any such attempts met resistance not only from the policyholders affected but also from politicians across the aisle, especially in large coastal states with a high number of potential flood victims such as Florida and New York. So far, all attempts to change the NFIP with the goal of making it reflect more accurately the real flood risk have faced intense bipartisan opposition. In September 2021, for example, 38 members of Congress petitioned House Speaker Nancy

Pelosi to block changes to the NFIP that would have considerably increased the premiums for many homeowners. With just three exceptions, all of them represented coastal states (Flavelle 2021). It is interesting to note that in a highly divided country, geography, exposure to natural hazards, and the prospect of financial help from the federal government seem to provide bipartisan common ground.

Lost in Translation?

Apart from the concrete difficulties of the attempts described above to get the "behavior" of nature rationally, statistically, and thus financially under control, a more fundamental problem in the relationship between nature and society comes into view when looking at the history of flood insurance. From the early efforts to insure against river floods and severe weather in the late nineteenth century (Unwetter-Ausschuss 1899) to the NFIP's struggle after the most recent hurricanes in the United States, the limits of insuring against natural catastrophes are evident. The natural dynamics of watercourses and oceans, low-pressure systems, and tropical storms could not (and still cannot) be fully translated into the modern rationality of weather and climate forecasts, mathematical models, insurance statistics, and national budgets. Nature and numbers are not a perfect match, it seems.

Like all insurance against natural hazards, the NFIP ultimately struggled with defining the type of hazard and distinguishing it from other uninsured natural phenomena. "What is a breath of fresh air to one person may already appear as a storm to another," the Association of German Private Fire Insurance Companies already recognized in 1898. That such differences were by no means of a purely academic nature, but could even lead to international entanglements, was most clearly demonstrated after the 1906 earthquake in San Francisco. Whether the destruction was caused directly by the earthquake or indirectly by fire made a considerable difference to the insurance companies involved. While many homeowners were insured against fire, most policies lacked an earthquake clause. This was true not least for the policies of Munich Re, which had a strong presence in the US market (Steinberg 2000, 34, Röder 2006). After Katrina, similar "interpretation problems" arose, only now the question was not earthquake or fire, but wind or water? Since most of those affected were insured against wind, but not flood, under regular property insurance provisions, there was a strong interest in attributing causation to the storm—a conflict that led to quite a few, mostly unsuccessful lawsuits (Grace and Klein 2007).

Conclusion

The instrument of insurance, which had been used so successfully in all sorts of social areas, seemed to offer an ideal approach to deal with the steadily increasing amount of flood damages in the United States. However, purely private-sector solutions have historically failed time and again, and where they exist today, they tend to be cosmetic in nature. As long as the individuals and businesses truly at risk are charged premiums that they are unwilling or unable to pay, flood insurance will not be a real alternative to government assistance. With Ulrich Beck, one can even argue that the "absence of adequate *private* insurance coverage (…) is *the* institutional indicator of the transition to the uncontrollable risk society of the Second Modern Age" (2007, 239, emphasis U.B.), in which the promise of insurance coverage is maintained only in Potemkin villages and the risks are ultimately borne by those directly affected by the disaster. The problem is aggravated by the substantial and constantly changing challenges of climate change. It is doubtful whether, under these conditions, the flood hazard can be modeled in a way that accurately reflects the real flood risk of homeowners, municipalities, and, in the end, the federal government. Thus, as the *New York Times Magazine* has pointed out, both insurance companies and residents "are playing a game of risk and timing" (Jarvis 2017).

References

Abbott, Ernest B. 2014. Flood Insurance and Climate Change: Rising Sea Levels Challenge the NFIP. *Fordham Environmental Law Review* 26 (1): 10–55.

American Insurance Association. 1966. Brief an HUD, New York, 24. Juni 1966, National Archives and Records Administration (NA), College Park, Maryland/Record Group (RG) 311/ Entry 3, B4.

Annalen des gesamten Versicherungswesens. 1899a. 30 (2. November 1899), Versicherung gegen Unwetterschäden.

Annalen des gesamten Versicherungswesens. 1899b. 30 (7. Dezember 1899), Versicherung gegen Unwetterschäden.

Arps, L. 1965. *Auf sicheren Pfeilern: Deutsche Versicherungswirtschaft vor 1914*. Göttingen: Vandenhoeck & Ruprecht.

Barry, J. M. 1997. *Rising Tide: The Great Mississippi Flood of 1927 and How It Changed America*. New York: Simon & Schuster.

Beck, U. 2007. *Weltrisikogesellschaft: Auf der Suche nach der verlorenen Sicherheit*. Frankfurt a.M: Suhrkamp.

Bernstein, Peter L. 1996. *Against the Gods. The Remarkable Story of Risk*. New York: Wiley.

Bonß, Wolfgang. 2013. Risk. Dealing with Uncertainty in Modern Times. *Social Change Review* 11 (1): 7–36.

Bramwell, Lincoln. 2023. Understanding Wildfire in the Twenty-First Century: The Return of Disaster Fires. *Environmental History* 28 (3): 467–494.

Castonguay, Stéphane, and Matthew Evenden, eds. 2012. *Urban Rivers. Remaking Rivers, Cities, and Space in Europe and North America*. Pittsburgh, PA: Pittsburgh University Press.

Dauber, Michele Landis. 2013. *The Sympathetic State: Disaster Relief and the Origins of the American Welfare State*, 2013. Chicago et al.: University of Chicago Press.

Deutsche Versicherungs-Zeitung. 1900. Ein Epilog zur Unwetterversicherung, 28, April 8, 1900.

Elliott, Rebecca. 2021a. *Underwater: Loss, Flood Insurance, and the Moral Economy of Climate Change in the United States*. New York: Columbia University Press.

Elliott, Rebecca (2021b) "Plan B: The Collapse of Public-Private Risk Sharing in the US National Flood Insurance Program," in: Critical Disaster Studies, Jacob A.C. Remes and Andy Horowitz eds. Philadelphia: University of Pennsylvania Press, 116–129.

Ewald, F. 1989. *Die Versicherungs-Gesellschaft. In: Kritische Justiz* 22:385–393.

Ewald, F. 1991. Insurance and Risk. In *The Foucault Effect*, ed. G. Burchell, C. Gordon, and P. Miller. Chicago: Chicago University Press.
Fischer, M. 2004. Solidarität schafft Sicherheit. Die Kantonalen Gebäudeversicherungen als System. In *Katastrophen und ihre Bewältigung: Perspektiven und Positionen*, ed. Christian Pfister and Stephanie Summermatter. Bern, Stuttgart, Wien: Haupt.
Flavelle, Christopher. 2021. *The Cost of Insuring Expensive Waterfront Homes is About to Skyrocket*, 24. Sept: New York Times.
Frost, Lionel E., and Eric L. Jones. 1989. The Fire Gap and the Greater Durability of Nineteenth-Century Cities. *Planning Perspectives* 4:333–347.
Gerdes, V. 1963. Insuring the Flood Peril. *The Journal of Insurance* 30 (4): 547–553.
Grace, M. F., and R. W. Klein. 2007. Facing Mother Nature. *Regulation* 30 (3): 28.
Grutzner, C. (1955) *Flood Insurance: Pros and Cons*. In: New York Times, 28. August.
Hacking, Ian. 2014. *The Taming of Chance*. Cambridge: Cambridge University Press.
Harvey, David. 2005. *A Brief History of Neoliberalism*. Oxford: Oxford University Press.
Hinshaw, Robert E. 2006. *Living with Nature's Extremes: The Life of Gilbert Fowler White*. Boulder, CO: Johnson Books.
Hood, J. 2000. Who Insures Against Floods, and Why. *In: Consumers' Research Magazine* 83 (10): 16–20.
HUD. 1966. *Insurance and other Programs for Financial Assistance to Flood Victims*. A Report from the Secretary of the Department of Housing and Urban Development (HUD) to the President, 8. August 1966, 20, NA/RG 311/Entry 2, B2.
Hutchinson, J. 2000. Disasters and the International Order: Earthquakes, Humanitarians, and the Ciraolo Project. *International History Review* 22 (1/March 2000), 1–36.
Jarvis, Brooke. 2017. When Rising Seas Transform Risk into Certainty. *New York Times Magazine*, April 18.
Johner, W.H. 1958. *Die Elementarschadenversicherung in der Schweiz. Der heutige Stand der Versicherung und die Möglichkeiten eines weiteren Ausbaus* (Inaugural-Dissertation, Universität Bern), Wintherthur: P.G. Keller.
Kalenberg, C. 1998. *Zur Versicherbarkeit von Hochwasser- und Überschwemmungsschäden*. Karlsruhe: Verlag der Versicherungswirtschaft.
Kildow, Judith, and Jason Scorse. 2012. *End Federal Flood Insurance*. New York Times, November 29.
Washington Post. 2005a. *Editorial: Rethinking Flood Insurance*, 21. September 2005.
King, R. 2005. *Federal Flood Problem: The Repetitive Loss Problem*. Congressional Research Service Report for Congress. http://fas.org/sgp/crs/misc/RL32972.pdf (30/07/08).
Knowles, Scott Gabriel, and Howard C. Kunreuther. 2014. Troubled Waters: The National Flood Insurance Program in Historical Perspective. *Journal of Policy History* 26 (3): 327–353.
Langbein, W. B. 1953. Flood Insurance. *Land Economics* 29 (4): 323–330.
Lanz-Stauffer, H., and C. Rommel. 1936. *Elementarschäden und Versicherung*. Bern: Selbstverlag des Rückversicherungsverbandes.
Luebken, U. (2007) 'Der große Brückentod': Überschwemmungen als infrastrukturelle Konflikte im 19. und 20. Jahrhundert. In: *Saeculum. Jahrbuch für Universalgeschichte* 58(1): 89–114.
Luhmann, N. 1996. *Das Risiko der Versicherung gegen Gefahren*. In: Soziale Welt 47: 273–83.
Makropoulos, M. 1990. Möglichkeitsbändigungen: Disziplin und Versicherung als Konzepte zur sozialen Sicherung von Kontingenz. *Soziale Welt* 41:407–423.
Mohr, H. H. 1963. *Grundlagen der Sturmversicherung*. Karlsruhe: Verlag der Versicherungswirtschaft.
Moss, D. A. 2002. *When All Else Fails: Government as the Ultimate Risk Manager*. Cambridge, MA, London: Harvard University Press.
Münchner Rück. 1927. *Münchener Rückversicherungs-Gesellschaft an den Präsidenten des Reichsaufsichtsamtes für Privatversicherung* (Abschrift), München, 20. September 1927, Bundesarchiv Berlin, R3101/17116.
New York Times. 1927a. *Flood Insurance Halted*, 26 April, 1927.

New York Times. 1927b. *Suffering in Hartford Increased by Cold*, November 9, 1927.
New York Times. 1884. *The Ohio Floods*, February 9.
Neue Preussische Zeitung. 1903. *Versicherung gegen Hochwasserschäden*, 25. August 1903.
Oberholzner, F. 2006. Ein Novum der landwirtschaftlichen Risikovorsorge. Die Gründung der Bayerischen Landeshagelversicherungsanstalt 1884. *Jahrbuch für Geschichte des ländlichen Raumes* 3:46–71.
Otte, Jedidajah. 2023. Florida rocked by home insurance crisis. *New York Times*, July 15.
Power, F.B., and E.W. Shows. A Status Report on the National Flood Insurance Program–Mid 1978. *The Journal of Risk and Insurance* 46 (2/1979), 61–76 (61).
Quante, P. 1926. *Hochwasser und Überschwemmungsschäden*, Preussisches Statistisches Landesamt, Berlin, 15. März 1926, GStA PK, I. HA, Rep. 77, Tit. 52, Nr. 124.
Röder, T. J. 2006. *Rechtsbildung im wirtschaftlichen "Weltverkehr". Das Erdbeben von San Francisco und die internationale Standardisierung von Vertragsbedingungen (1871–1914).* Frankfurt am Main: Vittorio Klostermann.
Rommel, C. 1950. *Probleme der Hochwasser- und Überschwemmungsversicherung (Zeit- und Forschungsfragen der Versicherungswirtschaft, Heft 6).* Weißenburg: Mathematischer Tabellen-Verlag.
Rotman Zelizer, V. A. 1979. *Morals and Markets: The Development of Life Insurance in the United States*. New York: Columbia University Press.
Schwarze, Reimund, and Wagner, Gert G. 2009. *Natural Hazards Insurance in Europe?* Tailored Responses to Climate Change Needed, Working Papers in Economics and Statistics, No. 2009–06, University of Innsbruck, Department of Public Finance, Innsbruck.
Steinberg, Th. 2000. *Acts of God: The Unnatural History of Natural Disaster in America*. Oxford: Oxford University Press.
Strother, Logan. 2018. The National Flood Insurance Program: A Case Study in Policy Failure, Reform, and Retrenchment. *Policy Studies Journal* 46 (2): 452–480.
United States Senate, Banking and Currency Committee. 1956. *Federal Disaster Insurance*. Staff Study, 84th Congress, 2nd Session, Report No. 1313, 09/01/56.
United States Task Force on Federal Flood Control Policy. 1966. *A Unified National Program for Managing Flood Losses*. Washington, DC: U.S. Government Printing Office.
Unwetter-Ausschuss. 1899. *Sitzungsniederschrift*, 24. März 1899, Bundesarchiv, R3101/17116.
Verband Deutscher Privat-Feuerversicherungs-Gesellschaften. 1898. an von der Recke von der Horst, "Versicherung gegen Unwetterschäden", Berlin, 6. Dezember 1898, Bundesarchiv, R3101/17116.
Washington Post. 2005b. *Editorial: Rethinking Flood Insurance*, 21. September 2005.
Wermiel, Sara E. 2012. *Did the Fire Insurance Industry Help Reduce Urban Fires in the United States in the Nineteenth Century?* in: Bankoff/Lübken/Sand (2012), 235–53.
White, G.F. 1945. *Human Adjustment to Floods: A Geographical Approach to the Flood Problem in the United States*. University of Chicago: Research Paper 29, Department of Geography.

All of the Above: Energy, Environmental Politics, and US Climate Change Policy Since the 1970s Oil Crises

Andrew C. Isenberg and James M. Turner

Abstract Since the energy crises of the 1970s, the United States' commitment to addressing climate change has been plagued by inaction. Ironically, during the 1970s, when the United States adopted landmark environmental laws in response to the modern environmental crisis, it also expanded its commitment to what is now known as an "all-of-the-above" energy strategy in response to the energy crises. Since then, whether led by Republicans or Democrats, the United States has put concerns about energy ahead of the environment. This commitment was most apparent during President Ronald Reagan's administration in the 1980s, when Reagan dismissed environmentalists' emphasis on conservation and restraint and, instead, began to champion energy abundance as a key plank in the modern conservative agenda. During his administration, the United States expanded offshore oil drilling and western coal mining. That pattern persisted through subsequent Republican administrations, culminating in President Donald Trump's "America First" energy agenda. Although Democrats have positioned themselves as more concerned about environmental issues, they too have been complicit in the "all-of-the-above" energy strategy. In recent decades, even as Democrats have advanced key climate initiatives, they hedged their bets, pairing such initiatives with continued support for fossil fuel development, including fracking of natural gas and expanded offshore oil drilling. In a country addicted to energy, the "all-of-the-above" energy strategy forged in the 1970s has been common ground in the fractious landscape of American politics, much to the detriment of the planet.

A. C. Isenberg (✉)
University of Kansas, Lawrence, KS, USA
e-mail: isenberg@ku.edu

J. M. Turner
Wellesley College, Wellesley, MA, USA
e-mail: jturner@wellesley.edu

N. Rauscher, W. Werner (eds.), *Natural Disasters in the United States*, Risk, Governance and Society 22, https://doi.org/10.1007/978-3-031-96436-7_4

Introduction

On the night of April 20, 2010, a massive explosion occurred on the *Deepwater Horizon*, an oil rig operating about 80 km off the coast of Louisiana in the Gulf of Mexico. The blast killed eleven and injured sixteen workers. Initially, as the fire aboard the rig raged, members of the United States Coast Guard and the National Oceanic and Atmospheric Administration's Office of Response and Restoration assumed that the source of the fire was the 2.6 million liters of diesel fuel the rig carried. By daybreak, however, it was apparent to emergency responders that oil from the exploratory well that *Deepwater Horizon* had been drilling 1500 meters below sea level was fueling the blaze. Two days after the explosion, as the Coast Guard continued to try to extinguish the fire, the drill rig sank into the Gulf of Mexico. Oil flowed from the well for five months before the leak was capped. Between four and five million barrels of oil escaped the well during that time, making the *Deepwater Horizon* disaster the largest marine oil spill on record.[1] The spill affected 150,000 square kilometers of the Gulf—roughly the size of Austria and the Czech Republic combined.[2] Oil lapped up on over 1000 km of shoreline in Louisiana, killing wetland vegetation and millions of waterfowl. At sea, the spill devastated plankton, killed thousands of dolphins and other marine mammals, and decimated populations of fish and shellfish such as oysters and crabs. Since most of the oil never left the deep sea or returned to the deep sea from the surface, the worst damage was to the deep pelagic environment and the seafloor.[3]

The *Deepwater Horizon* blowout was a spectacular catastrophe. Yet, even if everything had gone as intended—the well successfully drilled, the oil pumped to the surface, and a new source of fossil fuel brought into production—the consequences would still have been catastrophic, only they would have occurred at a pace and a scale that rendered those impacts unexceptional and unnewsworthy. The four or five million barrels of oil that flowed into the ocean would have been refined into gasoline, diesel, heating oil, or jet fuel and become part of the roughly seven billion barrels of oil the United States consumes every year. In the United States, gasoline and diesel used for transportation account for 29% of US carbon dioxide emissions

---

[1] See "The Early Days and Hours of Deepwater Horizon," NOAA Office of Response and Restoration, April 20, 2020. https://blog.response.restoration.noaa.gov/early-days-and-hours-deepwater-horizon. *United States v. BP Exploration & Production*, US District Court, Eastern District of Louisiana, 15 December 2010. www.epa.gov/sites/default/files/2013-10/documents/deepwater-cp121510.pdf. Accessed 22 May 2022.

[2] Tamay M Özgökmen, et al., "Over What Area Did the Oil and Gas Spread During the 2010 *Deepwater Horizon* Oil Spill?" *Oceanography*, 29 (September 2016), 96–107.

[3] Nancy N Rabalais and R. Eugene Turner, "Effects of the *Deepwater Horizon* Oil Spill on Coastal Marshes and Associated Organisms," *Oceanography*, 29 (September 2016), 150–159; Steven A. Murawski, et al., "A Synthesis of *Deepwater Horizon* Impacts on Coastal and Nearshore Living Marine Resources," *Frontiers in Marine Science* (1 February 2021); Charles R. Fischer, Paul A. Montagna, and Tracey T. Sutton, "How Did the *Deepwater Horizon* Spill Impact Deep-Sea Ecosystems?" *Oceanography*, 29 (September 2016), 182–195.

(another quarter comes from burning coal and natural gas to produce electricity). Through its profligate consumption of fossil fuels, the US's historic carbon dioxide emissions far outpace any other country and have played a key role in driving up atmospheric carbon dioxide levels from 315 parts per million in 1960 to 350 parts per million in 1990 to over 410 parts per million in 2019. As a consequence of the consumption of fossil fuels by the United States and other nations, the average global temperature has risen 1.1 °C since 1880, with the majority of that rise since 1975.

As Bill McKibben, the environmental writer and activist, observed in the wake of the *Exxon Valdez* disaster in 1989 (a spectacular environmental disaster in its own right, spilling 257,000 barrels of crude oil into the Prince William Sound), "there is another oil slick." The metaphorical slick to which McKibben pointed was not oil in the ocean, visible for all to see; it was the invisible accumulation of carbon dioxide emissions in the atmosphere. "It spreads out above us every second of the day." As McKibben observed, "The greenhouse effect…is not the result of something going wrong." The problem was not a faulty well (as with *Deepwater Horizon*) or an intoxicated sailor (as with the Exxon *Valdez*). "It's harder to deal with than that, because it's just a result of normal life."[4]

In contrast to the spectacular disaster of *Deepwater Horizon*, climate change is a manifestation of what the environmental scholar Rob Nixon has called "slow violence." Despite its lethality, in an increasingly spectacle-driven media, politics, and academia, the slow violence of climate change is comparatively disregarded in the United States.[5] Climate change is, seemingly, in everyone's thoughts—a quick search of the term on Google Ngram Viewer shows that the term's usage has risen exponentially since the 1970s. Yet, relative to citizens of other countries, Americans remain relatively unconcerned about it. According to a poll conducted in 2021 by the Pew Research Center, only 60% of Americans believe that climate change will harm them personally, compared with 72% of respondents to the poll in other countries. Americans are also far more sanguine about their government's response to climate change: 47% of Americans think the United States government is doing a good job of responding to climate change; only 36% of poll respondents in the rest of the world think the US government is doing a good job. Above all, while most Americans support the development of clean, renewable energy sources, they are reluctant to give up fossil energy sources such as coal and oil. While 69% of those polled by the Pew Research Center in August 2023 thought that the United States should prioritize the development of energy sources such as wind and solar, in the same poll, almost an equal number (68%) said that the United States should continue to rely on a mix of energy sources that includes fossil fuels. A Pew poll earlier in 2023 found that Americans rank "dealing with climate change" seventeenth in their list of twenty-one national priorities; only 37 % of respondents ranked climate

[4] Bill McKibben, "The *Exxon Valdez* as Metaphor," *The New York Times*, April 7, 1989, A31.

[5] Rob Nixon, *Slow Violence and the Environmentalism of the Poor* (Cambridge, Ma.: Harvard University Press, 2013).

change as a top national priority. The top priority, according to the poll, was "strengthening the economy"; 75% of respondents ranked the economy as a top national priority.[6] In short, climate change has drawn increasing public attention in the past decade, as extreme weather phenomena become more common, and scientists are better able to attribute droughts, storms, and other weather events to global warming. Nevertheless, relative to the citizens of other countries, fewer Americans believe the problem is urgent.

By contrast, when the *Deepwater Horizon* exploded, the response was immediate. The Coast Guard responded with fireboats, helicopters, and crews aimed at rescuing survivors and extinguishing the fire. In the weeks that followed, the National Oceanic and Atmospheric Administration coordinated hundreds of staffers, dozens of ships, and aircraft to manage the spill. The response to the *Deepwater Horizon* was an example of what the political scientist Roger A. Pielke, Jr. calls "tornado politics." According to Pielke, in an emergency (such as an approaching tornado), people put aside their political differences and heed the advice of experts (such as a meteorologist who advises seeking shelter from the tornado). By contrast, in situations that are not seen as emergencies (as, for instance, the way that most Americans regard climate change), people's values interject themselves into their sense of the scale and urgency of the problem; the advice of experts is more easily ignored.[7] It is thus that Americans put sixteen other problems ahead of climate change in their ranking of national priorities.

In the United States, over the past decade, conservative Republicans have made a disregard for climate change a basic premise of their popular politics. When Donald Trump characterized climate change as a "hoax" in 2015, he consolidated a position that conservative Republicans had been gravitating toward for over a decade.[8] Only 8 years before, Senator John McCain had included climate action among his policy priorities as the Republican candidate for president. Yet it is more

---

[6] "Americans are less concerned—but more divided—on climate change than people elsewhere," Pew Research Center, September 14, 2021. https://www.pewresearch.org/short-reads/2021/09/14/americans-are-less-concerned-but-more-divided-on-climate-change-than-people-elsewhere/. "What the data says about Americans' views of climate change," Pew Research Center, August 9, 2023. https://www.pewresearch.org/short-reads/2023/08/09/what-the-data-says-about-americans-views-of-climate-change/#:~:text=Overall%2C%2037%25%20of%20Americans%20say,a%20Center%20survey%20from%20January.

"Economy Remains the Public's Top Policy Priority; COVID-19 Concerns Decline Again," Pew Research Center, February 6, 2023. https://www.pewresearch.org/politics/2023/02/06/economy-remains-the-publics-top-policy-priority-covid-19-concerns-decline-again/

[7] Roger A. Pielke, Jr., "Forests, Tornadoes, and Abortion: Thinking about Science, Politics, and Policy," in *Forest Futures: Politics and Policy for the Next Century*, ed. Karen A. Arabas and Joe Bowersox (Lanham, MD: Rowman & Littlefield, 2004), 145. See also Pielke, *The Honest Broker: Making Sense of Science in Policy and Politics* (Cambridge: Cambridge University Press, 2007), 40–41.

[8] Consider Sen, James Inhofe of Oklahoma, who in 2003 described climate change as the "greatest hoax ever perpetrated on the American people," or Rep. Dana Rohrabacher of California, who in 2013 said that "global warming is a total fraud." "No Climate Tax Pledge," NoClimateTax.com. Accessed 7 November 2017.

than the unspectacular nature of climate change or this rightward shift in American politics that has stymied US action on climate change. The reluctance stems from the political tumult of the decade of the 1970s—ironically, the same decade that produced most of the landmark US environmental laws. In the 1970s, the United States also intensified its commitment to what is now known as an "all-of-the-above" energy strategy. For four decades since, every time the United States, whether led by Republicans or Democrats, has faced a choice between environmental reform and energy security (provided in the form of profligate fossil fuel extraction), it has chosen the latter.

The Environmental Consensus of the 1960s and 1970s

That energy concerns would supersede environmental imperatives was not obvious at the start of the modern environmental era, however. In the 1960s and 1970s, a bipartisan consensus prevailed in the United States on environmental issues. Democratic Party leaders such as Edmund Muskie and Hubert Humphrey were prominent advocates of environmental protection, but such protection was not a source of partisan division. In 1964, Humphrey introduced into the Senate the bill that became the Wilderness Act. The House version of the bill was introduced by a conservative Republican from Pennsylvania, John Saylor. Between 1969 and 1973, President Richard Nixon, a Republican, signed into law the National Environmental Policy Act, the Clean Air Act, and the Endangered Species Act. He created the Environmental Protection Agency by executive order. Barry Goldwater, arguably the most conservative Republican in the Senate in the 1960s and 1970s, was a cosponsor of the Clean Air Act. In the House of Representatives, 95% of Republicans voted for the Clean Air Act of 1970, the Clean Water Act of 1972, and the Endangered Species Act of 1973. In the Senate, 100% of Republicans voted for those three laws.[9]

In the 1960s and 1970s, elected officials in both parties understood that American voters strongly supported environmental reforms. 20 million Americans attended events held across the United States in the days before and after the first Earth Day on April 22, 1970. Though billed as a "teach-in," it is misleading to think of Earth Day as an artifact of the 1960s youth counterculture. Many supporters of Earth Day were middle-class suburbanites of voting age who got the attention of elected officials in the early 1970s through relentless political campaigning. The organizers of

[9] Saylor supported prayer in school and opposed the 1966 Supreme Court decision in the *Miranda* case that had decreed that suspects must be apprised of their rights when arrested. See Thomas G. Smith, *Green Republican: John Saylor and the Preservation of American Wilderness* (Pittsburgh, Pa.: University of Pittsburgh Press, 2006); Peter Iverson, "'This Old Mountain Is Worth the Fight': Barry Goldwater and the Campaign to save Camelback Mountain," *Journal of Arizona History*, 38 (Spring 1997), 41–56; J. Brooks Flippen, *Nixon and the Environment* (Albuquerque: University of New Mexico Press, 2000). For Goldwater's views on the environment, see Goldwater, *The Conscience of a Majority* (Englewood Cliffs, N.J.: Prentice Hall, 1970), 212, 217.

the first Earth Day founded a group, Environmental Action, that branded twelve gubernatorial and congressional incumbents who most opposed environmental reform—a group that included both Republicans and Democrats—as the "Dirty Dozen." Seven of those twelve were defeated at the polls in the 1970 election. Every two years, Environmental Action targeted a new Dirty Dozen. In 1972, five of that year's Dirty Dozen lost their races. In 1974, when six of that year's Dirty Dozen were voted out of office, the *New York Times* declared that environmentalism was American politics' "inescapable issue."[10]

The political strength of environmentalism not only drew ideologically pliable Republicans such as Richard Nixon into supporting environmental protections. It forced one of the party's leading conservatives, Ronald Reagan, who was elected governor of California in 1966 and reelected in 1974, to accommodate himself to voters' views about the environment. In 1968, Reagan, out of concern for the environmental costs, blocked the building of a massive dam in Northern California.[11] In May 1971, he created an "Ecology Corps," modeled on Franklin Roosevelt's Civilian Conservation Corps, made up of conscientious objectors to the military draft who were put to work in state parks.[12] In 1972, in a calculated publicity stunt, he donned a cowboy hat and rode a horse in Yosemite National Park to announce his opposition to a federal highway that would have cut through the Sierra Nevada near the park.[13] By the end of his second term as governor, he had added 145,000 acres to the state park system.[14] Early in his term as governor, Reagan created the California Air Resources Board, and throughout his tenure, he supported regulations aimed at curbing automobile pollution and reducing smog.[15]

Governor Reagan was a willing protector of the environment. Yet, there were limits. Reagan remained committed to the potential of free enterprise, and even as he agreed to California environmental protections, he voiced concerns that such

---

[10] Adam Rome, *The Genius of Earth Day: How a 1970 Teach-In Unexpectedly Made the First Green Generation* (New York: Hill and Wang, 2014), 211–213; Gladwin Hill, "Conservationists Count Successes: See Their Role as Vital in Many Election Contests," *New York Times* (5 November 1970), 37; "Inescapable Issue," *New York Times*, (24 November 1974). For suburban Americans' participation in the environmental movement, see Christopher C. Sellers, *Crabgrass Crucible: Suburban Nature and the Rise of Environmentalism in Twentieth-Century America* (Chapel Hill: University of North Carolina Press, 2012).

[11] Tom Goff, "Environment Stressed in Reagan Water Plan," *Los Angeles Times* (15 November 1969), B12; Lou Cannon, *Governor Reagan: His Rise to Power* (New York: Public Affairs, 2003), 309–313.

[12] "Ecology Corps Created for C.O.'s by Reagan," *New York Times* (2 May 1971), 30.

[13] George Skelton, "The Man in the White Hat Who Saved the Sierra," *Los Angeles Times* (28 June 1997).

[14] Jordan Rau, "A Seasoned Style, Green Record as Gov." *Los Angeles Times* (7 June 2004).

[15] Jerry Gillam, "Reagan Presents Clean Air Law to Legislature," *Los Angeles Times* (23 January 1970), A1. See also John S. Saloma, "California's Reagan tests politics of the environment," *Boston Globe* (8 February 1970), A23; Cannon, *Governor Reagan*, 307. Daniel A. Farber, "The Conservative as Environmentalist: From Goldwater and the Early Reagan to the twenty-first Century," *Arizona Law Review*. vol 59, no. 1005 (2017), 1021–1022.

regulations might go too far. He revealed those concerns when running for governor in 1966, when he famously told a lumber consortium in San Francisco, "We've got to recognize that where the preservation of a natural resource like the redwoods is concerned, there is a common sense limit. I mean, if you've looked at a hundred thousand acres or so of trees—you know, a tree is a tree, how many more do you need to look at?"[16] After taking office, his opposition to the creation of Redwood National Park succeeded in reducing the park to half of its proposed size.[17] He repeatedly criticized environmentalists for their "exaggerated rhetoric"; his preferred term for environmentalists was "doom-cryers."[18] And yet through the early 1970s Reagan remained a part of the political consensus on the environment and continued to position California as a forerunner on environmental policy. Though conservatism was ascendant in US politics beginning in the mid-1960s, Reagan, Goldwater, and other conservatives joined the larger political consensus on environmental protection for several reasons[19]: the popular strength of environmentalism in the 1960s and early 1970s, the urgency of environmental issues, such as Los Angeles's smog crisis and a 1969 Santa Barbara oil well spill, and because no one fully anticipated the constraints that state laws, such as the California Environmental Quality Act, or federal laws, such as the Clean Air Act, would place on free enterprise.

The environmental consensus of the late 1960s and early 1970s contained within it a tenuous compromise on restraining the production of domestic fossil fuels from federal lands. The federal government is the largest landowner in the United States; it owns 96.5 million hectares or 27% of the United States; it also holds the rights to coal, oil, and natural gas beneath the surface of federal lands and to oil beneath the Outer Continental Shelf up to a distance of 370 km from shore. Since the 1930s, the federal government has opened small areas of its offshore oil reserves to exploitation. In the late 1950s and early 1960s, rising oil prices increased oil companies' pressure on the federal government to lease more offshore areas. In March 1962, the federal government obliged: the Bureau of Land Management leased 411 tracts that

---

[16] Cannon, *Governor Reagan*, 177.

[17] Susan Schrepfer, *The Fight to Save the Redwoods: A History of Environmental Reform, 1917–1978* (Madison: University of Wisconsin Press, 1979), 146.

[18] William Endicott, "Reagan Unveils Environmental Protection Plan But Hits 'Doomcriers,'" *Los Angeles Times* (April 8, 1972), 10A. For environmentalists as "doom-cryers," see also Reagan's letter to a group of graduating high school students in New Jersey in Helene von Damm, *Sincerely, Ronald Reagan* (Ottawa, Il.: Green Hill Publishers, 1976), 156.

[19] For the rise of conservatism, see Lisa McGirr, *Suburban Warriors: The Origins of the New American Right* (Princeton, N.J.: Princeton University Press, 2001); Donald T. Critchlow, *The Conservative Ascendancy: How the GOP Right Made Political History* (Cambridge, Ma.: Harvard University Press, 2007); David Farber, *The Rise and Fall of Modern American Conservatism: A Short History* (Princeton, N.J.: Princeton University Press, 2010); Kim Phillips-Fein, *Invisible Hands: The Businessmen's Crusade against the New Deal* (New York: Norton, 2010); Darren Dochuk, *From Bible Belt to Sunbelt: Plain-Folk Religion, Grassroots Politics, and the Rise of Evangelical Conservatism* (New York: Norton, 2011); Kevin M. Kruse, *One Nation under God: How Corporate America Invented Christian America* (New York: Basic Books, 2016).

month; the 800,000 hectares leased totaled more than all previous federal offshore leases combined. The hasty expansion of offshore drilling in the 1960s outpaced regulation. The US Geological Service Conservation Division did not inspect rigs regularly and routinely issued waivers to rigs found in violation of regulations intended to prevent spills.

The result of rapid expansion and lax oversight was predictable: in January 1969, a rig off the coast of Santa Barbara, California, that had been granted a waiver blew out. For eleven days, the rig spewed oil. Between 80,000 and 100,000 barrels escaped altogether, creating a 200,000-hectare oil slick that affected 50 km of California coastline.[20] The Santa Barbara spill occurred only a week after Nixon's inauguration; the accident not only encouraged Nixon's effort in his first term to position himself as an advocate of environmental protection, but it cast a shadow over federal offshore oil leasing for the next ten years. In response to the spill, Nixon's secretary of the interior issued new safety guidelines prohibiting drilling in environmentally sensitive areas. In October 1972, Congress passed the Coastal Zone Management Act, which authorized states to develop plans to balance the interests of industries and coastal environments. Just a few weeks later, California voters approved by referendum Proposition 20, the Coastal Zone Conservation Act, which created a commission to establish restraints on development on both land and sea along California's coast. Although Governor Reagan opposed Proposition 20, it passed with the support of 55% of California voters.[21] Most American voters had little desire to return to the under-regulated offshore drilling of the 1960s.

A similar compromise reined in coal mining on federal lands in the 1970s. In the early 1970s, less than 2% of coal mined in the United States came from federal lands. The United States had issued hundreds of leases to mine coal on federal lands but 90% of federal coal leases were inactive.[22] The federal government had little interest in increasing coal production from federal lands. In 1973, Nixon's secretary of the interior even issued a moratorium on new leases. A 1976 law, the Federal Coal Leasing Amendments Act, discouraged the development of coal mining on federal lands by requiring competitive bids for leases and sales at fair market values. New environmental regulations further discouraged strip mining, a mining strategy that accounted for the majority of US coal production by the mid-1970s. When Jimmy Carter assumed the presidency in 1977, one of his first acts was to sign into law a bill imposing regulations on strip mining. The law imposed added taxes on

---

[20] National Commission on the BP Deepwater Horizon Oil Spill and Offshore Drilling, "The History of Offshore Oil and Gas in the United States (Long Version)," Staff Working Paper No. 22, 1–13.

[21] James D. Dilts, "In California, the coast is conservation-land," *Baltimore Sun* (December 28, 1972), A1; Cannon, *Governor Reagan*, 314–315.

[22] David Howard Davis, "Energy on Federal Lands," in Charles Davis, ed., *Western Public Lands and Environmental Politics*, second ed. (Boulder, Co.: Westview Press, 2001), 147.

strip-mined coal and required mining companies to restore lands that had been stripped to their prior condition.[23]

The Oil Crises of the 1970s and the Expansion of Domestic US Energy Production

The oil crises of the 1970s undermined the emerging consensus around restraining and regulating domestic energy production on federally owned lands. In October 1973, in retaliation for US support for Israel during the Yom Kippur War, the Organization of Arab Petroleum Exporting Countries implemented an embargo on oil exports to the United States, driving up the price of oil from $3 to nearly $12 per barrel by January 1974. The result in the United States was rapid and devastating inflation: the average annual inflation rate rose from 3.2% in 1972 to 6.2% in 1973 to 11% in 1974. By the summer of 1974, about 20% of American gas stations had no fuel to sell. At remaining stations, motorists endured long lines and government-imposed rationing. Frustrated by the limited supplies of fuel, truckers embarked on a brief strike in December 1973. In 1974, to curb gasoline consumption, the federal government imposed a deeply unpopular national highway speed limit of 55 miles per hour (88 km/hour).[24]

Initially, these measures enjoyed bipartisan support. In November 1973, President Richard Nixon urged Americans to embrace a "spirit of sacrifice" and take steps to conserve energy, including lowering home heating, driving more slowly, and turning off unnecessary lights.[25] It was the last time a Republican would urge such austerity on the American people. At the same time, Nixon announced another set of initiatives, which he billed Project Independence. Nixon championed a new domestic agenda to promote federal research, public-private partnerships, and regulatory reforms aimed at rapidly expanding the production of a wide range of energy sources, including fossil fuels and nuclear power. The goal—one that the United States would fall far short of achieving—was to meet US energy needs with domestic sources by 1980. Congress agreed: in the fall of 1973, as the oil crisis was unfolding, Congress exempted the construction of the Trans-Alaska Pipeline, a 1200-km-long pipeline designed to bring oil from Alaska's Prudhoe Bay on the Arctic Ocean to the port of Valdez on the Pacific, from the requirement of an environmental impact statement. Such a statement was mandated by the National

[23] Ben A. Franklin, "House Democrats Renew Drive for Strip-Mine Bill," *New York Times* (August 25, 1976); "Carter signs strip-mine control bill but says he will try to strengthen it," *Chicago Tribune* (August 4, 1977), 6. Chad Montrie, *To Save the Land and People: A History of Opposition to Surface Coal Mining in Appalachia* (Chapel Hill: University of North Carolina Press, 2003).

[24] See David Frum, *How We Got Here: The 70 s* (New York: Basic Books, 2000); Meg Jacobs, *Panic at the Pump: The Energy Crisis and the Transformation of American Politics in the 1970s* (New York: Hill and Wang, 2016).

[25] https://www.presidency.ucsb.edu/documents/address-the-nation-about-national-energy-policy

Environmental Policy Act of 1970, the first of the major 1970s environmental laws that is sometimes called the "Magna Carta" of environmental laws. Congress then rushed through an authorization act for the pipeline. Construction began in 1974. The hasty congressional action effectively made clear that the hope of lower energy costs outweighed environmental concerns.

The spike in oil prices in 1973–1974 did not immediately cause the 1970s environmental consensus to collapse. But as concerns about the constraints imposed by environmental regulations grew and the energy crisis persisted, the consensus became increasingly fragile. Jimmy Carter, elected to the White House in 1976, did not anticipate how quickly concerns over energy and the economy had begun to override concerns for the environment. During his time in office, Carter set a tone of frugality and environmentally sensitive retrenchment: he installed solar panels on the roof of the White House and, during the winter, turned the thermostat in the White House down to 65 ° F (18 °C). In 1979, Carter announced an energy plan that he and his advisers had taken years to develop: a phased deregulation of oil prices, gas rationing, and the development of alternative energy sources. In many respects, it built on the agenda that Nixon had set forth in 1973. Carter wanted Americans to pay more for the energy they consumed—and consume less of it.[26] Yet Carter's ambitious plans were overtaken by the Iranian Revolution in 1979, which disrupted oil production in one of the world's largest oil-producing nations. The subsequent cut in Iranian oil production caused the price of a barrel of oil to more than double to nearly $40. Inflation in the United States rose to 11.3% in 1979 and 13.5% in 1980.[27]

The energy crises of the 1970s accelerated the ascendancy of conservative Republicans, strengthening their calls to unshackle industry from all forms of regulation—particularly the far-reaching environmental laws that had been enacted in the late 1960s and early 1970s. In this context, running for the White House, Ronald Reagan positioned himself in sharp contrast not only to Jimmy Carter, but the consensus politics that had been important to laying the policy foundation of US environmentalism. When Reagan announced his intent to seek the Republican nomination for president, one of his chief talking points was the energy crisis. Appealing to his audience, he reminded them "Last summer you probably spent hours sitting in gasoline lines." He warned that, "This winter, some will be without heat and everyone will be paying much more simply to keep home and family warm." In Reagan's view, conservation was not an energy policy. "At best, it means we will run out of energy a little more slowly." Instead, Reagan championed a vision of energy abundance, not one of scarcity and limits. The first step toward realizing this vision was committing to more domestic oil and gas production in the United

[26] Jacobs, *Panic at the Pump*, 182 ff.

[27] Burton I. Kaufman, *The Presidency of James Earl Carter, Jr.* (Lawrence: University Press of Kansas, 1993), 140–148; Otis L. Graham, Jr., *Presidents and the American Environment* (Lawrence: University Press of Kansas, 2016), 252–271.

States.[28] The policy that Reagan articulated as a presidential candidate in 1980 would become Republican orthodoxy over the next several decades.

That commitment changed the geography of energy production in the United States. In western states, where the federal government owned much of the land, ranging from 30% of Montana to 90% of Alaska, numerous Republican politicians called for loosening regulations on federal lands.[29] When Ronald Reagan, on the campaign trail in 1980, said that the federal government was "making it impossible for people to have any access or enjoyment from [public] lands, unless, maybe, you're a backpacker," his statement was, in point of fact, utterly untrue.[30] But it resonated with conservatives in western states. Laws meant to protect and regulate resource use on federal lands appeared to westerners—and the oil and mining companies allied with them—like unnecessary restrictions that held back the ability to drill for oil and dig coal on federal lands and thus free the United States from dependence on foreign energy supplies. Among right-leaning politicians in the American West who adhered to these views, a loose-knit political movement, known as the Sagebrush Rebellion, emerged. Their central claim: the western states, not the federal government, should hold title to western lands. If states, rather than the federal government, controlled public lands, the Sagebrush Rebels believed, they could lift federal restrictions on land use, producing more timber, more coal, more natural gas, and more jobs. In April 1979, the Nevada state assembly passed a law asserting state control over the federally owned public domain within Nevada's boundaries. The territory, most of which was administered by the federal Bureau of Land Management, amounted to roughly 79% of the state. Arizona, New Mexico, Utah, Wyoming, and Colorado followed suit. Legally, these state laws had no effect. Yet, as symbolic politics, the Sagebrush Rebellion captured the rising tide of frustration in the West, among citizens, industry, and politicians. That tide helped carry Reagan to the White House in the election of 1980.[31]

---

[28] Ronald Reagan, "Official Announcement," November 13, 1979, https://www.4president.org/speeches/1980/reagan1980announcement.htm and Sabin, *The Bet: Paul Ehrlich, Julian Simon, and Our Gamble over Earth's Future* (New Haven: Yale University Press, 2013), 138

[29] In 1978, the *Conservative Digest* characterized wilderness preservation laws as a "Big Federal Land Grab." "The Big Federal Land Grab," *Conservative Digest*, 4 (December 1978), 7–11. See James M. Turner, "'The Specter of Environmentalism': Wilderness, Environmental Politics, and the Evolution of the New Right," *Journal of American History*, 96 (June 2009), 123–148.

[30] Reagan, quoted in *Field & Stream* (October 1980); in Mark Green and Gail MacColl, *Reagan's Reign of Error* (New York: Pantheon, 1987), 98.

[31] Reagan's election came at the end of a tumultuous period in US politics. Between 1970 and 1980, one US president resigned (something that had never happened before) and two incumbent US presidents were turned out of office (something that had not happened since 1932). US Congressional representatives tend to be reelected at high rates, but between 1964 and 2020, two of the three elections in which the lowest percentage of Congressional representatives were reelected came in 1970 (85%) and 1974 (88%). Sitting US senators tend to win reelection as well, but between 1964 and 2020, the three elections in which the lowest percentage of senators were reelected were in 1976 (64%), 1978 (60%), and 1980 (55%). Environmental Action and the Sagebrush Rebellion were not the only reasons presidents, senators, and Congressional representa-

Reagan's first secretary of the interior, James Watt, did not cede federal lands to state control as the Sagebrush Rebels had hoped, but he enthusiastically opened up domestic energy production on federal lands. In 1973, when the American energy crisis began, American coal production was centered in Appalachia and the Ohio Valley. Watt set out to change that—with astonishing success. When he took office, a 1976 law governing the leasing of federal land for mining coal outlined a selective and deliberate pace of leasing. Effectively disregarding that law, Watt directed the Department of the Interior to offer extensive tracts of public lands to coal companies. The first parcels to go up for bid were in the Powder River Basin on the Wyoming-Montana border. It contains the world's largest known reserve of fossil fuels—even more than the natural gas and oil fields of the Middle East. Unlike Appalachian coal fields, where the thickness of coal seams is measured by the meter, in Wyoming, coal seams are as tall as eight-story office buildings.[32] At the direction of Watt, the Department of the Interior put up the largest federal leases in history in 1982, offering tracts holding an estimated 1.45 billion metric tons of recoverable coal reserves.

Watt opened up offshore oil drilling in the same way that he opened strip mining in the Powder River Basin. When Watt assumed office in 1981, he inherited a federal offshore oil program that had, over the course of the previous decade, tried to expand domestic oil production in response to the 1970s oil crises without risking the kind of poorly regulated drilling that had led to the 1969 oil spill off the coast of Santa Barbara, California. Beginning in 1978, Interior secretaries were authorized to propose five-year plans in which they spelled out which offshore acreage would be available for leases. In 1982, when Watt had his first opportunity to propose a plan, he abandoned the balanced approach, gleefully announcing at a National Ocean Industries Association meeting, "We will offer one billion acres for leasing in the next five years."[33] Rather than having the Department of the Interior select particular tracts for oil companies to bid on, Watt threw open large areas and allowed oil companies to select their own tracts. In seven sales between 1983 and 1985—many of which took place after Watt had been forced from office but adhered to the leasing plan he had set out—the Department of the Interior leased over 2600 tracts, nearly as much as all the lease sales since 1962 combined. Those leases set the stage for a dramatic expansion in oil production in the Gulf of Mexico.

---

tives were turned out of office between 1970 and 1980, but in many key races, those issues were critical.

[32] Timothy J. Considine, "Powder River Basin Coal: Powering America," Natural Resources 4, no. 8 (2013): 514–533.

[33] "Offshore Drillers Hail Billion-Acre Lease Plan," July 19, 1982, UPI Archives, https://www.upi.com/Archives/1982/07/19/Offshore-drillers-hail-billion-acre-lease-plan/6994395899200/

Steve Greenberg, *Los Angeles Daily News*, 5 June 1981

The Creation of a New Energy Consensus

In the 1970s, there had been a bipartisan consensus on environmental protection. In the early 1980s, the Reagan administration forged a new consensus: a commitment to developing federal oil and coal reserves to promote US independence from foreign energy supplies. Successive secretaries of the interior under Presidents Reagan, the first Bush, Clinton, and the second Bush continued the policy of extensive and largely uncompetitive coal leases on federal lands in the West, particularly the Powder River Basin. In 1973, 10% of the coal in the US had come from federal lands. By 2014, however, 40% of coal mined in the United States came from federal lands; almost 90% of that federal coal was mined in the Powder River Basin. Two hundred power plants in 38 states burn Powder River Basin coal. If the Powder River Basin were its own country, it would be the seventh-largest producer of greenhouse gases in the world, trailing China, the rest of the United States, India, Russia, Japan, and Germany.[34]

[34] Nidhi Thakar and Michael Madowitz, "Federal Coal Leasing in the Powder River Basin: A Bad Deal for Taxpayers," July 29, 2014. Center for American Progress. https://cdn.americanprogress.org/wp-content/uploads/2014/07/ThakarPowderRiver-brief.pdf. Accessed April 14, 2017.

The extent to which Republicans put fossil fuel production ahead of conservation became apparent in another debate over the public lands: the fate of the Arctic National Wildlife Refuge. In the wake of the Exxon *Valdez* oil spill in 1989, environmentalists pushed to protect the refuge's coastal plain as a wilderness area, which would have effectively barred future oil and gas development. Environmentalists argued that the United States could save more gasoline than the refuge would ever produce by raising fuel efficiency standards for vehicles—a measure that would also help address nascent concerns about climate change and promote national energy security (a concern amid the Persian Gulf War). Yet, since the Reagan administration, Republican presidents have championed opening up the refuge's coastal plain to oil and gas exploration. When Democrats nearly succeeded in passing legislation to protect the refuge's coastal plain and raise fuel efficiency standards in 1991, President George H. W. Bush threatened a veto.[35] He would not consign Americans to drive small energy-efficient cars; instead, he championed expanding fossil fuel production as the best way to secure the nation's domestic interests and protect it from dependency on the Middle East.[36] It would take two more decades for Republicans to succeed in opening the refuge to drilling, which finally came to pass (at least temporarily) during the first Donald J. Trump administration.

Even while Democrats held the line on drilling in the Arctic National Wildlife Refuge and gave new attention to other environmental issues, especially global warming, a shared commitment to expanding domestic oil and gas production remained a bipartisan priority. Expanded offshore oil leases, like expanded coal leases, continued not just through Republican administrations but through both of Bill Clinton's terms and through Barack Obama's first term. By 2008, when BP acquired a lease for Mississippi Canyon Block 252, which became the site of the *Deepwater Horizon* disaster, the expanded offshore oil lease system that James Watt had inaugurated a quarter of a century earlier had become standard. Despite Obama's strong stance on climate change (a position that helped earn him the Nobel Peace Prize in 2008), from the start, his administration focused on what he came to describe as an "all-of-the-above" energy strategy. That strategy included renewable energy sources. The administration spearheaded the American Recovery and Reinvestment Act, which prioritized investments in solar, wind, and battery technology, to help pull the United States out of the 2008–2009 economic recession and lay the foundations for a clean energy industry. Despite those commitments, however, Obama remained supportive of fossil fuel developments. During most of his first term, Obama's support for offshore drilling remained relatively strong. On April 2, 2010, speaking in Charlotte, North Carolina, Obama explained his all-of-the-above approach to energy policy, noting specifically that "oil rigs today generally don't

[35] James Morton Turner, *The Promise of Wilderness: American Environmental Politics since 1964* (Seattle: University of Washington Press, 2010), 290–94.

[36] George H. W. Bush, "Teleconference Remarks with the American Gas Association," October 28, 1991, *The American Presidency Project.* https://www.presidency.ucsb.edu/documents/teleconference-remarks-with-the-american-gas-association. October 28, 1991.

cause spills." Eighteen days later, the blowout occurred on the *Deepwater Horizon*. Although in the aftermath of the *Deepwater Horizon* blowout, the Obama administration began to roll back the extent of offshore oil leases in the Gulf of Mexico, his administration continued to pursue its all-of-the-above energy strategy. The most consequential energy development during Obama's presidency was the dramatic expansion of fracking for natural gas. In his 2014 State of the Union Address, Obama championed natural gas production. "If extracted safely, it's the bridge fuel that can power our economy with less of the carbon pollution that causes climate change."[37] Even as the Obama administration adopted policies that strengthened oversight of fossil fuel extraction, domestic production of both natural gas and oil increased substantially between 2010 and 2018.[38]

The Limits of the Environmental Consensus

So, what happened to the environmental consensus of the 1970s? In important respects, it persisted. Many Republican politicians have remained committed to many of the core ideas of 1970s environmentalism: clean air, clean water, and regulating chemical pollution. In 1988, Reagan signed the Montreal Protocol, an international agreement that committed the United States to phasing out chlorofluorocarbons, the chemicals that were contributing to the depletion of the stratospheric ozone layer.[39] In 1990, George H. W. Bush signed an amendment to the Montreal Protocol, which accelerated the phase-out of chlorofluorocarbons. That same year, he signed into law amendments that strengthened the US Clean Air Act to reduce acid rain. In 1993, however, when the administration of Bill Clinton tried to move from these core issues of pollution to address climate change with a proposed Climate Change Action Plan, opposition from both parties whittled his ambitious plan down to a 4.3 cents-per-gallon tax on gasoline. His efforts to secure a US commitment to the Kyoto Protocol, which would have committed the United States to reducing its greenhouse gas emissions—following the model of the Montreal Protocol and the ozone hole—faltered when it ran up against united opposition of Democrats and Republicans in the US Senate. When Clinton did secure environmental policy victories, it was on more traditional issues: he had success restricting logging on public lands and protecting endangered species and roadless wilderness areas—issues that resonated with the core of the 1970s environmental consensus. But addressing climate change ran afoul of the consensus on keeping energy prices low that emerged in the United States following the oil crises of the

[37] Barack Obama, "State of the Union Address," January 28, 2014. https://obamawhitehouse.archives.gov/the-press-office/2014/01/28/president-barack-obamas-state-union-address. Accessed 13 September 2023.

[38] US Energy Information Administration, *Annual Energy Outlook* 2018, February 6, 2018, 32.

[39] Ronald Reagan. "Statement on Signing the Montreal Protocol on Ozone-Depleting Substances," 5 April 1988. The American Presidency Project. http://www.presidency.ucsb.edu/ws/?pid=35639

1970s. In recent years, environmentalists have argued in court that the Clean Air Act of 1970 ought to extend to regulating carbon dioxide, methane, and other heat-trapping gases. In court, the argument has met with success; in the political realm, however, it has been a different story.[40]

As part of the environmental consensus of the early 1970s, the United States started down the road toward restricting fossil energy production and reckoning with the end of abundance—until the oil crises detached energy from other environmental issues. In the context of American politics, the oil crises of the 1970s made clear that despite the strong concern for the environment in the United States, policies that would limit fossil fuel production, increase the cost of energy, or compromise US energy abundance or security were nonstarters. In short, energy concerns overrode environmental reform. And that bipartisan consensus around energy production has endured: using federal lands to produce fossil fuel energy to keep energy prices in the United States low and avoid a repetition of the inflation of the 1970s has, to varying degrees, been championed by both Democrats and Republicans. Thus, many Republican officials simultaneously present themselves as in favor of a clean environment and high levels of domestic fossil energy production. In 2018, for instance, the Interior Department during the administration of Donald Trump relaxed federal rules regulating the venting of methane while drilling for oil and gas on federal lands. David Bernhardt, an official in the Interior Department, said, "We're for clean air and clean water. At the same time, we're for reasonable regulations" of domestic energy production.[41] While Republicans have led the drive for exploiting federal lands for coal and oil, Democrats, cognizant of their loss of the White House in 1980 owing to energy-driven inflation—have often been junior partners, as Barack Obama's embrace of an "all-of-the-above" energy strategy illustrated. The 1970s have thus cast a long shadow in US politics: elected officials of both parties have concluded—rightly or wrongly—that opposing environmental protection is bad for one's electoral chances, but it is even worse to be held responsible for high energy prices.

Even President Joseph Biden, who promised to ban new oil and gas permits on federal lands and made climate action a centerpiece of his agenda, has not been able to escape the durability of this consensus. The strong bipartisan commitment to domestic fossil fuel production sharply constrained the Biden administration's climate agenda. Despite passing the most far-reaching climate policy in US history, the Inflation Reduction Act in August 2022, the policy does little to restrict the production of fossil fuels. Instead, its most substantive policy measures are incentives aimed at encouraging investments in renewable energy technologies and electric vehicles. With its passage came the tacit admission that the United States, even in the face of the climate emergency, still could not muster the political will to adopt a climate policy that imposed restrictions, taxes, or caps on fossil fuel production or

[40] *Massachusetts v. EPA*. S. Ct. 05–1120 (2007).

[41] Kurtis Alexander, "Another greenhouse gas regulation to be weakened by Trump administration," *San Francisco Chronicle* (18 September 2018).

greenhouse gas emissions.[42] In fact, the Inflation Reduction Act, harkening back to an all-of-the-above energy strategy, actually included several provisions requiring the administration to expand fossil production on federal lands—provisions that were included to appease politicians and secure the law's passage in the US Senate. According to those requirements, the Biden administration auctioned off almost 30 million hectares of the Gulf of Mexico for deep-water oil and gas production in March 2023.[43] In these respects, the Inflation Reduction Act mirrors the results of the August 2023 Pew poll that showed that while 69% of Americans wanted the development of renewable energy sources, 68% believed that the United States should continue to rely on a mix of renewables and fossil fuels.

Conclusion

In one respect, the 40-year-long effort to generate more fossil energy from federal lands has succeeded. The United States now produces more energy than it consumes. The vast majority of that energy comes from fossil fuel sources (petroleum 35%, natural gas 34%, and coal 10%). Yet, because the price of a barrel of oil is determined by an international market, energy independence has not freed the United States from the same problem of crisis-driven high energy prices that it faced in the 1970s. In 2022, the war in Ukraine drove the average price of a gallon of gasoline in the United States to $4.59. Adjusted for inflation, that is about what Americans paid for gasoline in 1980. The monthly inflation rate for June 2022 was 8.6%—the highest rate for any month since December 1981.[44] In short, the forty-year experiment in maximizing fossil energy production from public lands led the United States back to where it started. Less spectacularly but more ominously, the bipartisan consensus on domestic energy production has left the United States incapable of addressing the challenge of climate change.

For all the successes of US environmental policy in measurably reducing air and water pollution and protecting human health, the US's unswerving commitment to an energy politics predicated on abundance has limited the US's response to climate change. Over the last five decades, while climate change has grown increasingly important to US environmental politics, it has never been treated as an environmental emergency in the same way that Americans have responded to such disasters as the *Exxon Valdez* or *Deepwater Horizon*. The closest the United States has come to

[42] Kate Aranoff, "Why is the Fossil Fuel Industry Praising the Inflation Reduction Act," *The New Republic,* March 10, 2023. https://newrepublic.com/article/171086/fossil-fuel-industry-praising-inflation-reduction-act

[43] Bureau of Ocean Energy Management, "Lease Sale 259," March 29, 2023. https://www.boem.gov/oil-gas-energy/leasing/lease-sale-259

[44] US Energy Information Administration, "U.S. Energy Facts Explained," https://www.eia.gov/energyexplained/us-energy-facts/. Accessed 15 June 2022. Ella Koeze and Clifford Krause, "Why Gas Prices Are So High," *New York Times* (14 June 2022).

addressing climate change, with the American Recovery and Investment Act in 2009 and the Inflation Reduction Act in 2023, has also been predicated on a politics of more. Instead of affecting an energy transition—in which renewable energy sources *replace* fossil fuels, the United States has adopted policies that *add* renewable energy investments even as fossil fuel production continues to expand. While that approach may be politically expedient and while it has played a key role in scaling up the deployment of clean energy technologies in the United States, without policies that significantly reduce fossil fuel consumption and greenhouse gas emissions, the United States will have failed in the face of the greatest environmental challenge of all.

Letting Nature Do Its Thing: Early Flood Control and Environmental Disaster Along the Los Angeles River

Jan Hansen

Abstract Los Angeles, often praised for its nearly perfect weather, grapples with a delicate balance between utopia and dystopia. Despite the idealized image promoted by city boosters in the nineteenth and early twentieth centuries, the region faced significant dangers such as wildfires, earthquakes, and periodic flooding of the Los Angeles River. This essay delves into the pre-1940 history of flooding in Los Angeles, investigating why addressing the flood risk of the Los Angeles River took until the 1930s and exploring alternative solutions dismissed by economic and political leaders. The essay contributes to existing scholarship by highlighting early flood control efforts initiated "from below" by citizens, business owners, and private railroad companies. These initiatives, predating federal intervention in the 1930s, underscore a grassroots approach characteristic of responses to environmental challenges in many cities in the far American West. The analysis extends to the 1930 Olmsted-Bartholomew plan and examines proposed flood-risk management strategies that could have shielded the region from river overflow. Despite their potential effectiveness, these alternatives were not pursued, illustrating how economic priorities took precedence over alternative urban settlement patterns. In essence, the essay explores the interplay between the perception of environmental threats, civic responses, and societal values associated with these responses. The frontier myth played a crucial role in shaping these values, as it encouraged self-help approaches, defining Los Angeles as a do-it-yourself city. By examining these diverse themes, the essay uses the case of the Los Angeles River to uncover the management of environmental disasters in the far American West.

J. Hansen (✉)
Humboldt-University, Berlin, Germany
e-mail: jan-eric.hansen@geschichte.hu-berlin.de

N. Rauscher, W. Werner (eds.), *Natural Disasters in the United States*, Risk, Governance and Society 22, https://doi.org/10.1007/978-3-031-96436-7_5

Introduction

The summer of 1935 was just as scorching and dry as most summers used to be in Southern California. Bathed in relentless sunlight, Downtown Los Angeles recorded an average temperature of 98 °F (37 °C). No rain was in sight. At the height of summer, when the heat was most intense, heavy machinery, including bulldozers, construction trucks, and cranes, roared to life. Their mission was to confront a river that, for most of the year, lay dormant but transformed into an unpredictable force during the winter months. Earlier in the year, in recognition of the persistent threat posed by the unruly river, President Franklin D. Roosevelt had sanctioned a substantial allocation of $13.9 million from Works Progress Administration (WPA) funds. This financial injection aimed to implement flood control measures in Los Angeles. Overseeing the colossal task was the US Army Corps of Engineers. By October, a local workforce of 2000 individuals was engaged in the monumental project, channelizing the river from Lankershim Boulevard in North Hollywood to Elysian Park just shy of downtown, covering 11.5 miles. A year later, Congress passed the Flood Control Act, allocating an additional $70 million to further bolster the Corps of Engineers.[1]

Fast forward to March 1938, and Los Angeles found itself grappling with the most devastating flood in its history, surpassing the calamitous floods of 1914 and 1934. The river's unleashed fury wreaked havoc, particularly in areas excluded from the protective measures built by the Corps of Engineers. Countywide, the toll was staggering: 87 lives lost, 108,000 acres submerged, and property damage exceeding $78 million. Countless residents were left homeless, the motion picture industry along the riverfront suffered substantial losses, and at its Long Beach mouth, the river's volume exceeded even that of the mighty Mississippi near St. Louis. Postdisaster, engineers reevaluated and revised flood control plans. The new vision that they came up with called for a deeper river channel and higher paved banks resembling fortress walls. Construction efforts now extended into tributary systems and mountainous regions, incurring an additional cost of $268 million. Construction work persisted well into the late 1950s and early 1960s.[2]

Today, the Los Angeles River stands as a testament to the transformative power of human engineering over nature. It bears little resemblance to traditional rivers because it is confined within an expansive and entirely concrete channel mostly devoid of vegetation. The water coursing through it is predominantly treated sewage and street runoff. Only during the winter months does this artificial channel momentarily transform into a raging river.

The transformation of the nineteenth-century Los Angeles River reflects the perceived triumph of human ingenuity over nature, rendering it almost unrecognizable as a natural watercourse. However, this transformation was not swift. Despite

[1] Blake Gumprecht, *The Los Angeles River: Its Life, Death, and Possible Rebirth* (Baltimore: Johns Hopkins University Press, 1999), 205–15.

[2] Ibid., 215–33.

recurrent flooding, political and economic leaders, alongside engineers and property owners, did not manage to engage in flood-risk management that would tame the river until the 1930s. It was only in 1935 that the federal government intervened, initiating large-scale flood control measures. The question lingers: was the federal takeover inevitable, or were there alternative, unexplored solutions to the problem that haunted the region for decades?

This essay poses a set of three questions: Why did it take Los Angeles so long to find an appropriate way to address the frequent problem of flooding of the Los Angeles River? How was the chosen solution—the construction of a concrete riverbed—connected to the societal reaction to droughts that was taking shape in the region in the late nineteenth and early twentieth centuries? To what extent was the channelization of the river without alternatives? What were the paths not taken? These three questions are intertwined in many ways. Most importantly, they are bound together by the tendency of certain actors in the United States to avoid recognizing and addressing risks that endanger their society.[3] By extending the research question of this edited volume to the far American West, where the frontier was closing on the shores of the Pacific, this essay aims to understand how early settlers, property owners, farmers, City Council members, Chamber of Commerce leaders, and representatives of the federal government in distant Washington, DC, assessed Southern California's specific vulnerability to disasters, more specifically, floods and droughts.

While the region faced a higher susceptibility to natural disasters than most other areas in North America, grappling with significant losses from wildfires and earthquakes, this chapter delves into the intricate dynamics of water—its abundance and scarcity. Indeed, the complex interplay of floods and droughts posed a formidable challenge for early Los Angeles residents. Both were supposedly "natural" disasters, yet they evolved into crises due to specific human approaches to water.

In doing so, the essay contributes to the extensive body of research on water in Southern California, which has predominantly concentrated on major infrastructures such as the Los Angeles and Colorado River aqueducts.[4] These aqueducts played a pivotal role in importing water from distant sources to Southern California, driving unprecedented population growth and economic expansion in the region. Other works have focused on water within the city. Exploring the everyday history of water in Los Angeles, Blake Gumprecht delves into the Los Angeles River's

[3] See the introduction to this edited volume.

[4] William L. Kahrl, *Water and Power: The Conflict over Los Angeles' Water Supply in the Owens Valley* (Berkeley: University of California Press, 1982); Donald Worster, *Rivers of Empire: Water, Aridity, and the Growth of the American West* (New York: Oxford University Press, 1985); Marc Reisner, *Cadillac Desert: The American West and Its Disappearing Water* (New York: Viking, 1986); Norris Hundley, *The Great Thirst: Californians and Water, 1770s–1990s* (Berkeley: University of California Press, 1992); Steven P. Erie, *Beyond Chinatown: The Metropolitan Water District, Growth, and the Environment in Southern California* (Stanford: Stanford University Press, 2006).

historical significance for the area's development,[5] while Jared Orsi provides a comprehensive account of the region's destructive floods, revealing that technologically engineered solutions to flooding are as susceptible to disasters as natural waterways.[6]

Venturing beyond the scope of the Los Angeles River, the exploration of other rivers also yields valuable insights. Richard White's seminal analysis of the Columbia River in the Pacific Northwest demonstrates how the interplay of energy and labor intertwined white settlers, indigenous communities, and the river, eliminating the conventional divide between humanity and nature.[7] In a more recent work, Jim Clifford investigates the industrialized marshland along London's River Lea, highlighting how environmental challenges like pollution and water scarcity triggered significant political transformations in Greater London at the turn of the twentieth century.[8]

The essay builds on these works and expands on them carefully by contending, firstly, that property owners' and urban planners' delay in instituting comprehensive flood prevention measures in the nineteenth century was due to their do-it-yourself ethos, driven by the prevalent frontier spirit. In Southern California, where financial and economic interests often took precedence over meticulous urban planning, property owners hesitated to embrace measures preventing the Los Angeles River from overflowing due to a limited understanding of the factors causing inundations. Most importantly, a dominant self-help imperative stood in the way of an overarching regional solution to the flood danger. The essay contends, secondly, that public resistance to the 1930 Olmsted-Bartholomew plan, which advocated for sustainable flood prevention by restricting development on flood-prone land, underscores how residents, business owners, and politicians prioritized economic opportunities over fostering sustainable urban settlement patterns. Consequently, Los Angeles's vulnerability to disaster was fundamentally human made, and the very essence of vulnerability was self-inflicted.

Exploring historical solutions to Los Angeles's flooding problem, the essay interweaves various threads of the city's history, incorporating both well-known aspects and original research on the contributions of everyday citizens to the region's social, economic, and political history. Initially, the essay situates the topic within a brief history of California's approach to water and land, focusing on the theme of extraction. It then proceeds to closely examine why Angelenos were quick to forget

---

[5] Gumprecht, *Los Angeles River*.

[6] Jared Orsi, *Hazardous Metropolis: Flooding and Urban Ecology in Los Angeles* (Berkeley: University of California Press, 2004).

[7] Richard White, *The Organic Machine* (New York: Hill and Wang, 1995).

[8] Jim Clifford, *West Ham and the River Lea: A Social and Environmental History of London's Industrialized Marshland, 1839–1914* (Vancouver: University of British Columbia Press, 2017). Other works include Craig E. Colten, *Unnatural Metropolis: Wrestling New Orleans from Nature* (Baton Rouge: Louisiana State University Press, 2005); Uwe Lübken, *Die Natur der Gefahr: Überschwemmungen am Ohio River im 19. und 20. Jahrhundert* (Göttingen: Vandenhoeck & Ruprecht, 2014); Sara B. Pritchard, *Confluence: The Nature of Technology and the Remaking of the Rhône* (Cambridge, MA: Harvard University Press, 2011).

recent flood experiences when dry years occurred. Subsequently, the essay delves into the history of early flood control initiatives undertaken by private individuals, portraying early Los Angeles as a do-it-yourself city. These individuals offered individualistic solutions to community or regional problems that aligned with the prevailing frontier spirit in Los Angeles at the time. After outlining the contributions of private railroad companies to flood protection measures in the 1880s and 1890s, the essay briefly traces the path toward a centrally controlled approach to flood protection that emerged after the flood of 1914. The brevity of this section is justified by the sheer number of publications investigating these developments.

Finally, the essay more extensively explores alternative flood control solutions discussed in the 1920s, notably the Olmsted-Bartholomew plan of 1930, which emphasized the significance of urban settlement regulation for flood protection. Despite its failure, this plan raises questions about how the reluctance of historical actors in the United States to address a region's vulnerability to natural disasters played out in Southern California. This question is discussed in the conclusion.

Extraction as Leitmotif of California History

The history of California unfolds as a narrative of relentless extraction, a tale of diverse groups exploiting natural resources to an unparalleled extent in comparison to other North American regions. Indigenous communities had thrived for millennia in what today we call California, adapting to the semiarid environment by managing and transforming the landscape into meticulously tended gardens, employing techniques like pruning to foster new plant growth.[9] The Spanish Franciscans exploited indigenous lives in their missions, forcing them to become Christians and slave laborers in their agricultural conquest of California.[10] The character of California's history as a narrative of extraction becomes unmistakably pronounced in the early waves of American settlement. The initial pioneers were gold miners, drawn by the allure of extracting minerals during the gold rush.[11] Subsequently, a land rush unfolded, marked by the usurpation of indigenous territories—stripped of soil, plants, and animals.[12] "Extraction," writes Mark Arax in his brilliant exploration of agriculture in California's Central Valley, "was how California was conceived in myth by sixteenth-century Spain and then born as a real place."[13]

---

[9] M. Kat Anderson, *Tending the Wild: Native American Knowledge and the Management of California's Natural Resources* (Berkeley: University of California Press, 2006).

[10] Steven W. Hackel, *Children of Coyote, Missionaries of Saint Francis: Indian-Spanish Relations in Colonial California, 1769–1850* (Chapel Hill: University of North Carolina Press, 2005), 127ff.

[11] Benjamin Madley, *An American Genocide: The United States and the California Indian Catastrophe, 1846–1873* (New Haven: Yale University Press, 2016), 67–73.

[12] Ibid., 157–72.

[13] Mark Arax, *The Dreamt Land: Chasing Water and Dust across California* (New York: Knopf Doubleday Publishing Group, 2019), 35.

The history of water in Los Angeles during the nineteenth and early twentieth centuries reflects an ongoing struggle to exploit nature through technological interventions.[14] Angelenos diverted the waters of the Los Angeles River through open irrigation ditches to their farmland, built an aqueduct to steal the waters of the Owens River some 215 miles northeast, effectively ending agriculture in the Owens Valley, and drilled holes in the ground to suck the last drop of water out of the bone-dry soil.[15] This endeavor unfolded in two distinct but interrelated phases: the first involved reshaping the semiarid landscape into a vision of a vast Mediterranean garden, while the second entailed attempts to mitigate the constant threat of natural disasters.

The allure of Los Angeles as a sunny utopia, propagated by influential city builders like Charles Nordhoff, Charles Dwight Willard, Charles Fletcher Lummis, Harrison Gray Otis, Harry Chandler, and Abbot Kinney, played a pivotal role in attracting settlers and tourists to Southern California.[16] These newcomers, enticed by the "myth of climate,"[17] brought financial capital, business ideas, and specific expectations for the city's appearance. Their influence transformed the landscape into an agricultural powerhouse, turning the once dusty town into a rapidly developing American metropolis, recognized for its vineyards, orchards, and green lawns amid a landscape parched by drought.[18] The settlers extracted resources ad libitum, fostering a growing divide between human aspirations and the region's limitations. Perhaps, at the heart of Southern California's civilization woes lay a profound misunderstanding—the desire for more than the land could sustain.

In Southern California, rain is a rarity, but when it does come, it can fall in torrents. Precipitation typically ranges between four and forty inches per year, although the variability can be significant. The public has come to know this region as the "Land of Little Rain," the beautiful title of a book written by Mary Hunter Austin.[19] The amount of rainfall depends on factors such as water temperature in the equatorial Pacific and atmospheric pressure. El Niño, a weather phenomenon associated with above-average water temperatures in the equatorial Pacific, often results in increased precipitation in Southern California. Conversely, higher-than-average air pressure in the Gulf of Alaska can alter the strength of the jet stream during winter,

---

[14] For an in-depth exploration of how this process unfolded nationwide, David E. Nye, *American Technological Sublime* (Cambridge, MA: MIT Press, 1994).

[15] James Tejani, "Dredging the Future: The Destruction of Coastal Estuaries and the Creation of Metropolitan Los Angeles, 1858–1913," *Southern California Quarterly* 96, no. 1 (2014), 5–39: 19.

[16] Lawrence Culver, *The Frontier of Leisure: Southern California and the Shaping of Modern America* (Oxford: Oxford University Press, 2010), 15–51.

[17] Norman M. Klein, *The History of Forgetting: Los Angeles and the Erasure of Memory* (London: Verso, 2020), 19.

[18] Douglas C. Sackman, *Orange Empire: California and the Fruits of Eden* (Berkeley: University of California Press, 2005).

[19] Mary Hunter Austin, *The Land of Little Rain* (Boston: Houghton Mifflin Company, 1903).

redirecting cold fronts and rainfall into the region.[20] As air from the Pacific Ocean flows over the Los Angeles basin, it encounters the surrounding mountain wall, causing it to rise. The ascending air becomes less capable of holding moisture, leading to precipitation primarily in the foothills and mountains. Rain in the mountains gives rise to rivulets and streams that converge and grow more potent. The water ultimately trickles into a vast underground lake beneath the San Fernando Valley—the origin of the Los Angeles River.[21]

Before the 1930s channelization of the Los Angeles River, its course was undefined, merging with open marshes and tulip fields. The river kept shifting course, at times flowing west toward present-day downtown and at other times heading directly south to San Pedro Bay.[22] During heavy rains or river flooding, the high alkali content in the soil, resulting in caliche formations, prevented water absorption. Therefore, rainfall led to devastating floods in various years, wreaking havoc on farmlands, buildings, and infrastructure.[23] Over 336 square miles were susceptible to flooding before municipal authorities implemented effective flood protection programs.

Forgetting About Past Floods

It was the interplay of water temperature and atmospheric pressure in the Pacific that produced precipitation patterns in Southern California, and the high alkali content in the arid soil that contributed to the river's tendency to overflow, but these factors were not primarily responsible for the destruction done to human civilization. As Gumprecht asserts, it was the rapid urban and industrial development in Los Angeles that created "an environment ripe for disaster."[24] Because in Los Angeles, the value of land rested on its proximity to water, a crucial element for cultivating the semiarid landscapes, settlers expanded their farmland into floodplains, vying for access to higher groundwater tables and the ability to divert water without city permits. They sometimes even constructed their houses in the dry

---

[20] Mike Davis, *Ecology of Fear: Los Angeles and the Imagination of Disaster* (New York: Metropolitan Books, 1998), 5.

[21] Orsi, *Hazardous Metropolis*, 3–8.

[22] United States Army, Corps of Engineers, Engineer Office, "Los Angeles County Flood Control Research," 1914–1915, compiled by J. W. Reagan, 2 Vols., Huntington Library, San Marino, CA, 12–13.

[23] The river became a raging torrent that flooded the land for miles in 1825, 1832, 1842, 1851–52, 1858–62, 1867–68, 1875–76, 1884, 1886, 1889, 1891–92, and 1914. See the narratives compiled in "Los Angeles County Flood Control Research," Huntington Library; J. M. Guinn, "Exceptional Years: A History of California Floods and Drought," *Historical Society of Southern California, Los Angeles* 1, no. 5 (1890), 33–35.

[24] Gumprecht, *Los Angeles River*, 198.

riverbed, colloquially described as being "on the sand."[25] No one hindered them from developing the floodplains, as zoning regulations were to come much later.[26]

The very tracts positioned for prosperity were the first to be inundated when the river overflowed during winter and spring. This vulnerability to floods was exacerbated by the removal of trees, bushes, and smaller plants from the riverbanks for use as building materials. The practice weakened the riverbanks, making them less resistant to swelling water flow. Intensive farming further contributed to soil erosion and increased runoff, diminishing the soil's ability to absorb water. Some farmers saw flooding as beneficial for the enrichment of their soil, but they usually changed their minds when they saw what the river could do to their land.[27]

Not only was there remarkably little concerted effort by the municipal government to address the danger of river flooding, but it is also noteworthy how swiftly the residents of Los Angeles tended to forget the perils of heavy rains and inundations. Even after devastating floods in 1884, 1886, 1889, and 1891–1892, urban dwellers continued to build houses on the floodplain and cultivate land near the river. The dramatic flood of 1884, for example, did not lead to sustainable flood control measures because Los Angeles experienced a burgeoning real estate boom a few years later.[28] Many new settlers who had never experienced a flood before arrived in the region. They were hungry for cheap land and, with no memory of past floods, bought tracts of land where they thought they could realize their business dreams.

Similar stories can be told about other floods. Even in 1914, when Los Angeles found itself underwater again, residents appeared surprised by the fact that the river could flood and inundate parts of the city. That year, the *Los Angeles Times* admitted that the city, having never anticipated such a forceful flood, was unprepared for the disaster.[29] Who could blame them? Between episodes of river flooding, there lay extended periods of droughts, causing the Los Angeles River to dwindle to a dry creek.[30] For instance, the years between the floods of 1875–1876 and 1884 were marked by below-average rainfall, reinforcing this sense of false security.[31] Most years of the 1890s were again so exceptionally dry that even Harry F. Stafford, who served as city engineer at the time, urged his fellow citizens to "give considerable

[25] "Los Angeles County Flood Control Research," Huntington Library, Vol. 1, 36

[26] Gumprecht, *Los Angeles River*, 209.

[27] "Los Angeles County Flood Control Research," Huntington Library, Vol. 2, 459.

[28] Joseph Netz, "The Great Los Angeles Real Estate Boom of 1887," *Annual Publication of the Historical Society of Southern California, Los Angeles* 10, no. 1/2 (1915).

[29] "No Unemployed," *Los Angeles Times*, February 25, 1914. Also quoted in Orsi, *Hazardous Metropolis*, 15.

[30] Guinn, "Exceptional Years," 36–37.

[31] Only the winter of 1884 brought intensive downpours caused by El Niño conditions in the Pacific. According to H. Greenewalt, who witnessed the flood of 1884, about forty to fifty houses were destroyed, and several residents died in the floods. "Los Angeles County Flood Control Research," Huntington Library, Vol. 1, 27.

attention to the storm drain problem."[32] Most Angelenos were quick to forget about past floods. As Norman M. Klein convincingly shows, the erasure of lived experiences was central to the creation of the Los Angeles myth.[33]

The demographics of Los Angeles's population and the long periods of drought between floods caused an enduring disconnect between the reality of natural disasters and the idealized vision propagated by boosters. In the booster era, disasters like floods, wildfires, and earthquakes were inconceivable and belonged to a world left behind by migrants in the East or Midwest. Boosters sold the promise of sunshine, not the potential for dystopia. Early residents either perceived floods as exceptions or conveniently forgot about them, focusing on obtaining more water for building their homes and maintaining their green lawns or gardens in a semidesert environment rather than dealing with an abundance of floodwater or conserving what they had.

The lack of environmental knowledge about the history of flooding in Southern California and precipitation patterns in desert-like ecosystems was intricately tied to the marginalization of Hispanic residents who held valuable knowledge about past floods. Ethnic prejudice, as Orsi contends, hindered the dissemination of this knowledge and the construction of flood protection infrastructure.[34] Stories of flooding were well-known in Mexican communities, passed down through generations, yet Euro-American settlers, eager for real estate and business opportunities, ignored this knowledge.[35] This negligence, driven by a desire to displace Mexican culture and traditions, contributed to what William Deverell calls the "whitewashing" of Los Angeles.[36] The city thus manifested an asymmetry of knowledge and power, where Hispanic culture sought to coexist with nature, giving water the necessary space, while American settlers aimed to conquer and exploit nature through technological means.

Do-it-Yourself Flood Control

If there was flood protection in nineteenth-century Los Angeles, it was organized by everyday citizens, private landowners, and railway companies, and only rarely by the city administration itself. Due to a weak municipal government in Los Angeles,

[32] Harry F. Stafford, Annual Report of City Engineer, City of Los Angeles, Cal., For the Year Ending Nov. 30, 1901, December 16, 1901, Los Angeles City Archives, Box B-2294, Folder "City Engineer 1901."

[33] Klein, *The History of Forgetting*, 31–38.

[34] Orsi, *Hazardous Metropolis*, 16.

[35] See the interviews conducted after the 1914 flood, documented in "Los Angeles County Flood Control Research," Huntington Library.

[36] William Deverell, *Whitewashed Adobe: The Rise of Los Angeles and the Remaking of Its Mexican Past* (Berkeley: University of California Press, 2004), 121.

urban development significantly relied on private initiatives.[37] This was as true for flood control as for constructing other infrastructures, such as roads, bridges, and irrigation ditches.[38]

Los Angeles was a do-it-yourself city. Until the end of the nineteenth century, there was arguably little urban planning efforts by municipal authorities.[39] Archival evidence reveals a dominant pattern: migrants from other parts of North America or Europe arrived in Southern California, purchased a large or small tract of land, erected a private residence, developed their land by planting crops like citrus fruits, wheat, or alfalfa, and opened businesses in the Downtown area of Los Angeles. Observing that Los Angeles's urban development lacked behind those places that they had left for Southern California, these middle-class, white residents frequently wrote letters to municipal authorities demanding urban improvements: the grading of a street, the laying of a sewer, or the piping of an open irrigation ditch. In addition, urban dwellers alarmed the mayor or the Common Council about stagnant water in low-lying land, which an increasing number of residents toward the end of the century decried as a hygienic concern. Private individuals became the first community planners of Los Angeles.

Urban infrastructure formed an intricate network: where sewers were laid and streets were graded; where streets were improved, storm drains became an issue. This was not necessarily about large-scale flooding of the Los Angeles River. Instead, it was more about small-scale inundations that involved privately held property. In many cases, the poor condition of urban infrastructure caused residents' properties to flood. J. C. Zahn, a notorious letter writer to the Council, was among those who were particularly alert to the flood risk of his property and frequently petitioned the Council to address issues like bursting pipes or broken ditches. In January 1890, he complained about his land along Flower Street between Third and Fourth Streets being flooded due to a burst conduit after heavy rainfall in Los Angeles at the end of the previous year. Zahn blamed the mismanagement of a reservoir's gates for the flooding:

> For we old residents have made our observation for the last 10 years or more, whenever the rainstorm is at its highest downpouring, then the floodgates of the above-named reservoir are let loose, in addition to the heavy downpour, and then woe to the residents below, who are at the mercy of the floods caused by the opening of the floodgates, in addition to the raging rainstorm, which tears up and washes away graded streets, which has, and will cost the City larger amounts of money for repairs etc.[40]

---

[37] Robert M. Fogelson, *The Fragmented Metropolis: Los Angeles, 1850–1930* (Berkeley: University of California Press, 1993), 28–42.

[38] As a source of inspiration from a different historical context, Arne Kaijser, "System Building from Below: Institutional Change in Dutch Water Control Systems," *Technology and Culture* 43, no. 3 (2002), 521–48.

[39] This only changed after around 1900, Greg Hise, *Magnetic Los Angeles: Planning the Twentieth-Century Metropolis* (Baltimore: Johns Hopkins University Press, 1997).

[40] J. C. Zahn, January 27, 1890, 99, Los Angeles City Archives, Box B-4, Vol. 73.

According to Zahn, the flooding of his property resulted from the interplay between natural events, such as rainstorms, and deficiencies of human-made infrastructure, specifically the sizes of the reservoir and stormwater drains. In a subsequent letter penned about two months later, Zahn elaborated on the extent of the damages. One of his two houses on the lot "fell in consequence down with a crash."[41] Zahn admitted that he was able to repair the damage "but it will never be as good as it was before." In total, Zahn claimed $1250 in damages from the city. After 6 months, Zahn wrote a third letter to the Council, requesting repairs at the intersection of Flower and Third Streets and the addition of gutters to prevent water runoff onto his property. He pointed out that the brick conduit, stretching around 300 feet between Fourth and Fifth Streets, was filled with sand, leaving only about 6 inches of clearance for water flow. Consequently, he expressed concern that the conduit would be ineffective in handling the upcoming winter storms.[42]

What is important here is that private individuals took the lead in demanding that the municipal administration build effective flood control measures. In other words, the city government relied on private parties to identify where urban stormwater infrastructure was dysfunctional or needed improvement. In 1889, resident Charles Bauer found himself facing a similar issue as Zahn did. He lived on the corner of Temple and Beaudry Streets and complained that his property had been seriously damaged by water from the street during recent rains. He stated that his cellars were flooded, and the water had caused damage to the walls, the building in general, as well as the sidewalks and streets. The necessary repairs would be a significant hardship for him, he said, particularly considering the substantial amount of money he had invested in improvements on his lot and street grading. He believed that a municipal catch basin or storm drain was necessary at the corner fronting his property.[43]

Property owners often engaged in this kind of water activism, urging the city to address issues with malfunctioning storm drains or poorly designed street grades that led to stormwater accumulating in front of, around, or on their properties. Identifying themselves as taxpayers, a group of residents at 15th, Dorchester, Normandie, and Dewey Streets alerted the Council that vacant lots in their neighborhood were flooded when heavy rains came.[44] Property owners at the intersection of Fifth and Wall Streets revealed that, due to street grade adjustments, Wall Street's level fell below that of Fifth Street. Consequently, rainwater from the north of Fifth Street collected at the northeast intersection, forming a large pool on Wall Street. This obstructed pedestrians, forcing them to wade through or detour across Fifth Street to the east of Wall Street.[45]

---

[41] J. C. Zahn, March 22, 1890, 338, Los Angeles City Archives, Box B-5, Vol. 74.

[42] J. C. Zahn, October 25, 1890, 892, Los Angeles City Archives, Box B-6, Vol. 77.

[43] Charles Bauer, October 22, 1889, 831, Los Angeles City Archives, Box B-4, Vol. 72.

[44] Petition to Mayor and City Council, January 3, 1910, Los Angeles City Archives, Box B-1272, Folder "Health Commission 1913."

[45] Chas W. Baldwin, October 13, 1890, 869, Los Angeles City Archives, Box B-6, Vol. 77.

Importantly, almost all residents who placed their signatures under letters or petitions to the Council had names that likely originated in Western European countries such as Great Britain, Ireland, Germany, France, Belgium, and The Netherlands. There were also some Italian and Portuguese surnames, but they constituted a clear minority. Furthermore, the number of Mexican or Spanish family names was surprisingly low, given the fact that California, some four decades earlier, had been Mexican. The dominance of White, Euro-American residents among the letter writers testifies to the power structures in Los Angeles toward the end of the nineteenth century.

Non-White residents of Los Angeles suffered from racial discrimination and, quite tangibly, from the poor condition of or lack of access to infrastructure in their neighborhoods. Streets were first paved in front of businesses and neighborhoods inhabited by Euro-American newcomers. Chinese neighborhoods, one of the most significant minorities in the city, were systematically excluded from infrastructure improvements because the city charter required landowners to pay a property line fee to fund the paving of streets or the construction of underground water mains.[46] Most Chinatown residents near the old town square neither owned their homes nor had the financial means to pay the frontage assessment. Ethnic prejudice and social conceit intersected and were inscribed into the city's laws.[47] Moreover, the location of the Chinese quarter in a low-lying portion of the city close to the river led to frequent inundations of Chinese homes after rainstorms. In 1898, about 90% of Chinatown was owned by two respected members of the Los Angeles business and political elite, Sepulveda and Schafer.[48] The unsanitary conditions and the unwillingness to mitigate the flood risk thus have to be attributed to systematic neglect.

Unlike Chinese residents, White Angelenos were deeply committed to safeguarding their properties against inundation, advocating for the construction of additional storm drains to manage excess water effectively.[49] In many instances, they went beyond mere advocacy, providing detailed suggestions for the design and implementation of stormwater pipes. This proactive involvement showcased their willingness to take responsibility for the maintenance of essential infrastructure.[50] It highlights that the traditional boundary between those who make public services available and those who utilize them—essentially, the dichotomy between producers and consumers of urban infrastructure—was blurred in this do-it-yourself city at the fringes of Euro-American civilization. Property owners, far from seeing themselves solely as entitled consumers, embraced a sense of responsibility for enhancing flood control measures in their newly adopted hometown.

---

[46] *Revised Charter and Compiled Ordinances and Resolutions of the City of Los Angeles. Compiled and Indexed by Wm. M. Caswell*, (Los Angeles: Evening Express Steam Printing Establishment, 1878), 132–35.

[47] David Samuel Torres-Rouff, *Before L.A.: Race, Space, and Municipal Power in Los Angeles, 1781–1894* (New Haven: Yale University Press, 2013), 219–27.

[48] Health Officer to City Council, June 26, 1898, Los Angeles City Archives, Box B-1272, Folder "Health Officer 1899."

[49] Joseph Hyans, December 16, 1889, 1031, Los Angeles City Archives, Box B-4, Vol. 72.

[50] W. H. Workman, September 16, 1889, 672, Los Angeles City Archives, Box B-4, Vol. 71; Capitol Milling Co., February 15, 1890, 210, Los Angeles City Archives, Box B-4, Vol. 73.

An Individualistic Solution to a Community Problem

Until the major flood of 1884, flood control was perceived as an individual responsibility rather than a task for the government. Remarkably, in the 1860s and 1870s, there were only a few calls for the government to protect citizens from flood dangers. When a private individual wrote an angry letter to the Common Council, they sought reimbursement for the damage that floods had caused to their property. Only rarely did they express expectations that flood protection was the responsibility of municipal authorities. How did Angelenos then organize this individualistic approach to flood control without coercion from above?

In the view of members of this settler society, these individualistic solutions to a community problem seemed justified in the face of the extraordinary circumstances they confronted in the far American West. Property owners flocked to Los Angeles with dreams of success, aspiring to cultivate large vineyards or orchards, cattle, or dairies, or make a living as butchers, bakers, mill owners, constructors, entrepreneurs in garments and textiles, or pioneers in the burgeoning oil industry. The city's promise of opportunity attracted a diverse array of individuals seeking prosperity in various fields. Part of Los Angeles's promise was the absence of state regulation, or so they thought.

Settler societies in the American West during the nineteenth century were characterized by a unique set of norms and values shaped by westward expansion and justified by the Manifest Destiny ideology. While it is important to note that these tropes related to the frontier varied among different groups and individuals in Southern California, many residents of early Los Angeles shared the belief in some general characteristics. They emphasized masculine self-reliance and independence. Most importantly, they expected their fellow settlers to be capable of handling the harsh challenges of the frontier environment, such as frequent floods.

The rules and institutions that were derived from these frontier tropes to coordinate this individualistic approach to flood control were important for the reproduction of power. In the absence of a strong central state, self-governance through local institutions was common. Local institutions in Los Angeles relied on informal mechanisms of self-help, and individuals were expected to take responsibility for their well-being, including effective flood control. This is why most urban dwellers in nineteenth-century Los Angeles accepted flood protection to be the responsibility of the individual citizen rather than a state function. They did not approach flood control as a regional problem requiring a concerted effort, allocation of financial resources, and the establishment of powerful agencies. Instead, flood protection was localized and focused on safeguarding private properties from overflowing pipes or ditches.[51]

Municipal authorities did play a role in local flood-risk management, but only to the extent that individual property owners asked for their assistance. The state's role was reactive, not proactive. Consequently, flood protection measures were not necessarily implemented along the river but as near as possible to private properties.

---

[51] Gumprecht, *Los Angeles River*, 174.

Eventually, flood control until 1884 was not about constructing levees or prohibiting settlements in floodplains.[52] It involved enlarging reservoirs, replacing storm drains with larger ones, and filling up low-lying land. However, these solutions were short-lived, limited in scope, and relied on manipulating existing urban infrastructure.

Confining the River

In nineteenth-century Los Angeles, where the provision of public services was lacking or was often not even wanted by residents, the city government was delegating the responsibility of safeguarding the community from floods to private parties. An illustrative example of this outsourcing occurred in the aftermath of the 1884 flood. Following the destruction of many buildings and farmland, as well as an unknown number of deaths, the Common Council approved the granting of a 50-foot right of way on the west bank of the river to the Southern California Railway.[53] The condition for this privilege was the construction and maintenance of a levee from Mission Street south to First Street. Eventually, the railroad extended the levee along the city's southern boundary, covering a total of 3.7 miles, and completed the project in March 1888. While the city undertook the task of constructing a levee on the opposite riverbank, it did not match the same length. Both levees suffered severe damage in the flood of 1889–90.[54]

During the last two decades of the nineteenth century, railway companies frequently undertook levee construction, laying elevated tracks on railroad embankments. This arrangement proved advantageous for the city, as the elevated tracks served the dual purpose of facilitating mobility and preventing flood damage. Private railroad companies were, to some extent, the pioneers in Southern California to adopt a large-scale approach to flood control. However, their strategy came with inherent problems. By confining floodwaters to a smaller area than the natural floodplain, it intensified the force of the water. The use of trestle bridges by the railway companies also obstructed the unrestricted flow of water.[55] These structures,

---

[52] However, in December 1870, fifteen property owners along the river urged the Council to "take immediate steps [...] to avert the threatened danger." The Council minutes suggest that the city installed some form of flood protection, which was destroyed soon afterward. "Minutes of the Common Council," December 22, 1870, Los Angeles City Archives.

[53] "An ordinance to provide for the construction of a levee upon the westerly side of the Los Angeles river, for the protection of property of inhabitants of said city from the high waters of said river, and for sale of city lands and grant of right-of-way to the Riverside, Santa Ana and Los Angeles Railway Company, for the construction of such levee," adopted December 8, 1886, *Compiled Ordinances and Resolutions of the City of Los Angeles, Vol. III, Compiled and Indexed by Freeman G. Teed, Clerk of the Council*, (Los Angeles: E. H. Freeman, Book and Job Printer, 1887), 224–31: 227.

[54] Gumprecht, *Los Angeles River*, 161–62.

[55] As one resident recounted, "The greatest cause of the overflow and damage is caused by the way the railroad bridges and wagon bridges are built. If you take all of the piles and obstruction out of

supported by closely spaced wooden piles driven into riverbeds, often acted as dams during heavy storms. The accumulation of debris behind the trestles exacerbated the issue, forcing water out of stream channels and onto adjacent lands.[56]

While the lack of knowledge about adequate flood-risk management infrastructure posed a significant challenge in safeguarding the community from inundations, it was not impossible to identify the roots of the problem. In 1891, a concerned citizen addressed the Common Council, expressing frustration over the city's disruption of railroad service with the rest of the country and the halt in mail delivery after floods. The petitioner then urged the city to compel railroad companies to install proper bridges, citing the current issue of small spans of trestle work trapping floating trees and logs. The solution he proposed included requiring railroad companies to adopt larger spans of the trestle, provide ample headroom, construct long-span bridges instead of short ones, establish deep and stable foundations, and utilize iron skew bridges.[57]

There are also traces in the archives showing that some residents were not only aware of why railroad embankments and bridges over floodplains exacerbated the flood risk but also why the river's overflowing damaged the built environment. In an article published in the *Los Angeles Times* in the summer of 1882, J. J. Warner, the head of the Canal and Reservoir Company that provided households with drinking water, identified settlement patterns in the floodplains as the primary cause of flood damage.[58] In a letter to the newspaper just a few weeks later, he referred to a lack of knowledge about past floods as the biggest obstacle to effective flood control measures.[59] However, Cassandra's calls from citizens like Warner went unheard or were forgotten in the next dry spell.

Therefore, problems kept persisting. In 1899, the city engineer admitted that both the urban storm drain system and flood control infrastructure along the river line remained in their infancy. In general, railroad companies were doing great in maintaining the levees, but "a large section in Los Angeles" during heavy downpours was "totally unable to handle the runoff without serious inconvenience and loss to many people."[60] Even when the city had a proper storm drain system, it would still be difficult for city workers to handle all the debris from the hill waters.

---

the river, you have done away with most of the cause of the damage." "Los Angeles County Flood Control Research," Huntington Library, Vol. 2, 407

[56] Gumprecht, *Los Angeles River*, 151–53.

[57] Petition No. 135, February 24, 1891, Los Angeles City Archives, Box B-6, Vol. 78.

[58] J. J. Warner, "A Warning," *Los Angeles Times*, July 30, 1882.

[59] J. J. Warner, "Letter to the Editor of The Times," *Los Angeles Times*, August 6, 1882. Also quoted in Gumprecht, *Los Angeles River*, 154.

[60] City Engineer, Annual Report 1899, Annual Reports 1898–1900, 8, Los Angeles City Archives, Box B-110, Vol. 237.

Toward a Centrally Controlled Approach to Flood Protection

The turning point in the history of flood control in Los Angeles came in 1914. Caused by an El Niño event, heavy rains fell over Southern California. As a result, about 2000 acres were flooded.[61] In the wake of the flooding, more and more residents became convinced that the city needed better flood protection. The emerging insurance industry also pressed Angelenos hard to protect their property better against flood damage. Finally, business leaders, engineers, and politicians decided that it was time to move away from individualized flood protection solutions (undertaken by landowners and privately owned companies) to create a centrally controlled authority to develop a countywide plan based on scientific knowledge. Ultimately, the establishment of the Los Angeles County Flood Control District in 1915 reflected the progressive spirit of the early twentieth century, with its reliance on the problem-solving abilities of objective experts and the creation of new political forms of urban governance.[62]

Under the direction of Chief Engineer James W. Reagan, the Los Angeles County Flood Control District prepared comprehensive flood control plans in the years that followed. Among other things, engineers conducted extensive interviews with (Euro-American and Mexican) residents who had witnessed flooding in Los Angeles in the nineteenth century.[63] They also traveled to Europe to learn about flood control in the Alps from French, German, Swiss, and Austrian engineers, transferring their knowledge to Southern California.[64] As so often before, however, implementation of measures proved difficult, and several construction projects remained unfinished or were never begun. Reasons included disagreement among flood control engineers, members of the City Council, and the influential Chamber of Commerce over who should be required to pay for the construction work—those who lived in flood-prone areas and would benefit or the entire community.[65] In addition, with the years between 1914 and 1934 being dry, voters were increasingly unwilling to approve more tax money for flood protection.

On New Year's Day 1934, after three days of heavy rain and preceding wildfires, masses of water, debris, and mud rushed through La Cañada Valley, Glendale, and Los Angeles. On that day, approximately 39 people died, 198 homes were destroyed, and more than 400 homes became uninhabitable.[66] This event drew national media

---

[61] Gumprecht, *Los Angeles River*, 167–71; Davis, *Ecology of Fear*, 16–17.

[62] Gumprecht, *Los Angeles River*, 179; Orsi, *Hazardous Metropolis*, 37.

[63] The engineers interviewed numerous Mexican residents. Some were of advanced age and recounted their memories; others recalled stories they had heard from parents, grandparents, neighbors, or friends. "Los Angeles County Flood Control Research," Huntington Library.

[64] Barrington Moore, "Checking Floods in the French Alps," Reports of the Board of Engineers Flood Control to the Board of Supervisors, Los Angeles County, California, Submitted July 27, 1915, [Los Angeles: publisher not identified, 1915], 81–84, Huntington Library.

[65] Gumprecht, *Los Angeles River*, 178; Orsi, *Hazardous Metropolis*, 47.

[66] Harold C. Troxell and John Q. Peterson, "Flood in La Canada Valley, California, January 1, 1934," in *Water Supply Paper 796C* (Washington, D.C.: U.S. Geological Survey, 1937), 68–69.

attention to Los Angeles's flooding problem, and it convinced the federal government to appropriate significant financial means to channelize the Los Angeles River, erect dams and catch basins in the San Gabriel Mountains, and construct storm drains across town. As the extent and magnitude of these measures are concerned, this can be read as an overcompensation of previous missed opportunities to protect the region's communities from flood damage.

Parks, Playgrounds, Parkways

The federal government's solution to the flood risk of the Los Angeles River was as radical as it could be. In confining the river to a 51-mile concrete riverbed, the Corps of Engineers tamed this stream with scientific knowledge and the brutish force of landscape engineering. However, in doing so, as Mike Davis convincingly shows, they brought about the "death of the Los Angeles River."[67] They destroyed the river's ecosystem and killed fish cultures, birds, and plants. They also opened floodplains to commercial development, sacrificing nature for profit. However, such an approach was not inevitable, as the Olmsted-Bartholomew plan of 1930 shows.[68]

The roots of this plan lay in the rapid urban growth following the completion of the Los Angeles Aqueduct in 1913, convincing city leaders that some form of urban planning was necessary. As was so often the case in this region characterized by weak political institutions but strong economic aspirations, the Chamber of Commerce took the lead.[69] For the Chamber's board members, urbanization in the past decades had overwhelmed the region. More precisely, what Los Angeles was lacking was a meaningful number of recreational spaces—parks, playgrounds, and simplified access to the mountains and beaches of the region.[70]

It was correct that the City of Los Angeles had fewer parks and playgrounds than other cities in the nation. But was that a problem? Los Angeles was a sprawling city with horizontal growth and only a few multistory buildings in the downtown area. The Santa Monica or San Gabriel Mountains were not far away, and beaches were within commutable distance for those who could afford to buy an automobile. However, providing urban dwellers with green spaces for recreational activities in their neighborhoods followed the City Beautiful idea that informed urban planning in many American cities at the onset of the twentieth century. A city should not only be functional but also aesthetic; it should offer the structures that facilitate social

[67] Davis, *Ecology of Fear*, 69.

[68] The plan is reprinted in Greg Hise and William Deverell, *Eden by Design: The 1930 Olmsted-Bartholomew Plan for the Los Angeles Region* (Berkeley: University of California Press, 2000), 65–283.

[69] On the role of the Chamber of Commerce in the history of Los Angeles, Sara S. Elkind, *How Local Politics Shape Federal Policy: Business, Power, and the Environment in Twentieth-Century Los Angeles* (Chapel Hill: University of North Carolina Press, 2011).

[70] Greg Hise and William Deverell, "Private Power, Public Space," in *Eden by Design*, 1–63: 2–3.

life, while also uplifting the morals of its residents. From this angle, the urban built environment was supposed to function as a means of social engineering, and the parks and playgrounds should "improve public health, reduce delinquency, and promote citizenship," as Hise and Deverell write.[71]

After extensive deliberations, the Chamber engaged the services of two prominent landscape and planning firms, namely Olmsted Brothers and Harland Bartholomew and Associates, to formulate a comprehensive plan. Building on their experience in cities like New York, Boston, Chicago, and Washington, D.C., and their knowledge of the spatial expanse of Los Angeles gained through a mobility study in 1924, Olmsted and Bartholomew were certainly well prepared for this task.[72] What they ultimately came up with was, according to Hise and Deverell, a "remarkable document."[73] They not only identified the deficiencies in urban space but also presented suggestions for establishing new parks, playgrounds, and parkways.[74] Their report breathed the spirit of the early twentieth-century belief in the moral order of social uplift through access to nature. Car-dependent mobility was not contrary but an intrinsic part of the proposed solution.[75]

Roads Not Taken

What makes the Olmsted-Bartholomew plan particularly intriguing is its innovative approach to flood control, emphasizing land use regulation and urban settlement patterns. The proposal recommended a prohibition on constructing buildings within floodplains, designating these areas for recreational parks and playgrounds instead, thus creating "a web of greenbelts that would be used to absorb swollen rivers during floods and devoted to recreation, auto touring, and transport the rest of the time."[76] An illustrative example was the Arroyo Seco, running from Altadena at the foothills of the San Gabriel Mountains to Downtown Los Angeles. The report identified the Arroyo Seco as an ideal route for a pleasure drive, accompanied by associated parkland that could serve as floodland after heavy rains.

---

[71] Ibid., 3.

[72] "A Major Traffic Street Plan for Los Angeles. Prepared for the Committee on Los Angeles Plan of Major Highways of the Traffic Commission of the City and County of Los Angeles by Frederick Law Olmsted, Harland Bartholomew, Charles Henry Cheney" (Los Angeles, May 1924), University of Southern California Libraries Special Collections, California Historical Society collection of Los Angeles Area Chamber of Commerce records and photographs, 0245–1, Box 73.

[73] "Preface to a Master Plan," in *Eden by Design*, vii.

[74] "Parks, Playgrounds, and Beaches for the Los Angeles Region: A Report Submitted to the Citizens' Committee on Parks, Playgrounds, and Beaches, by Olmsted Brothers and Bartholomew and Associates, Consultants [1930]," in *Eden by Design*, 83–91.

[75] Hise and Deverell, "Private Power, Public Space," in *Eden by Design*, 35.

[76] Orsi, *Hazardous Metropolis*, 105; Hise and Deverell, "Private Power, Public Space," 14–16.

The section titled "Parks and the Drainage Problem" in the report explicitly outlined the risks associated with uncontrolled population growth and the subdivision of flood-prone areas into house lots, business parcels, and industrial sites.[77] According to the authors, low-lying areas perpetually carried the risk of inundation, cautioning that "no matter how innocent it may look in dry weather, lowland must always be far less valuable for building purposes than other land."[78] The report highlighted the potential pitfalls of financial greed, ignorance, and forgetfulness among those who developed low-lying areas. Instead, it advocated understanding "what storm water can do in this country," referencing the region's flood history.[79] Knowing this history, the report argued, was crucial for making informed decisions about settlement patterns, particularly in regions with extended periods of drought between floods, such as Southern California.

However, the report emphasized that floodplains were not devoid of value; on the contrary, they were "just as good as any other for providing spaciousness of open scenery for parks and parkways."[80] Furthermore, acquiring low-lying land for municipalities was economically advantageous. The recommendation put forth was to prohibit "costly improvements in floodways unless and until adequate spaces have been set apart for handling the maximum floods and the floods have been confined to them by permanent channels, reservoirs, and reserved areas for percolation into the ground."[81] This indicated that the report did not outright oppose the implementation of flood protection measures, which were eventually introduced by the federal government a decade later. Rather, it displayed the foresight to caution against the development of floodplains without the presence of such essential flood control infrastructure.

While the report's proposals were innovative, it is important not to overstate their novelty. Establishing recreational spaces on flood-prone land was already a common practice in other American cities. For instance, Boston restricted urban development in several flood-prone areas. Notably, in 1926, the Olmsted Brothers had suggested a similar plan for the "Nigger Slough" in Los Angeles, an area near sea level with insufficient drainage. In their letter, they clearly articulated that low-lying land should only be used for agriculture and parks, emphasizing the dangers of uncontrolled private development near sea level.[82]

The Olmsted-Bartholomew plan not only provided an alternative to the channelization of the Los Angeles River but also presented a potential vision for the overall development of the Los Angeles region.[83] However, this alternative path was not

[77] "Parks, Playgrounds, and Beaches for the Los Angeles Region," 96–98.

[78] Ibid., 96.

[79] Ibid.

[80] Ibid.

[81] Ibid.

[82] Appendix No. III: Copy of Letter on Nigger Slough and Other Lands Lying Below Possible Drainage Levels, Submitted by Olmsted Brothers on May 21st, 1926, to the Board of Supervisors of Los Angeles County, in *Eden by Design*, 253–255.

[83] Hise and Deverell, "Private Power, Public Space," 52.

pursued. By February 1929, even before the planners could finalize their report, the Chamber's board of directors decided not to follow their suggestions. Reasons remain unclear, but they likely stemmed from concerns about the financial burden on Los Angeles taxpayers imposed by the proposed measures.[84] Furthermore, the report's proposal to establish new government agencies for overseeing parks and parkways was dismissed by the Chamber's board due to its resistance to political regulation.[85]

According to Olmsted and Bartholomew, public control should have played a decisive role in addressing the region's flood problem and park shortage. Specifically, they argued that only public control could prevent real estate speculation and unwise decisions about construction activities in floodplains.[86] This would have marked a stark departure from the previous approach to flood-risk management in Los Angeles, where private initiative was the driving force behind flood protection. Unsurprisingly, the Chamber showed little interest in creating institutions in Los Angeles that could potentially rival its political power.

The irony of the Chamber's success in suppressing the report's suggestion is that only a few years later, in the mid-1930s, the federal government did assume a prominent role and became a powerful player in the region, which the Chamber had dreaded in the 1920s. This involved creating the very institutions that the business elite had attempted to sidestep. The federal government's approach differed in many ways from the comprehensive planning envisioned by Olmsted and Bartholomew in their report. However, the federal government's involvement discouraged private initiatives in flood control and infused large-scale engineering into Southern California.[87]

Conclusion

If it is correct that certain actors in the history of the United States were hesitant to properly acknowledge and work toward preventing natural risks, then the example of Los Angeles reveals how this reluctance played out on the ground, and how the measures finally taken can be read as a kind of overcompensation for the missed opportunities to manage natural disasters (which, in fact, were human-made disasters) in the past.

When Euro-American settlers flocked to Southern California, building communities around present-day Los Angeles, they sought economic and personal success. Blindsided by the unnerving sunshine, they erased the memory of the raging waters of the Los Angeles River soon after the next drought set in. In their view, it was

[84] Ibid., 43.

[85] Ibid., 38.

[86] "Parks, Playgrounds, and Beaches for the Los Angeles Region," 127–28.

[87] Hise and Deverell, "Private Power, Public Space," 47–48.

inconceivable that Los Angeles was particularly vulnerable to disasters because this was the Mediterranean paradise that magazine articles, newspaper ads, and travel guides had sold them. Only because early migrants to Southern California did not want to acknowledge what was obvious to Hispanic residents who had lived in the area for many years could the citrus industry and other agricultural ventures prosper in the San Fernando or San Gabriel valleys.

Early city leaders did not proactively manage the flood risk of the river. Casualties due to flood catastrophes had to be accepted in the name of the region's economic growth and stunning success. Therefore, the mayors and council members of Los Angeles were hesitant to pass any ordinances prohibiting residents from buying tracts of land adjacent to the river or from erecting their houses in the floodplains, and no incentives for the population to move away from floodplains were provided. The reason why the indigenous people in the area, upon the arrival of the White man, had confined their settlements to elevated areas at some distance from the river was lost in the collective memory of the boomtown. The municipal government of Los Angeles did not realize the problems caused by railroad embankments because they were all too relieved that private companies had taken care of the flood problem.

Controlling the waters of the Los Angeles River was understood as the task of individual citizens, and not the government or some other public authority. Therefore, early Los Angeles, this do-it-yourself town, in the late nineteenth century saw a myriad of individual flood control solutions. Only after the devastating flood of 1914 did Council members and business leaders begin to understand that Los Angeles was indeed vulnerable to environmental forces beyond a single landowner's control. The first attempts to establish flood control measures failed spectacularly, resulting only in more deaths and property damages. Finally, the federal government, between the mid-1930s and the early 1960s, managed to construct infrastructure that prevented further deaths and damages from inundations.

The channelization of the Los Angeles River was a success story to the extent that it probably saved the lives of many people. However, it came with a significant price tag because it rendered obsolete any alternative plans that existed for a more sustainable future of the river. The Olmsted-Bartholomew plan had failed before the Corps of Engineers could ship massive construction equipment into the region. But with the river being confined in concrete walls, the vision that nature did not have to be conquered to make it compatible with human civilization also disappeared. Just as Los Angeles solved its water shortage problem in the 1910s by diverting a distant river to the San Fernando Valley, it solved its flood problem by extinguishing the Los Angeles River.

Los Angeles's approach to flood control is a lens into broader trends in American history in the nineteenth and early twentieth centuries. From the importance of private individuals and economic interests in the development of the city to the belief in apolitical technocrats and their expertise; from the erasure of Mexican flood memories to the exclusion of non-White and poor residents from flood protection, Los Angeles's infrastructure was a laboratory for several themes in US history. These strands are held together by a contemporaneous attempt to get a grip on what people at the time perceived as (environmental) disorder and replace it with a particular version of American civilization. The alleged victory of engineering over nature and its imminent dangers, however, should be understood as just another example of the prevalent reluctance to manage "natural" disasters in a way that was both effective and sustainable.

More recent efforts to revitalize the Los Angeles River and reimagine the city–river relationship, such as simplifying river access, establishing parks and trails, and enhancing social equity, while concurrently continuing to minimize flood risk, demonstrate that the solutions chosen in the early and mid-twentieth century no longer align with societal needs and norms in the twenty-first century.[88] With approximately half of all Los Angeles residents living within a short driving distance from the river, it seems crucial to seek a solution that addresses not only flood-risk management but also the needs of the diverse communities residing alongside the river. In essence, this river holds one key to Los Angeles's future.

References

Unpublished Primary Sources

"A Major Traffic Street Plan for Los Angeles. Prepared for the Committee on Los Angeles Plan of Major Highways of the Traffic Commission of the City and County of Los Angeles by Frederick Law Olmsted, Harland Bartholomew, Charles Henry Cheney" (Los Angeles, May 1924), University of Southern California Libraries Special Collections, California Historical Society Collection of Los Angeles Area Chamber of Commerce Records and Photographs, 0245–1, Box 73.

"Minutes of the Common Council," December 22, 1870, Los Angeles City Archives.

Petition No. 135, February 24, 1891, Los Angeles City Archives, Box B-6, Vol. 78.

Petition to Mayor and City Council, January 3, 1910, Los Angeles City Archives, Box B-1272, Folder "Health Commission 1913."

Baldwin, Chas W., to Council, October 13, 1890, 869, Los Angeles City Archives, Box B-6, Vol. 77.

Bauer, Charles, to Council, October 22, 1889, 831, Los Angeles City Archives, Box B-4, Vol. 72.

Capitol Milling Co. to Council, February 15, 1890, 210, Los Angeles City Archives, Box B-4, Vol. 73.

City Engineer, Annual Report 1899, Annual Reports 1898–1900, Los Angeles City Archives, Box B-110, Vol. 237.

Health Officer to City Council, June 26, 1898, Los Angeles City Archives, Box B-1272, Folder "Health Officer 1899."

Hyans, Joseph, to Council, December 16, 1889, 1031, Los Angeles City Archives, Box B-4, Vol. 72.

Moore, Barrington, "Checking Floods in the French Alps," Reports of the Board of Engineers Flood Control to the Board of Supervisors, Los Angeles County, California, Submitted July 27, 1915, [Los Angeles: publisher not identified, 1915], 81–84, Huntington Library.

[88] For a slightly outdated overview, Gumprecht, *Los Angeles River*, 274–301.

Stafford, Harry F., Annual Report of City Engineer, City of Los Angeles, Cal., For the Year Ending Nov. 30, 1901, December 16, 1901, Los Angeles City Archives, Box B-2294, Folder "City Engineer 1901."

United States Army, Corps of Engineers, Engineer Office, "Los Angeles County Flood Control Research," 1914–1915, compiled by J. W. Reagan, 2 Vols., Huntington Library, San Marino, CA.

Workman, W. H., to Council, September 16, 1889, 672, Los Angeles City Archives, Box B-4, Vol. 71.

Zahn, J. C., to Council, January 27, 1890a, 99, Los Angeles City Archives, Box B-4, Vol. 73.

Zahn, J. C., to Council, March 22, 1890b, 338, Los Angeles City Archives, Box B-5, Vol. 74.

Zahn, J. C., to Council, October 25, 1890c, 892, Los Angeles City Archives, Box B-6, Vol. 77.

Published Primary Sources

Compiled Ordinances and Resolutions of the City of Los Angeles, Vol. III, Compiled and Indexed by Freeman G. Teed, Clerk of the Council, (Los Angeles: E. H. Freeman, Book and Job Printer, 1887).

"No Unemployed," *Los Angeles Times*, February 25, 1914.

Revised Charter and Compiled Ordinances and Resolutions of the City of Los Angeles. Compiled and Indexed by Wm. M. Caswell, (Los Angeles: Evening Express Steam Printing Establishment, 1878).

Troxell, Harold C., and John Q. Peterson, "Flood in La Canada Valley, California, January 1, 1934," in *Water Supply Paper 796C* (Washington, D.C.: U.S. Geological Survey, 1937), 68–69.

Warner, J. J., "A Warning," *Los Angeles Times*, July 30, 1882a.

Warner, J. J., "Letter to the Editor of The Times," *Los Angeles Times*, August 6, 1882b.

Secondary Literature

Anderson, M. Kat, *Tending the Wild: Native American Knowledge and the Management of California's Natural Resources*. Berkeley: University of California Press, 2006.

Arax, Mark. 2019. *The Dreamt Land: Chasing Water and Dust across California*. New York: Knopf Doubleday Publishing Group.

Austin, Mary Hunter. 1903. *The Land of Little Rain*. Boston: Houghton Mifflin Company.

Clifford, Jim. 2017. *West Ham and the River Lea: A Social and Environmental History of London's Industrialized Marshland, 1839–1914*. Vancouver: University of British Columbia Press.

Colten, Craig E. 2005. *Unnatural Metropolis: Wrestling New Orleans from Nature*. Baton Rouge: Louisiana State University Press.

Culver, Lawrence. 2010. *The Frontier of Leisure: Southern California and the Shaping of Modern America*. Oxford: Oxford University Press.

Davis, Mike. 1998. *Ecology of Fear: Los Angeles and the Imagination of Disaster*. New York: Metropolitan Books.

Deverell, William. 2004. *Whitewashed Adobe: The Rise of Los Angeles and the Remaking of Its Mexican Past*. Berkeley: University of California Press.

Elkind, Sara S. 2011. *How Local Politics Shape Federal Policy: Business, Power, and the Environment in Twentieth-Century Los Angeles*. Chapel Hill: University of North Carolina Press.

Erie, Steven P. 2006. *Beyond Chinatown: The Metropolitan Water District, Growth, and the Environment in Southern California*. Stanford: Stanford University Press.

Fogelson, Robert M. 1993. *The Fragmented Metropolis: Los Angeles, 1850–1930*. Berkeley: University of California Press.

Guinn, J. M. 1890. Exceptional Years: A History of California Floods and Drought. *Historical Society of Southern California, Los Angeles* 1 (5): 33–35.

Gumprecht, Blake. 1999. *The Los Angeles River: Its Life, Death, and Possible Rebirth*. Baltimore: Johns Hopkins University Press.

Hackel, Steven W. 2005. *Children of Coyote, Missionaries of Saint Francis: Indian-Spanish Relations in Colonial California, 1769–1850*. Chapel Hill: University of North Carolina Press.

Hise, Greg. 1997. *Magnetic Los Angeles: Planning the Twentieth-Century Metropolis*. Baltimore: Johns Hopkins University Press.

Hise, Greg, and William Deverell. 2000. *Eden by Design: The 1930 Olmsted-Bartholomew Plan for the Los Angeles Region*. Berkeley: University of California Press.

Hundley, Norris. 1992. *The Great Thirst: Californians and Water, 1770s–1990s*. Berkeley: University of California Press.

Kahrl, William L. 1982. *Water and Power: The Conflict over Los Angeles' Water Supply in the Owens Valley*. Berkeley: University of California Press.

Kaijser, Arne. 2002. System Building from Below: Institutional Change in Dutch Water Control Systems. *Technology and Culture* 43 (3): 521–548.

Klein, Norman M. 2020. *The History of Forgetting: Los Angeles and the Erasure of Memory*. London: Verso.

Lübken, Uwe. 2014. *Die Natur der Gefahr: Überschwemmungen am Ohio River im 19. und 20. Jahrhundert*. Göttingen: Vandenhoeck & Ruprecht.

Madley, Benjamin. 2016. *An American Genocide: The United States and the California Indian Catastrophe, 1846–1873*. New Haven: Yale University Press.

Netz, Joseph. 1915. The Great Los Angeles Real Estate Boom of 1887. *Annual Publication of the Historical Society of Southern California, Los Angeles* 10, no. 1/2.

Nye, David E. 1994. *American Technological Sublime*. Cambridge, MA: MIT Press.

Orsi, Jared. 2004. *Hazardous Metropolis. Flooding and Urban Ecology in Los Angeles*. Berkeley: University of California Press.

Pritchard, Sara B. 2011. *Confluence: The Nature of Technology and the Remaking of the Rhône*. Cambridge, MA: Harvard University Press.

Reisner, Marc. 1986. *Cadillac Desert: The American West and Its Disappearing Water*. New York: Viking.

Sackman, Douglas C. 2005. *Orange Empire: California and the Fruits of Eden*. Berkeley: University of California Press.

Tejani, James. 2014. Dredging the Future: The Destruction of Coastal Estuaries and the Creation of Metropolitan Los Angeles, 1858–1913. *Southern California Quarterly* 96 (1): 5–39.

Torres-Rouff, David Samuel. 2013. *Before L.A.: Race, Space, and Municipal Power in Los Angeles, 1781–1894*. New Haven: Yale University Press.

White, Richard. 1995. *The Organic Machine*. New York: Hill and Wang.

Worster, Donald. 1985. *Rivers of Empire: Water, Aridity, and the Growth of the American West*. New York: Oxford University Press.

Representation and Perception

Hopeless Optimism: Framing Early Twentieth-Century Earthquakes as News and History in the Making

Jacob Birken

Abstract At the beginning of the twentieth century, disasters were thoroughly mediatized international events—they were experienced, retold, and represented against a backdrop of burgeoning modernization and ideologies stressing entrepreneurship and innovation. After an earthquake and fires destroyed large parts of San Francisco in April 1906, the survivors rallied behind visions of their city rebuilt grander, more beautiful, and prosperous than ever. To maintain this vision, Californians had to strategically tweak reality—including an interpretation of the disaster in which the earthquake's role in the city's destruction was severely downplayed. Rather than being the catastrophic effect of an unpredictable natural force, the events were reframed as a large-scale but ultimately manageable accident that conveniently created new business opportunities and a tabula rasa for urban renewal. As writers like Kevin Rozario argue, this attempt to make (economic) sense of disaster belongs to a larger "culture of calamity" in the United States, where catastrophic events are incorporated into a national narrative of creative destruction. While such optimistic inscriptions of disaster into an ongoing history of progress and growth might help survivors to come to terms with loss and trauma, they also hamper processes of disaster mitigation and prevention, effectively rendering a society more vulnerable to future calamity.

In this chapter, I want to expand upon this argument by juxtaposing the 1906 San Francisco disaster with US reactions to the 1908 Messina earthquake. Here—as already in narratives surrounding the 1906 disaster—Anglo-Saxon heroism and entrepreneurship are contrasted with the Southern Europeans' alleged backwardness and irrationality. Using examples from newspapers and magazines, but also insights from disaster sociology and psychology, I discuss how such narratives are informed by "Western" ideology on the one hand but also arise from (and amplify) general misconceptions about human behavior during times of crisis on the other.

J. Birken (✉)
University of Cologne, Cologne, Germany
e-mail: jacob.birken@uni-koeln.de

N. Rauscher, W. Werner (eds.), *Natural Disasters in the United States*, Risk, Governance and Society 22, https://doi.org/10.1007/978-3-031-96436-7_6

Introduction

Disasters might result from earlier disaster response, or, rather, a response to disaster that is wrongly understood to be conclusive. A society that not only mitigated and survived disaster but also successfully turned the event into history might willfully neglect future dangers—creating a narrative of *invulnerability* instead of taking its own vulnerability into account. In this chapter, I want to discuss some cultural, political, and psychological means used to frame catastrophe as a paradoxically optimistic event. For this, I will look at an American disaster—the 1906 San Francisco earthquake and fires—and at the American view of a Southern European catastrophe, the 1908 Messina earthquake. While the two events and the experiences of the survivors are hardly comparable, comparisons were indeed drawn by contemporaries and used to juxtapose not only the events but also the victims' and survivors' respective behavior. Often, irrational behavior is attributed to the 'Other'. This is a psychological issue—of experiencing one's own and others' behavior during a crisis—and a political one—of differentiating alleged local, national, or ethnic character. In the United States in the early 1900s, such distinctions were made against the backdrop of an ideology of "Western" progress, of the New World being propelled into the future by technological improvement and entrepreneurship, leaving behind a Europe mired in superstition and inefficiency.

The chapter is loosely structured around three topics: The media's interpretation of local and distant disaster in newspapers and magazines; the integration of catastrophe into an optimistic narrative of growth and improvement, with slight detours into turn-of-the-century aesthetics and urbanism; finally, a short attempt to reexamine the discussed issues from the perspective of disaster sociology and psychology.

"Nothing But a Large Cemetery": Reporting from and about Disaster Sites

About two and a half years after an earthquake and subsequent fires had reduced major sections of San Francisco to rubble, seismic disaster struck Messina in southern Italy on the morning of December 28, 1908. For a US audience, this was international or, due to the scope and nature of the disaster, *global* news. Daily papers printed early reports from Sicily, even though it was "impossible to obtain even an approximate idea of the vast damage done" due to "the fact that telegraphic and telephonic communications were almost completely destroyed", as the *New York Times* wrote on December 29.[1] During the next weeks and months, more detailed descriptions, eyewitness reports, and scientific or sociopolitical analyses appeared; daily papers and magazines first (re-)used archival images of Messina's vistas or the 1905 Calabria earthquake to illustrate their articles, as it took days or even weeks for

[1] "Italy Shaken; Thousands Die", *The New York Times* (29. December 1908), 1.

photographs of the actual events to reach the United States via ship. Nevertheless, photos were already established as useful or even preferred evidence. On January 4, the *Times* wrote of "[a]uthentic photographs of earthquake scenes which have arrived" in London and "indicate that the telegraphed descriptions are not a whit exaggerated", thus validating earlier descriptions with a secondhand reference to images not yet available for print.[2]

Editors and writers were clearly shocked by the event: *Collier's* called it "the most appalling disaster in the history of Christianity", *Harper's Weekly* the "most appalling disaster known in modern times, perhaps the most awful demonstration of the forces of nature in the history of the world."[3] Yet, the earthquake's newsworthiness was fleeting, soon to be replaced with other (relatively) relevant 1909 news items like aviation, the suffrage movement, William Taft's inauguration, or Teddy Roosevelt's African expedition.

In this sense, the Sicilian earthquake is paradigmatic for the representation of disaster and distant suffering in US media at the turn of the century: credible information is delayed due to logistical reasons; text precedes images; despite its emotional impact, the event's news cycle remains short, even though substantial reporting by experts might still be published months after the event—in April 1909, the *National Geographic Magazine* ran an article by Charles W. Wright of the US Geological Survey, while the *Century Magazine* printed reports by novelist Robert Hichens and by volcanologist Frank E. Perret, who had also submitted a much more technical "Preliminary Report" on the topic for the *American Journal of Science.*[4] *The Outlook* had already published "The Greatest Disaster in History" in late March, a long article by popular novelist Francis Marion Crawford, illustrated with photos and moody sketches of survivors by Italian-born painter Joseph Stella. Crawford's text was meant to be the "first paper" in a series on the disaster, but the chronically ill writer himself passed away on April 9.[5]

The disaster had thus been transformed from a shocking news item to a scientific or cultural subject. This affected both the quality and kind of information. Early news was speculative, spectacular, and often false. "The Rock of Charybdis now blocks the entrance to the Strait of Messina", the *Times* stated matter-of-factly on December 30, the tidal wave having "wrecked the lighthouses in the strait, including Faro Beacon", which then "crashed into the sea."[6] Arriving the next day, Frank

[2] "Photographs Confirm Worst", *The New York Times* (5. January 1909), 2.

[3] "The Earthquake in Sicily", *Collier's* 42.18 (23. January 1909), 12f; "The Tragedy on the Mediterranean", *Harper's Weekly* (9. January 1909), 7.

[4] Charles W. Wright: "The World's Most Cruel Earthquake", *National Geographic Magazine* 20.4 (April 1909): 373–396; Robert Hichens, "After the Earthquake", *The Century Magazine,* 77.6 (April 1909): 928–939; Frank E. Perret, "The Messina Earthquake", ibid.: 921–927; Frank E. Perret, "Preliminary Report on the Messina Earthquake of December 28, 1908", *The American Journal of Science* 160 (April 1909): 321–34.

[5] Francis Marion Crawford, "The Greatest Disaster of History", *The Outlook* 91.13 (27. March 1909): 673–690.

[6] "Dead in Quake may be 100,000", *The New York Times* (30. December 1908), 1.

Perret found the lighthouse still standing, as well as the "famous rock of Scylla" on the opposite coast.[7] "What is quite certain is that the first descriptions of the earthquake were pure invention", Francis Marion Crawford concluded, offering a simple explanation: "No one 'saw' anything" as the earthquake struck at night, and once "daylight revealed the truth, [...] there were no means of communication between the fallen city and the outside world."[8]

These are seemingly trivial, yet relevant observations. Extreme events like earthquakes will not only disrupt everyday life but also destabilize perception and memory, be it due to trauma or simply to the surroundings having become unrecognizable. Modern mass media, while promising near-immediate and objective news, is no less susceptible: Networks are disrupted, messages delayed, and locals much rather concerned with survival than with media production. This can be understood as a logistical issue, strongly related to technological affordance. In the early twentieth century, national and transatlantic telegraphy made it easy to transmit short texts over long distances. Still, information depends on context, and might—metaphorically speaking—deteriorate over distance, being then 'repaired' or supplemented with figments of a cultural imagination of disaster.

Reporting on the 1906 San Francisco catastrophe in the German *Coburger Zeitung*—a four-page daily newspaper—offers some insight in such processes. On April 20, the *Zeitung* published first news from California based on several telegrams and dispatches. An estimate of "nearly thousand" dead had been relayed to the "Cansas" [sic!] City post office from Los Angeles; another telegram mentioned that the buildings of the "Coal" and "Exanild" newspapers had been destroyed—extreme misspellings or misreadings of the local "Call" and "Examiner".[9] While the short article sums up the estimated damage to infrastructure and specific buildings, not much is said about the inhabitants besides 'fearful' and 'agitated' people fleeing their homes in nightgowns. This changes in the first-page editorial on April 21, where survivors are described as "madly running to the left and right", while "gangs of looters, mostly comprised of Chinese, Japanese, and [African Americans], stalked the streets for spoils" with "even white Policemen" joining in.[10] As the African American community in San Francisco was no larger than half a percent of the city's population, this racialized description hints toward a more generalized (European) cliche of the United States as a site of ethnic conflict.[11] Indeed, a short news item on a later page adheres more closely to a Northern, urban US racist imaginary—here, bloody street fights between Chinese, Hispanics, Italians, and "Americans" allegedly ensued after their quarters' destruction forced the

[7] Perret, "Preliminary Report", 326.

[8] Crawford, "Greatest Disaster", 680.

[9] "San Franzisko von einem schweren Erdbeben heimgesucht!", *Coburger Zeitung* (20. April 1906), 1f.

[10] "Das Erdbeben in Kalifornien", *Coburger Zeitung* (21. April 1909), 1. My translation, omitting a German racial slur.

[11] On San Francisco's demographics, see William Issel and Robert W. Cherny, *San Francisco, 1865–1932* (Berkeley: University of California Press, 1986), 56.

inhabitants to evacuate to other neighborhoods.[12] This story can be found in other international newspapers and traced back to an account from Oakland printed in the New York *Sun* on April 19. The original passage reads as follows:

> The Chinese came out of their underground burrows like rats and tumbled into [Portsmouth] square, beating such gongs and playing such noise instruments as they had snatched up. They were met on the other side by the refugees of the Italian quarter. The panic became a madness. At least two Chinamen were taken to the morgue dead of knife wounds, given for no other reason, it seems, than the madness of panic. There are 10.000 Chinese in the quarter and there are thousands of Italians, Spaniards and Mexicans on the other side. [...] The two uncontrolled streams met in the centre of the square and piled up on the edges. There they fought all the morning, until the Regulars restored order with their bayonets.[13]

None of this can be confirmed by local or official sources; a photo from the morning of the 18th shows people waiting or wandering around in front of the Hall of Justice (Fig. 1). An often-reproduced series of photos taken by San Francisco artist Arnold Genthe on the 18th in Chinatown and his personal account of the day present a very different scenario, in which the city became the site of spectacle, the inhabitants being spectators instead of (violent) actors. "[T]here was no hysteria, no signs of real terror or despair", Genthe wrote in his 1936 memoir. The bohemian artist himself spent the morning breakfasting at a nearby hotel, taking pictures, and having drinks with friends until someone reminded him that, as most houses in the quarter, his studio would soon be dynamited to stop the flames—which it was, without Genthe being able to save his possessions, despite attempts to bribe the militia with a bottle of 1868 German wine.[14]

Genthe's near-delusional, carefree behavior and similar accounts by other San Franciscans stand in stark contrast to the descriptions of crazed and panicking disaster victims in American and international media. I will discuss contradictions between first-person and mediated narratives of disaster (and the general unreliability of either) in more detail later; here, I want to point out that these contradictions can be shaped by individual perception but also by a general cultural imaginary of disaster that is routinely invoked if concrete information is not available. This cultural imaginary is structured around tropes often associated with a tragic failure or desolation of modern institutions, and cliched depictions of allegedly vulnerable or menacing groups of people. The *Coburger Zeitung* writes of two hundred dead children in the ruins of an orphanage in its April 21 issue; the following issue mentions that "hundreds" ill people had "fled" from hospitals, many of them throwing themselves into the sea, while "700 madmen had escaped from the asylum".[15] The

[12] "Original-Telegramme und telephonische Nachrichten", *Coburger Zeitung* (21. April 1909), 3.

[13] "Earthquake Lays Frisco in Ruins", *The Sun* (19. April 1909), 1. The *Coburger Zeitung* follows an abridged version with similar wording to an article in *The Age* from Melbourne, Australia, April the 21st. The substitution of 'Mexican' with "amerikanisch" seems to be the *Coburger Zeitung's* mistake.

[14] See Arnold Genthe, *As I Remember* (New York: Reynal & Hitchcock, 1936), 88–92.

[15] "Das Erdbeben in Kalifornien", *Coburger Zeitung* (21. April 1909), 1; "Das Erdbeben in Kalifornien", *Coburger Zeitung,* (22. April 1909), 1. My translations.

Fig. 1 Bear Photo, [Chinese crowds in Portsmouth Plaza, 1906]. California History Room, California State Library, Sacramento, California

original source for the latter report is the *Oakland Tribune*, which published the following on the evening of April 18:

> A conductor on one of the San Jose local trains reports that the Insane Asylum at Agnews is a total wreck, killing many of the inmates. It is reported that the attaches of the institution who were about at the time were saved. The building is a total wreck. The ruins took fire shortly after the collapse. There were about 700 persons in the building. The above report has been officially confirmed, and, it is added, scores of the inmates who escaped death are roaming about the country in a state of panic.[16]

Other newspapers reprinted complete or abridged versions of the conductor's report. The—most likely false—impression that *any* number of inmates had escaped was conflated with the total number of inmates: "A conductor on one of the San Jose trains reports that 700 of the inmates of the St. Agnew [sic!] Insane Asylum are roaming about the country in a state of panic", the *Washington Post* wrote on April 19, the same misinformation as published by the *Coburger Zeitung*.[17]

---

[16] "Agnews Insane Asylum is Destroyed", *Oakland Tribune* (18. April 1906), 2.

[17] "Earthquake and Flames Bring Death and Ruin to City of San Francisco", *The Washington Post* (19. April 1906), 1. For a detailed account of the events at Agnews and an even more egregious example of misinformation in the press, see Philip L. Fradkin, *The Great Earthquake and Firestorms of 1906* (Berkeley: University of California Press, 2005), 150–153.

Similar stories (re-)appeared after the 1908 Messina disaster. English and US newspapers printed an account by Lily Wolffsohn, who in turn had transcribed an oral report by "a man employed by a German cotton firm":

> The Messina prison was destroyed and the wardens killed, but most of the convicts escaped. They prowled among the ruins, robbing and murdering. They cut off the fingers of dead and wounded to take the rings, some of them singing songs as they plied the knife.[18]

Looters cutting off fingers is a trope within the disaster imaginary that can be traced back at least to the 1889 Johnstown flood, and is often used as a racist or xenophobic motif. In that year, "a party of 13 Hungarians" was allegedly "out for plunder"; after one of them severed the finger of a dead woman, they were confronted by local farmers. "Nine of the brutes escaped, but four were literally driven into the surging river and their death."[19] In the 1900 Galveston flood, the trope was attached to African Americans, where, according to an article, ninety Black men were "shot by citizen soldiery while looting and mutilating the bodies of the dead for plunder."[20] In 1906, this trope was so well known that *The Sun* wrote that "certain rumours about men caught stealing jewels from the dead and cutting off the fingers of corpses to get at the rings are regarded as preposterous", as the "dead were among the very poor."[21] Nevertheless, it was effective enough to reappear in 1908, perhaps against the 'new' setting of southern Europe.

Lurid depictions of specific groups as susceptible to violent or irrational behavior in times of crisis shape the accounts of disasters on even more general levels. An editorial in the *Leipziger Neueste Nachrichten* uses the news from San Francisco for an unabashedly ethnonationalist agenda. Comparing the Californian disaster to the eruption of Mount Vesuvius in early April 1906, the article juxtaposes 'Germanic' and 'Romanic' national characters: While the 'Germanic' Americans rely on their own strength to combat the flames and—again—the looting mob, the Italians succumb to hysteria and idolatry, attempting to stop the volcano's "devastating force" with "dirges and litanies".[22] This chauvinist but also openly anti-Catholic agenda permeates US reporting, though in a usually more subdued manner. Reports on the Vesuvius eruption seldom omit mentioning "processions of villagers, carrying images of saints and the Madonna and praying for a cessation of the eruption", churches "crowded day and night with people praying for deliverance from impending peril", or the statue of St. Anne being "taken to the mountain side to confront the

[18] Lily Wolffsohn, "Refugees' Stories", *The Daily Record and Mail* (31. December 1908), 5.

[19] "Thousands Die!", *The Lancaster Intelligencer* (5. June 1889), 3.

[20] "Death List May Reach 10,000", *The Humeston New Era* (19. September 1900), 7.

[21] "Burned City Takes on Courage", *The Sun* (22. April 1906), 1. The 'finger cutting' trope as a notorious canard is also mentioned in an anecdote from the *Sun's* offices in 1906, in Haynes Johnson, "A Wound That Will Be Long in Healing and Never Forgotten", *The Washington Post* (19. April 1981).

[22] Quoted in "Die Katastrophe in Kalifornien", *Coburger Zeitung* (24. April 1906), 1.

lava" and "frequently moved backward as the tide advanced."[23] This is not to say that this—well-documented—religious framing of disaster in Italy is included as outright misinformation in US news media; nevertheless, it contributes to the representation of southern Italian disaster response as essentially different from, or even the opposite of Californian disaster response.

This framing might have influenced international interpretations of the disaster's aftereffects. "No real comparison between Messina and San Francisco is possible", a correspondent for the *London Times* wrote:

> I was in San Francisco five days after the earthquake, and already wooden structures were being put up, work had begun on the tramway lines, and the main streets were crowded with cheerful, hopeful people. The residential portion of the city had been spared, and just across the bay was Oakland, to serve as a temporary place of business. The case of Messina is altogether different. Soon there will be nobody left but troops. A small settlement may be established, but Messina has disappeared.[24]

Others shared this prognosis. "It will be regarded as certain that a considerable portion of the population will bow before the wrath of the gods and seek homes elsewhere", the *Chicago Tribune* claimed; in a later article, the *London Times* correspondent quotes a professor Paterno, who insists that "Messina will remain nothing but a large cemetery."[25] "A splendid little city and prosperous a week ago, now dead without hope of resurrection", the *New York Times's* own correspondent Mary Maclean wrote from the ruins of Reggio in Calabria.[26] This fatalistic—and, from today's perspective, unfounded—prediction needs to be complicated for several reasons.

First, the 1908 earthquake led to mass death and material destruction on an entirely different scale. Even though the 1906 disaster caused thousands of deaths, displaced many more, and leveled large parts of San Francisco's inner city, the effects were much easier to mitigate: After the quakes and before the fires, most of the affected San Franciscans were able to evacuate, while a tsunami struck the Italian cities shortly after the quake, and torrential rainfall and aftershocks made rescue attempts even more difficult. Current research estimates the loss of 80,000 lives.[27] In a report from early February 1909, a writer for the *Outlook* mentioned that the removal of corpses was still ongoing, the Royal Civil Commission estimating

---

[23] "Vesuvius Causes Terror", *The New York Times,* 6. April 1906, p. 1; "Vesuvius Threatens Destruction of Towns", *The New York Times* (7. April 1906), 1; "Ashes From Vesuvius Overwhelm a Town", *The New York Times* (8. April 1906), 1.

[24] "Ruined Cities First Pictured", *The New York Times: The Marconi Transatlantic Wireless Dispatches* (10. January 1909).

[25] "Chaotic Rocks Entomb Cities Forever Lost", *Chicago Daily Tribune* (30. December 1908), 1; "Messina to Remain Merely a Cemetery", *The New York Times* (15. January 1909), 3.

[26] Mary Maclean, "In the Wake of the Disaster", *The New York Times: Magazine Section* (31. January 1909).

[27] See the respective entry in the *Catalogue of Strong Earthquakes in Italy,* http://storing.ingv.it/cfti/cfti5/quake.php?21318IT.

"sixty thousand" remaining in the ruins while removing "at least one hundred and forty" per day.[28]

"We Will Rebuild": Disaster and Progress

The question of rebuilding is mostly ideological. There are obvious reasons for *not* rebuilding major settlements in seismic hazard zones; according to the *Chicago Times,* the Italian King Victor Emmanuel III was expected to "urge that human residence in the earthquake zone be abandoned" in the days after the disaster.[29] Yet, plans for rebuilding were made almost immediately. In the case of San Francisco, playing down seismic risk became a concerted strategic effort, including semiofficial regulations to only talk about the devastating fires and effectively shutting down Californian academic seismology.[30] Clearly distinguishing between these two types of risks—fire and earthquake—was useful for both economic and ideological reasons. As insurance policies did not cover damage caused by natural disasters, there was a clear financial incentive to stress the conflagration as a distinct event and to even remove the earthquake as a noteworthy cause of the destruction. Furthermore, an earthquake is unpredictable and beyond human control, while a fire can be—at least theoretically—contained and, in the future, prevented using the appropriate means. Ultimately, this is an epistemological distinction, establishing different hierarchies of power between humans and their hazardous environment. To Californians, the earthquake presented an existential risk on unknown terms but also an obvious, immediate risk to the image of their region. Virtually every city remains susceptible to fire hazards without taking countermeasures, while strong seismic activity is a regional issue—an association that Californians actively attempted to avert. Accordingly, the earthquake's unpredictability was strategically shifted into the realm of the entirely unintelligible, precluding further scientific inquiry and, paradoxically, transforming vulnerability into invulnerability. As Pierre N. Beringer, a regular contributor to San Francisco's media, claimed in the Californian *Overland Monthly* magazine in May 1906, "[n]o one can tell when an earthquake will occur nor is there any reason to believe that earthquakes will occur again in San Francisco."[31]

East Coast writers and even the regional press ridiculed the San Franciscans' strategic earthquake denialism and speech regulations—"[n]o set of men can prescribe the terms which shall be applied to a catastrophe having few parallels in

[28] W. E. Davenport, "The Situation in Messina", *The Outlook* 91.12 (20. March 1909), 630.

[29] "District May be Deserted", *The Chicago Tribune* (1. January 1909), 2.

[30] See Fradkin, *Great Earthquake and Firestorms*, 231f, 260–262.

[31] Pierre N. Beringer, "The Destruction of San Francisco", *Overland Monthly* 47.5 (May 1906), 392–406, here 397. For other examples, see Fradkin, *Great Earthquake and Firestorms*, 268.

history", the *Oakland Tribune* wrote.[32] Nevertheless, such culturally or officially enforced, optimistic narratives can be easily read against a US-American or specifically Western backdrop of boosterism and a 'rugged' frontier ideology: Disasters are interpreted as proving grounds for individuals and collectives and as an incentive for renewal. As historian Kevin Rozario pointed out in *The Culture of Calamity,* this can be traced back to Puritan moral concerns but ultimately feeds into capitalist notions of creative destruction, where a city in ruins becomes a business opportunity for savvy entrepreneurs.[33] Both interpretations do not contradict but much rather support each other. In his *Overland Monthly* article, Beringer wrote:

> San Francisco, visited by earthquake and fire, will rise once more chastened and, more beautiful than before, in a cleansed majesty, in robes of white, it will turn to a magnificent future.[34]

This highly aestheticized vision was inherently political. For Beringer, 'cleansing' the city not only meant using the *tabula rasa* to realize a 1905 plan by urban designer Daniel Burnham—proponent of the so-called City Beautiful movement—but also to remove Chinatown from its inner-city location (neither was implemented).[35] He, too, framed San Francisco's inevitable rise as a (White) American project:

> Carthage never rose again, Rome is glorious only in its decay, but the indomitable American spirit rebuilt Chicago; Baltimore, phoenix-like, sprang from its ashes.[36]

This distinction between Old and New World disasters and between the associated ideas of the respective continent's historicity also informed 1908 disaster reports. Articles often mentioned Messina's and Reggio's long history, which was portrayed as a history of violence and calamity.[37] In contrast, US history as settler-colonialist history was presented as fairly recent, strategically omitting the existence (and violent erasure) of precolonial histories. For urban aesthetes, the ruins remaining after nineteenth- and early twentieth-century disasters thus even provided a simulacrum of the ancient history the New World seemed to lack—local photographer Edgar A. Cohen chastised fellow artists for including wrecked scaffolding in their photos of the ruins of San Francisco's City Hall, thus destroying "the feeling that it might be a scene in ancient Greece."[38] These ruins could, of course, only be of sentimental and aesthetic interest as they would soon be removed, making room for modern new

---

[32] "Stop the Cinch Business", *Oakland Tribune* (25. April 1906).

[33] Kevin Rozario, *The Culture of Calamity. Disaster and the Making of Modern America* (Chicago: The University of Chicago Press, 2007).

[34] Beringer, "The Destruction of San Francisco", 393.

[35] Ibid., 398. For the local politics behind the 'City Beautiful', see Judd Kahn, *Imperial San Francisco. Politics and Planning in an American City, 1897–1906* (Lincoln: University of Nebraska Press, 1979).

[36] Ibid., p. 393.

[37] See, for example, Crawford, "The Greatest Disaster", 675, Perret, "The Messina Earthquake", 922f; Wright, "The World's Most Cruel Earthquake", 375.

[38] Edgar A. Cohen, "With a Camera in San Francisco", *Camera Craft* (June 1906), 182–194, here 193. For the role of ruins in US culture, see Nick Yablon, *Untimely Ruins: An Archaeology of*

buildings. They were the site of an emergent, future America, while European ruins remained testaments of the Old World's eternal pastness. This interpretation not only helps to make sense of calamity by turning it into spectacle but also contributes to nation-building narratives.

It is important to remember that—as irrational as they often seem—such framings are firmly set within Western concepts of modernity, and thus ideologically opposed to the allegedly premodern handling of disaster in Southern Europe. In 1906, San Francisco had entered a performative stage of risk society: environmental dangers were assessed or even acknowledged only in relation to profit and progress. Within this ideological framework, earthquake vulnerability simply did not compute. It took over eight decades and the dogged work of archivist Gladys Hansen to reassess the official number of less than 500 casualties to 1498 documented and at least 3000 estimated deaths.[39] Tellingly, only a single or low two-digit number of the thousands of photos of the 1906 disaster show corpses, most of which are described as killed looters rather than casualties of earthquake or fire.[40] Even for witnesses, it might have been easier to conceptualize the destruction as "a scene in ancient Greece" than one of mass death: The ruins of an imaginary American past offered a more agreeable vista than a present (and future) marred by contingent environmental forces. Such narratives, of course, undermine efforts to prevent or mitigate potential future disaster; with a correspondingly fatalistic attitude, Gladys Hansen dedicated her 1989 book *Denial of Disaster* to not only "the known and unknown victims who suffered and died in the last Great Earthquake" but also "to those who will suffer and die in the next."[41]

Beyond such displays of denialism, optimistic disaster narratives can also be understood—and told—as stories of communality and empowerment. This is an argument Rebecca Solnit makes in *A Paradise Built in Hell* (2009), quoting numerous accounts from San Francisco that describe an atmosphere of generousness and altruism after the disaster.[42] "Never, in all San Francisco's history, were her people so kind and courteous as on this night of terror", Jack London wrote in his "Story of an Eye-Witness" for *Collier's* in May 1906.[43] Within this political narrative, the city's destruction (temporarily) removes oppressive social hierarchies and norms, which—like material wealth—become irrelevant in the stricken community's struggle for survival. While Kevin Rozario does not disavow such accounts, he correctly adds that in many cases, this optimistic, egalitarian narrative remained a "fantasy of the privileged", who had both the social networks and financial means to weather

American Urban Modernity (Chicago: The University of Chicago Press, 2009), and Miles Orvell, *Empire of Ruins* (Oxford: Oxford University Press, 2021).

39 Fradkin, *Great Earthquake and Firestorms*, 189–191.

40 This is my own research; see Jacob Birken, *Die kalifornische Institution. Fernwestliche Weltbilder um 1906* (Heidelberg: arthistoricum.net, 2018), esp. 50–53.

41 Gladys Hansen and Emmet Condon, *Denial of Disaster* (San Francisco: Cameron and Company, 1989).

42 Rebecca Solnit, *A Paradise Built in Hell* (New York: Viking Penguin, 2009), esp. 23–33.

43 Jack London, "The Story of an Eye-Witness", *Collier's* (5. May 1906), 22f.

property loss.[44] "People who were unable to afford dependable shelter against the ravages of winter storms were hardly likely to place a high value on the primitive delights of a natural disaster", he writes.[45]

Concurrent with the fatalistic prognoses of Messina's and Reggio's demise and their juxtaposition with Anglo-Saxon industriousness, US writers also invoked optimistic narratives after the southern Italian catastrophe. In an article on the "Mammoth Reconstruction" of San Francisco three years after the earthquake, German American poet Herman Scheffauer—member of the same bohemian circles as London and Genthe—wrote:

> Let us not plume ourselves to proudly, for as surely as the sun shines and human hope and endeavor survive fire and earthquake, old and poverty stricken Messina with half her population dead will in her turn do all that young and wealthy San Francisco has already done, and if she do not accomplish it in three years she will in seven, and if with less haste, then with infinite more order.[46]

Other authors attempted to integrate the disaster into even larger historical or even metahistorical contexts. "Not even the earth itself is done, or nearly done, much less the peoples on it and their civilization", the editors of *Harper's Weekly* argued. "There is no visible net profit in earthquakes, but the perpetual destruction, of which they are a detail, is the price of progress." Yet, this price is worth paying, as in its fight against "all the destructive forces of nature", mankind "grows strong and wise" until "progress begins to gallop where it used to walk."[47] In an article on January 9, liberal Christian magazine *Outlook* put the "tragedy in southern Italy" into perspective: "Let us take our place a million miles away, and look at this awful scene in its due proportions in the universe." Here, humans toil away, "infinitely little" in their "relation to the universe", yet very much convinced of the greatness of their achievements. "But nothing they build endures", the author—most likely chief editor Lyman Abbott—writes in a tone reminiscent of Shelley's "Ozymandias"; "Babylon, Thebes, Corinth, disappear in ruin as irretrievable as Messina."[48]

Again, perpetual decay is just a backdrop for moral uplift:

> There is something awful in the succession of destructions which slowly or suddenly turn irresistibly to ruin man's greatest achievements. There is something sublime in the indestructible energy which follows every apparent failure with a new and nobler endeavor, and makes each new disaster minister to a larger life of the spirit.[49]

In this case, the progress set in motion by disaster is not material or social but spiritual: "The tidal wave and the trembling mountain are themselves making the city of

[44] Rozario, *Culture of Calamity*, 116.

[45] Ibid., 117.

[46] Herman Scheffauer, "The Soul of San Francisco", *The San Francisco Sunday Call* (18. April 1909), 5.

[47] "The Theology of Earthquakes", *Harper's Weekly* (9. January 1909), 5.

[48] "The Lesson of the Earthquake", *The Outlook* (9. January 1909), 55f, here 55.

[49] Ibid., 56.

God, which is built of undaunted souls schooled in disaster."[50] Yet, I want to argue that these narratives remain valid in the Progressive Era United States exactly because they move between different or even conflicting materialist and spiritual ideologies. An interesting example of this ambiguity are the works of Californian artist Maynard Dixon—another San Francisco bohemian—made in relation to the 1906 disaster.[51]

Dixon, a successful illustrator for various Western topics around the turn of the century, contributed covers for Californian monthly *Sunset* and the Sunday supplement of the *San Francisco Chronicle* in the aftermath of the earthquake and fires; he also illustrated *Francisca Reina,* a 1908 book by local poet Amelia Woodward Truesdell that collected various elegiac or humorous pieces on the disaster and its consequences. One of Truesdell's poems, again, uses the trope of the superstitious Catholic. An unnamed "Italian" —depicted by Dixon as the mustached, cap-wearing stereotype (Fig. 2)—attempts to stop the flames by praying to St. Francis, and waving a banner with the Saint's visage; only on the next day he realizes that the banner was showing the (Frenchman) Francis de Sales instead of (Italian) Francis of Assisi:

> So that's why the City of Francis was burned;/The wrong saint was called to defend./If Assisi'd been there he'd have heard the wild prayer,/And mayhap would have changed the end.[52]

More complex and interesting than this gag is a loose series of Dixon's illustrations showing San Francisco as an anthropomorphic allegory, usually a majestic, semi-nude, or robed woman. Two of these 'Franciscas' figure prominently on the covers of *Sunset,* a magazine funded by Southern Pacific Railroad in 1898 and, as historian Kevin Starr writes, "a key prism through which the people of the Far West were glimpsing the possibilities and futures of themselves and their region."[53] The 1906 disaster, of course, challenged such futures. Dixon's two *Sunset* covers show how the disaster was quickly integrated into a narrative of Western progress and prosperity. On the cover of the May 1906 *Sunset*—a makeshift, black-and-white issue of only eight pages—the allegorical figure rises from the smoke billowing above the burning city, its minuscule inhabitants barely visible at the bottom, and themselves not more than spectators (Fig. 3). One month later, Dixon used a very similar composition to turn calamity into a vision of a better future: Here, the figure is emerging from a sublime cloudscape, carrying a model of rebuilt San Francisco in her arms. Again, minuscule people are seen at the bottom of the image, but now, they are a

[50] Ibid.

[51] For Dixon's life and role in the Californian arts scene, see Donald J. Hagerty and Maynard Dixon, *The Life of Maynard Dixon* (Layton: Gibbs Smith, 2010).

[52] Amelia Woodward Truesdell, "The Reason Why", in *Francisca Reina* (Boston: The Gorham Press, 1908), 42f.

[53] Kevin Starr, "Sunset Magazine and the Phenomenon of the Far West", in *Sunset Magazine. A Century of Western Living, 1898–1998* (Stanford: Stanford University Libraries, 1998), 31–75, here 32.

Fig. 2 Maynard Dixon, „That martyr ablaze he wigwagged aloft". Illustration in Amelia W. Truesdell, *Francisca Reina, 42*

workforce moving toward the allegorical figure and the future city, through and past the ruins of destroyed, old San Francisco (Fig. 4).

In 1906, this is surely an affective image, evoking hope in a community that had just survived disaster. Yet, it is also an eminently political image that mirrors Pierre N. Beringer's vision of San Francisco rising in "cleansed majesty, in robes of white". In the previous decades, real estate developer and, in 1897–1902, mayor James Duval Phelan had struggled to implement the 'City Beautiful' concept in San Francisco, a holistic approach to transform American cities and their social conditions through top-down, aesthetically uplifting urban design. City Beautiful mastermind Daniel Burnham had demonstrated his concept of monumental, neoclassical architecture at the 1893 Chicago World's Columbian Exposition. Celebrating "the four hundredth anniversary of the discovery of America by Christopher Columbus

Fig. 3 *Maynard Dixon, Title illustration for Sunset, May 1906*

Fig. 4 *Maynard Dixon, Title illustration for Sunset, June–July 1906. The Bancroft Library, University of California, Berkeley*

by holding an international exhibition of arts, industries, manufactures, and the products of the soil, mine, and sea", this fair aimed to present the United States as the stage and driving force of progress.[54] While the main section—dubbed the 'White City'—showcased international exhibits in spacious, representative pavilions, a second section—the Midway Pleasance—included various entertainment venues and ethnographic exhibits, ranging from Native American tribes to a faux-historical German village. This spatial arrangement further strengthened the distinction between international, yet US-led progress and its historic 'Other'—be it colonized Natives or picturesque Europeans, which would be encountered as evolutionary steppingstones while the Exposition's visitors made their way toward the 'White City'.[55] Such ideas informed US-American concepts of aesthetic and social urban order around 1900 and culminated in Beringer's racist announcement that "[f]ire has reclaimed to civilization and cleanliness the Chinese ghetto and no chinatown will be permitted in the borders of the city."[56]

At the request of Phelan, Burnham delivered a *Report on a Plan for San Francisco* in 1905, a very general, top-down concept of restructuring the city around boulevards and so-called Places, round areas with civic or commercial functions.[57] Interestingly, the model character of both Burnham's 1905 plan and the successful 1893 'White City' made it difficult to implement. As historian Judd Kahn describes the conundrum, "[o]nly the new San Francisco could motivate the citizens to create the new San Francisco"—after all, the uplifting effect of a US-American 'City Beautiful' remained unproven, at least beyond the 1893 exhibition site that had been torn down after half a year.[58] Nevertheless, American fairs were explicitly mentioned as examples to be followed in San Francisco's rebuilding. Petitioning for financial aid for the city in Congress, Nevada senator Francis Newlands claimed that "in the great expositions of the country [...] great white cities have sprung up almost in a nighttime, with a perfection of plan and a beauty of detail unequaled in the history of the world."[59] In this sense, the spectral image of future San Francisco carried by the figure on Maynard Dixon's *Sunset* cover fits into the US discourse of the (White) model city emerging as a near-miraculous proof of Western progress.

Dixon's other 1906 images both further and complicate this narrative. A full-page cover illustration for the *San Francisco Chronicle Sunday Supplement* on May 20 can be read as an allegory for the City Beautiful's promise and dilemma. The motif of the model city and the majestic cloud-woman is juxtaposed with a ghoulish figure emerging from the smoke of an industrialized city; below, billboards advertise products like "Disgusto Cigars" or "Rot-n-Rye Whiskey". At the bottom of the

54 Benjamin Harrison, "Proclamation 301—World's Columbian Exposition", 24. December 1880.

55 See Chap. 2 in Robert W. Rydell, *All the World's a Fair* (Chicago: The University of Chicago Press, 1984), 38–71.

56 Beringer, "The Destruction of San Francisco", 398.

57 See Chap. 4 in Kahn, *Imperial San Francisco*, 80–102.

58 Kahn, *Imperial San Francisco*, 71.

59 Congressional Record – Senate, 2. May 1906, 6245. See also Kahn, *Imperial San Francisco*, 163f.

image, a man with a trowel in his hand pauses his work in the ruins of San Francisco. He looks to the left—to the model city—yet his pose is oriented to the right. "Beauty an Asset/Ugliness a Liability", large Art Deco letters declare at the top. While this might be understood to advertise the 'City Beautiful', it much rather relegated Burnham's and Phelan's project to the realms of the aesthetic and moral. Just a few days after the disaster in April, the *Chronicle* itself and its publisher, M. H. de Young had begun organizing local opposition to the Burnham plan; "[e]ven those who admit that the plans are pretty in theory admit that their practical application is impossible", the newspaper wrote in an editorial on May 24.[60] Ultimately, San Francisco (including Chinatown on its original site) was rebuilt quickly and pragmatically, while—as in Dixon's illustrations—the model city remained a model.

There is yet another layer to the implementation and representation of progress in 1906 San Francisco. In three illustrations for Truesdell's *Francisca Reina*, the allegorical figure of the Californian city is even more ambivalent in her relation to crisis and urban modernity. In the first two, 'Francisca' is seen during her struggles against disaster—tossing away a cup with "black wine of despair" handed to her by "fate", grieving over the burning city and protecting the displaced survivors—while in the third, she has been reinstated as a "queen among cities", the buildings of the neoclassical urban metropolis becoming quite literally her seat of power.[61] This narrative resembles the later optimist writing on Messina in *Harper's Weekly* or the *Outlook.* Of the catastrophe, Truesdell writes in her second 'Francisca' poem:

> Extremity like thine revealed to us/That thou wert of God's plan unto the world/To civilize. We saw that thou must rise/In evolution of His purposes/From thy baptism of fire to higher life.[62]

This violent concept of progress is carried over to the third 'Francisca'. Dixon's allegorical figure is once again shrouded in dark smoke, yet here, it rises from the chimneys of rebuilt factories rather than from the city's ruins. The two opposites on Dixon's *Chronicle* cover are synthesized to a sublime, possibly menacing but ultimately majestic vision. As an aspirational figure of progress, 'Francisca' belongs to a visual tradition established earlier in the nineteenth century, most prominently with John Gast's ubiquitous 1872 *American Progress,* on which a gigantic personification of 'Columbia' brings settler colonialism and various boons of industrialized civilization from the East to the West. Dixon himself provided a related image for the first issue of *Out West* (previously *The Land of Sunshine*) in 1902, a magazine meant to promote California, its culture and political and economic issues (Fig. 5). Here, the towering figure once again emerges from a beautiful cloudscape, directing a group of minuscule settlers through harsh Western territory. A corresponding poem by Sharlot Hall dramatizes the call of the West, which, as in Truesdell's later text, has to be understood as a trial by fire:

---

[60] Kahn, *Imperial San Francisco*, 177–183; "'Cobweb' Plan is Strongly Opposed", *San Francisco Chronicle* (24. May 1906), 1.

[61] Amelia Woodward Truesdell, *Francisca Reina,* 8, 44.

[62] Ibid., 13.

Fig. 5 Maynard Dixon, "The Genius of the West," *Out West, January 1902, 1*

> The unmapped seas took tribute of many a dauntless band, and many a brave hope measured but bleaching bones in the sand;/Yet for one that fell, a hundred sprang out to fill his place,/For death at her call was sweeter than life in a tamer race.[63]

Her reference to "race" does not have to be read in a context of White supremacist racism, but rather as a chauvinist attack on "the Old-World cities that drowse by the eastern main", as Hall writes several stanzas later—from a Western perspective, East Coast cities are now effectively *European*, left behind as the frontier was pushed westward.[64] The ideological affinity between such allegories of the frontier (and of 'manifest destiny') and the 'Franciscas' shows that either of them needs to be understood against the backdrop of an emergent, specifically Western society of risk—risk including the dangers of the trail and 'wilderness' but also the harsh urban conditions under industrialized capitalism. To people subscribing to the market as the new frontier, postdisaster San Francisco might have appeared as an "ideal Social Darwinist world", as Judd Kahn wrote, rife with business opportunities for aggressive entrepreneurs.[65] In this sense, 1906 disaster capitalism predates the later 'Californian Ideology' of Silicon Valley and Mark Zuckerberg's motto to "move fast and break things", even though in this case the things came prebroken by the earthquake and fires.

Disaster and the Irrational 'Other'

I want to conclude with a different perspective on 'understanding' disaster. My previous examples show how disasters were integrated into a—sometimes religious, sometimes secular—narrative of human progress, in many cases associated with a Western and/or Anglo-Saxon national or ethnic project. This is psychologically and politically useful: disaster is no longer experienced as contingent, but as a necessary means for the betterment of the species or a society; sometimes, it will also be used to distinguish between an 'us' capable of overcoming crisis and a helpless 'them', who will succumb to panic and disorder. Such distinctions can be employed to argue for a social order where the latter group is, in one way or the other, subjugated: In this narrative, disasters will inevitably cause *certain groups* to panic or turn violent, which makes it necessary to forcefully impose strict rules and hierarchies. Disaster, then, becomes synonymous with a state of emergency that suspends everyday social relations and democratic, procedural governance. This was the case in San Francisco, where mayor Eugene Schmitz issued a 'proclamation' that the military and police in the city "have been authorized by me to KILL any and all persons found engaged

[63] Sharlot M. Hall, "Out West", *Out West* (January 1902): 3–5, here 4.

[64] This concept, of course, had been previously popularized by Frederick Turner, who by no coincidence introduced his Frontier Thesis in the context of the 1893 Columbian Exposition. See Turner, "The Significance of the Frontier in American History", in *The Frontier in American History* (New York: Henry Holt, 1920), 1–38.

[65] Kahn, *Imperial San Francisco*, p. 158.

in Looting or the Commission of Any Other Crime", despite any actual evidence of widespread looting; a US army officer, Charles Morris, ordered the destruction of the liquor that might be found in the city's bars, and kept his men busy pouring beverage into the gutter while they could have been helping with the evacuation.[66] "[P]revailing intemperence would unquestionably have precipitated street fights and riots, involving the certainty of much bloodshed", he later wrote in a letter defending his actions, an assumption that is impossible to verify—following Morris' logic, the lack of evidence for drunken mayhem would simply prove his orders had successfully prevented it from happening.[67] Morris' orders angered local store owners and alienated his superiors, but were in line with the paternalistic or authoritarian streaks of US-American progressivism that also birthed the City Beautiful movement. "Prohibition, pure and simple, was my unqualified aim and object", he wrote.[68] Yet—just as in Pierre N. Beringer's article—such concepts of moral and social order are built upon the exclusion or penalization of specific groups. Morris refers to

> a committee of highly respected residents of the district, who stated they had learned on good authority that the poor who had been burned out, felt that the rich had not suffered losses as had the poor, and were therefore determined to invade the Western Addition and lay it waste by incendiarism and attendant acts of spoil, looting, and violence.[69]

On a political level, such narratives legitimize the suppression of, and even violence against groups 'identified' as potentially dangerous. In the case of the San Francisco earthquake and fires, such narratives were far less effective than in other disasters like— for example—the 1923 Kantō earthquake in Japan, where rumors about Koreans poisoning wells or setting fires led to the torture and murder of possibly thousands of members of this minority at the hands of soldiers and lynching mobs, or racist vigilantism in the aftermath of Hurricane Katrina in 2005.[70] Nevertheless, the assumption that panic was inevitable and that specific groups were more prone to irrational or criminal behavior was as prevalent in 1906 as during earlier and later disasters. Two illustrations from a so-called instant history, a hastily compiled book on a current topic, show how such views were adapted to different demographics. The first, titled "Camping in the Golden Gate Park", shows a crowd of evacuees, all in various states of panic, grief, or stupor. While this is a diverse group in regard to appearance and age, another image—"The Evacuation of Chinatown"—shows a

---

[66] For Schmitz' proclamation, see Torrah Giles, "San Francisco, 1906: The Law and Citizenship in Disaster", *UCLA Historical Journal,* 28.1 (2017): 23–35; discussed also in Fradkin, *Great Earthquake and Firestorms*, 64–69; Solnit, *A Paradise Built in Hell,* 36–39.

[67] Col. Charles Morris, letter to the Adjutant General, July 3, 1907, http://sfmuseum.org/1906/liquor.html

[68] Ibid.

[69] Ibid.

[70] See J. Michael Allen, "The Price of Identity: The 1923 Kantō Earthquake and Its Aftermath", Korean Studies 20 (1996): 64–93, and A. C. Thompson, "Post-Katrina, White Vigilantes Shot African-Americans With Impunity", *ProPublica*, 19. December 2008, https://www.propublica.org/article/post-katrina-white-vigilantes-shot-african-americans-with-impunity

violent mob of near-identical, stereotyped Chinese.[71] Both groups are presented as irrational; the 'unmarked' evacuees as helpless and the racialized Chinese as dangerous. Either strategically or by simply reproducing cultural misconceptions, such images support the argument that times of crisis demand for authoritarian control, even though evidence suggests that dangerous irrational behavior is rather uncommon in such situations. In disaster sociology, this phenomenon is described as 'elite panic'—a fear of mass panic that would upend the social order, and thus also undermine the elite's claim to power: much rather a fear of *masses* than of *panic*.[72]

In this sense, this is a clearly political issue, yet interestingly correlates with a psychological phenomenon. As even early disaster research shows, people tend to underestimate the rationality of their own behavior during crisis, and overestimate the other's irrationality. In a study made after a 1949 fire in a men's dormitory at Oklahoma University, *all* participants claimed to have acted rationally, while accusing most of their peers of irrational behavior.[73] "I'm still amazed at how calm I was," one of the survivors wrote in a letter found in 1995, while accusing his roommate of panicking—"I told him to shut up but he kept screaming [...]. I socked him in the face and he finally shut up."[74] Even in this restricted constellation, supposedly rational and irrational behavior is established as a power differential that makes authoritarian violence necessary. Journalist James Hopper provided a more introspective account of this projection of irrationality onto the 'others' in an article on his experience of the 1906 disaster:

> As for me I felt a strange elation. I was immensely proud of myself. I had gone through that hideous minute and a quarter with full command over myself, and now I was calm, absolutely calm. I threw my chest out and looked with amazement upon my dazed co-citizens. And yet when a few days after, I saw again a friend who had met me just at that time, he told me that I had been so exited I couldn't talk, that my arms trembled as I gesticulated, and that my eyes were an inch out of their sockets.[75]

This phenomenon ultimately being a cognitive issue by no means overrides its political relevance—acknowledging this could even help to accept that the 'irrational other' is in itself an irrational projection, and to critically dissect any attempt to utilize it for political means. For historians, narratives of the 'irrational other' in accounts of crises or natural disasters introduce a peculiar conundrum, as they undermine the witnesses' reliability—the most probable but unsatisfying conclusion from the Oklahoma study or Hopper's article would be that *everyone* panics, while simultaneously being under the impression that they are the *only ones* not

[71] Charles Morris, *The San Francisco Calamity. By Earthquake and Fire* (n.p.: W. E. Scull, 1906). This is not the Charles Morris mentioned previously.

[72] See Lee Clarke & Caron Chess, "Elites and Panic: More to Fear than Fear Itself", *Social Forces* 87.2 (December 2008): 993–1014.

[73] Leonard Logan, Lewis M. Killian and Wyatt Marrs, "A Study of the Effect on Social Disorganization" (University of Oklahoma Research Institute, 1952), 82. See also Martha Wolfenstein, *Disaster. A Psychological Essay* (Glencoe: Free Press, 1957), 86.

[74] Leonard Jackson, "Physician Recalls OU Fire", https://eu.oklahoman.com/story/news/1995/12/20/physician-recalls-ou-fire/62369978007/

[75] James Hopper, "Our San Francisco", *Everybody's Magazine* (June 1906): 760a–h, here 760b.

panicking. In news media and political propaganda, the suspicion that *everyone else* panics might make even far-fetched scenarios seem plausible, and be used to legitimize measures to suppress the 'irrational others'. As this narrative is expanded from a psychological phenomenon to social contexts, the *rational I* and the *irrational them* can be aligned with specific group identities, the *rational I* becoming—for example—the 'us' of the progressive, industrious White Anglo-Saxons, and the *irrational them* being identified in the allegedly superstitious Chinese or Italians. Describing the relief work done by the British after the 1908 earthquake, the *London Times* boasted of the "rescues effected at the imminent risk of the sailors' lives and in ways that seemed to Italians miraculous":

> The genius for organization of the Anglo-Saxon race, the one quality in which that race is supreme, as cultivated Latins cheerfully admit, is being well displayed at the present time.[76]

On the one hand, it is not surprising that the survivors of a horrifying and deadly event would be exuberantly thankful for *any* help; on the other, being praised as "miraculous" saviors might have been exactly what the Britons would have expected in order to ascertain their self-perceived racial superiority. But what about the disaster accounts that transcended such narratives of the 'irrational other' to propagate a humanist, universalized vision of progress? In his article for the *Century Magazine*, Robert Hichens wrote:

> But with the great resignation there is a mingled resolution to take up the threads of life, to start again, to form new ties, set to new tasks. In the Italian Parliament it was said, "Messina shall rise again!" and a roar of cheering followed the prophesy. And most of these poor, stricken fugitives, crushed down under the iron heel of a great calamity, hold deep in their hearts the murmur, "I will rise again." It is marvelous that it should be so, but human beings are marvelous.[77]

Compared to chauvinist and paternalistic descriptions of (ir)rational behavior, this is obviously a far more progressive—and constructive—approach. Yet, against the backdrop of multiple human-made environmental crises, we might wonder if this optimism about humanity rising again and again is not just a subtler chauvinism, a vision of marvelous survival which should rather be a critical reassessment of risk-taking, be it building cities and infrastructure on seismic fault lines, or the lack of any meaningful action against anthropogenic climate change.

References

Allen, J. Michael. 1996. The Price of Identity: The 1923 Kantō Earthquake and Its Aftermath. *Korean Studies* 20:64–93.

Beringer, Pierre N. 1906. The Destruction of San Francisco. *Overland Monthly* 47 (5): 392–406.

Birken, Jacob. 2018. *Die kalifornische Institution. Fernwestliche Weltbilder um 1906*. Heidelberg: arthistoricum.net.

[76] "The Earthquake. Incidents of the Relief Work", *The Times* (11. January 1909), 7.

[77] Hichens, "After the Earthquake", *The Century Magazine,* 939.

Clarke, Lee, and Caron Chess. 2008. Elites and Panic: More to Fear than Fear Itself. *Social Forces* 87 (2): 993–1014.
Cohen, Edgar A. 1906. With a Camera in San Francisco. *Camera Craft*182–194.
Collier's. 1909. *The Earthquake in Sicily* 42 (18): 12f.
Crawford, Francis Marion. 1909. The Greatest Disaster of History. *The Outlook* 91 (13): 673–690.
Davenport, W. E. 1909. The Situation in Messina. *The Outlook* 91 (12): 630.
Fradkin, Philip L. 2005. *The Great Earthquake and Firestorms of 1906*. Berkeley: University of California Press.
Genthe, Arnold. 1936. *As I Remember*. New York: Reynal & Hitchcock.
Giles, Torrah. 2017. San Francisco, 1906: The Law and Citizenship in Disaster. *UCLA Historical Journal* 28 (1): 23–35.
Hagerty, Donald J., and Maynard Dixon. 2010. *The Life of Maynard Dixon*. Layton: Gibbs Smith.
Hall, Sharlot M. Out West. *Out West* (January 1902): 3–5.
Hansen, Gladys, and Emmet Condon. 1989. *Denial of Disaster*. San Francisco: Cameron and Company.
Harper's Weekly. *The Theology of Earthquakes*, 9. January 1909a, 5.
Harper's Weekly. *The Tragedy on the Mediterranean*, 9. January 1909b, 7.
Harrison, Benjamin. *Proclamation 301—World's Columbian Exposition*, 24. December 1880.
Hichens, Robert. 1909. After the Earthquake. *The Century Magazine* 77 (6): 928–939.
Hopper, James. *Our San Francisco*. Everybody's Magazine (June 1906): 760a–h.
Issel, William, and Robert W. Cherny. 1986. *San Francisco, 1865–1932*. Berkeley: University of California Press.
Kahn, Judd. 1979. *Imperial San Francisco. Politics and Planning in an American City, 1897–1906*. Lincoln: University of Nebraska Press.
Logan, Leonard, Lewis M. Killian, and Wyatt Marrs. 1952. *A Study of the Effect on Social Disorganization*. University of Oklahoma Research Institute.
London, Jack. *The Story of an Eye-Witness*. Collier's (5. May 1906): 22f.
Morris, Charles. 1906. *The San Francisco Calamity. By Earthquake and Fire*. N.p: W. E. Scull.
Morris, Charles, *Letter to the Adjutant General*, July 3, 1907. http://sfmuseum.org/1906/liquor.html.
Orvell, Miles. 2021. *Empire of Ruins*. Oxford: Oxford University Press.
The Outlook. The Lesson of the Earthquake, 9. January 1909, 55f.
Perret, Frank E. 1909a. The Messina Earthquake. *The Century Magazine* 77 (6): 921–927.
Perret, Frank E. 1909b. Preliminary Report on the Messina Earthquake of December 28, 1908. *The American Journal of Science* 160:321–334.
Rozario, Kevin. 2007. *The Culture of Calamity. Disaster and the Making of Modern America*. Chicago: The University of Chicago Press.
Rydell, Robert W. 1984. *All the World's a Fair*. Chicago: The University of Chicago Press.
Solnit, Rebecca. 2009. *A Paradise Built in Hell*. New York: Viking Penguin.
Starr, Kevin. 1998. Sunset Magazine and the Phenomenon of the Far West. In *A Century of Western Living, 1898–1998*, ed. Sunset Magazine, 31–75. Stanford: Stanford University Libraries.
Thompson, A. C. Post-Katrina, White Vigilantes Shot African-Americans With Impunity. *ProPublica*, 19. December 2008. https://www.propublica.org/article/post-katrina-white-vigilantes-shot-african-americans-with-impunity
Truesdell, Amelia Woodward. 1908. *The Reason Why*. In Francisca Reina. Boston: The Gorham Press, 42f.
Turner, Frederick. 1920. The Significance of the Frontier in American History. In *The Frontier in American History*, 1–38. New York: Henry Holt.
Wolfenstein, Martha. 1957. *Disaster. A Psychological Essay*. Glencoe: Free Press.
Wright, Charles W. 1909. The World's Most Cruel Earthquake. *National Geographic Magazine* 20 (4): 373–396.
Yablon, Nick. 2009. *Untimely Ruins: An Archaeology of American Urban Modernity*. Chicago: The University of Chicago Press.

Hurricanes and Theodicy in Twentieth and Twenty-First Century American Protestantism

Edward Manger

Abstract Theodicy occupies a significant place in American Christianity and the broader culture of the United States. However, only a small amount of scholarship exists on the intersection of theology, popular religion, and natural disasters in American Christianity. The discussions that do exist around the theodicy of natural disasters focus on the theological content and potential applications of any given theodicy and are rarely historicized. I contend that looking at Christian responses to natural disasters alongside theodicy reveals the changing nature of American Christianity across the twentieth and twenty-first centuries. Furthermore, by looking at the reactions to hurricanes by different strands of Protestantism, we can begin to understand a hitherto underappreciated element in America's particular vulnerability to the effect of natural disasters: A religiously encouraged acceptance of risk. The ways Americans theologize natural disasters or do not, are historically contingent and reflect the movements and trends in American Christianity at any given time. I argue that protestant responses to natural disasters reveal the fears, anxieties, and preoccupations of any movement and moment within American Christianity. Each movement, by confronting or ignoring the causes of natural disasters, contributed in its own way to a cultural acceptance of vulnerability by a focus on the aftermath of any given event rather than future minimizing of risk.

Introduction

The Bible is full of weather. Rain, winds, and floods occur regularly throughout scripture and are depicted as under God's control. The Psalmist describes God riding on the clouds as one who reigns over the elements and commands them to do his will, either bringing life and growth or death and destruction (Ps. 78:26, 104:3, 107:25, 135:7, 147:18, 148:8). Furthermore, the Bible's most memorable handling

E. Manger (✉)
Heidelberg Center for American Studies, Heidelberg University, Heidelberg, Germany
e-mail: Edward.Manger@erskine.edu

N. Rauscher, W. Werner (eds.), *Natural Disasters in the United States*, Risk, Governance and Society 22, https://doi.org/10.1007/978-3-031-96436-7_7

of the question of suffering is book-ended by divinely ordained weather patterns. In the opening of the book of Job, the titular antagonist is afflicted by a violent tempest that destroys his property and kills his relatives. Then, toward the end of the narrative, God speaks to him from the whirlwind to answer Job's objections and reaffirm his sovereignty over creation (Job. 1:19, 37:9, 38:1, 40:6). This image of God presented in the Bible raises many questions regarding the meaning and purpose of natural disasters and the suffering they cause. These questions have been wrestled with repeatedly throughout the church's history in the form of theodicies, attempts to reconcile the goodness and power of God with the presence of evil and suffering in the world. As Peter Thuesen has commented, the weather can be seen as a primordial power that leads inexorably to theological reflection and attempts to frame our idea of the divine.[1] Historically, the most common way to reconcile the suffering caused by natural disasters with the goodness and power of God was through judgment theodicies. Simply stated, this is the view that God controls natural disasters and uses them to punish peoples, nations, or societies for their sins.[2] Many Christians in America have readily embraced this approach, even in recent years. In the wake of the devastating Hurricane Katrina, which hit New Orleans on June 18, 2005, the conservative Christian broadcaster Pat Robertson decried the sin of abortion as the root cause of the hurricane, which he believed was nothing more or less than the instrument of God's judgment against a sinful nation.[3]

This view of natural disasters has derived legitimacy from the belief, held by white Protestants throughout the country's history, that they are a covenanted people in a special relationship with God. As such, the people of America have been seen as held to certain expectations or requirements in that covenant. If these requirements were neglected or left unfulfilled, then chastisement and punishment would inevitably result.[4] A specific form of public lament over sin known as the Jeremiad developed in New England as a mode of discourse intended to remind hearers of their covenanted status and call them to repentance and obedience to the

[1] Peter J. Thuesen *Tornado God: American Religion and Violent Weather* (Oxford: Oxford University Press, 2020), 6.

[2] Mark S. M. Scott *Pathways in Theodicy: An Introduction to the Problem of Evil* (Minneapolis: Fortress Press, 2015), 6. Bruce R. Reichenbach *Evil and A Good God* (New York: Fordham University Press, 1982), 88.

[3] Meghan Friedman, "God's Wrath Caused Katrina," TIME, September 16th, 2011, http://content.time.com/time/specials/packages/article/0,28,804,1953778_1953776_1953771,00.html and "Pat Robinson Controversies," Last modified January 13th, 2010, https://www.cbsnews.com/pictures/pat-robertson-controversies/

[4] Although this has been highlighted by scholars from Perry Miller onward as a foundational belief of what became America, it is not a uniquely American belief, but rather developed out of European precedents. Furthermore, one must note that careful theologians have always distinguished between being 'The' covenanted people of God and 'A' covenanted people of God. See Perry Miller *The New England Mind: The Seventeenth Century* (New York: The Macmillan Company, 1939) and Mary Morrissey "Elect Nations and Prophetic Preaching: Types and Examples in the Pauls Cross Jerimiad," in *The English Sermon Revised: Religion, Literature and History 1600–1750*, ed. Laurie Ferrell and Peter McCulloch (Manchester: Manchester University Press, 2000), 43–58.

rules of the covenant. If they did this, they would receive blessings and avoid punishment. The Jeremiad has permeated American society and political discourse and serves as evidence of the deep resonance the idea of America as a chosen nation has to its people. Calls to repentance for specific sins were commonplace in the colonial period when Christians engaging in a literal reading of the Bible saw multiple examples of God punishing nations with temporal chastisement, sometimes with destructive weather patterns. This attitude extended through the Revolutionary War up through the Civil War and into the later nineteenth century. However, as the twentieth century approached, this mode of public discourse developed and evolved into separate strands that Andrew Murphey has termed "traditionalist" and "progressive."[5] These contrasting Jeremiads differed starkly in their view of what covenantal requirements the American people may or may not have been keeping. Additionally, with the solidifying of what Robert Bellah famously called American "civil religion"[6] in the twentieth century on the one hand and the emergence of the religious right on the other, the traditional notion of a covenanted nation has shifted to be allied with particular sociopolitical agendas and specific moral flashpoints. Thus, national temporal judgment and chastisement are less universally accepted and more contested among Christians and even self-professed conservatives and evangelicals. For example, some socially conservative ministers such as Southern Baptist Russell Moore have argued that the link between America and the Old Testament nation of Israel does not exist, and promises from the Old Testament, such as 2 Chronicles 7:14—a crucial verse in describing the relationship between God and the United States in the traditional covenantal view of the nation—must apply to the church, not to America.[7]

By the twenty-first century, therefore, despite American Christianity's history of associating natural disasters with divine judgment, the comments by Pat Robertson about Katrina—and others in their vein—caused outrage and consternation nationally and worldwide.[8] Nevertheless, in the popular imagination, reference to Christians and natural disasters will most likely revolve around one of two images:

[5] Andrew R. Murphey *Prodigal Nation: Moral Decline and Divine Punishment from New England to 9/11* (Oxford: OUP, 2008), 110–114.

[6] Bellah defined civil religion as 'a set of beliefs, symbols, and rituals' that express 'a religions dimension for the whole fabric of America life.' Robert N. Belllah, "Civil Religion in America," *Dædalus, Journal of the American Academy of Arts and Sciences* 96, No. 1 (Winter 1967): 1–21, 3. For the development of American Civil Religion and its relation to theological and religious belief, see Philip Gorski *American Covenant: A History of Civil Religion from the Puritans to the Present* (New Jersey: Princeton University Press, 2017), Peter Gardella, *American Civil Religion: What America Holds Sacred* (Oxford: Oxford University Press, 2014).

[7] The text in the KJV reads, 'If my people, which are called by my name, shall humble themselves, and pray, and seek my face, and turn from their wicked ways; then will I hear from heaven, and will forgive their sin, and will heal their land. 'Russell Moore "2 Chronicles 7:14 isn't about American Politics" RussellMoore.com, January 14th, 2016. https://www.russellmoore.com/2016/01/14/2-chronicles-714-isnt-about-american-politics/

[8] Matthew Yglesias "Hagee on Katrina," The Atlantic, arch 5, 2008, https://www.theatlantic.com/politics/archive/2008/03/hagee-on-katrina/45516/

Calls of judgment upon sin on the one hand or disaster relief efforts and charitable aid on the other. The reality is, however, that all Christian traditions give a significant place to the concept of suffering in both the experience of believers and the church. Additionally, beliefs about the relationship of humanity to divine providence and the role of God in the world are at the heart of the Christian life. Ideas of renewal, spiritual growth through death and suffering, and the question of good and evil all run through the Bible as consistent themes and, as such, are manifest in liturgy, prayers, and sermons that echo around the churches, chapels, and places of worship of the entire country weekly, even when there is no natural disaster in the headlines.

Consequently, when disaster strikes, the Christian response is far more nuanced and varied than the cry of judgment or calls for material aid (although these do form a well-publicized portion of the Christian response). Part of this stems from the fact that there is not one 'Christian' response: the Christian community is made up of different denominations and traditions, but more than that, different individual believers, congregations, ministers, para-church organizations, denominational structures as well as public intellectuals and prominent spokespersons. Each of these has its own priorities, messages, and views they wish to present when 'responding' to natural disasters, such that it is hard to speak of a single 'Christian response' meaningfully. However, the responses made by Protestant Christians in the wake of natural disasters are those of everyday Christian life given extra urgency and intensified by sudden public attention by an event about which Christians and churches feel qualified to speak. These concerns include apologetics, pastoral care, identity formation, spiritual growth, aid, and calls to repentance, but only rarely do they focus on reducing the vulnerability of communities to natural disasters in the future. As such, natural disasters can form an interesting lens through which to examine the ways in which theology, Christian practice, and belief filter down to the concrete realities of living in the real world with physical and emotional turmoil and suffering while at the same time not directly addressing the practical concerns of future risk.

Despite this, and the fact that theodicy occupies a significant place in American Christianity and the broader culture of the United States and is reflected in the significant points of contact and conflict between the two, only a small amount of scholarship exists on the intersection of theology, popular religion, and natural disasters in American Christianity. The discussions that do exist around the theodicy of natural disasters focus on the theological content and potential applications of any given theodicy and are rarely historicized.[9] I contend that looking at Christian responses to natural disasters alongside theodicy reveals the changing nature of American Christianity across the twentieth and twenty-first centuries. Furthermore, by looking at the reactions to hurricanes by different strands of Protestantism, we can begin to understand a hitherto underappreciated element in America's particular

[9] See, for example, D.P. O'Mathúna "Christian Theology and Disasters: Where is God in All This?" In: O'Mathúna, D., Dranseika, V., Gordijn, B. (eds) *Disasters: Core Concepts and Ethical Theories. Advancing Global Bioethics*, vol 11. 2018

vulnerability to the effect of natural disasters: A religiously encouraged acceptance of risk. The ways Americans theologize natural disasters or do not, are historically contingent and reflect the movements and trends in American Christianity at any given time. I have chosen four hurricanes spread across the twentieth and twenty-first centuries as a window into this dynamic to show that while the prevalence of a prominent civil religion persists in America, allowing the idea of America as a covenanted people to continue even without that specific language, the idea of judgment as a viable answer to the theodicy question is not the only or even the most significant Christian response in the wake of natural disasters. Instead, I argue that protestant responses to natural disasters reveal the fears, anxieties, and preoccupations of any movement and moment within American Christianity. Each movement, by confronting or ignoring the causes of natural disasters, contributed in its own way to a cultural acceptance of vulnerability by a focus on the aftermath of any given event rather than future minimizing of risk. In addition, I propose that judgment, as a cause of natural disasters, was recast by the Social Gospel movement, concerned with the social implications of natural disasters; sidelined by Fundamentalists, concerned with theology over society; ignored by Neo-Evangelicals, attempting a cultural consensus; resurrected by the religious right, afraid of the precarious nature of their relationship to mainstream culture; and rejected by recent movements within evangelicalism, focused on the apologetic and evangelistic potential of natural disasters. In each case, by explaining the theological significance of hurricanes according to the terms of their own religious priorities, protestants implicitly conceded the inevitability, and indeed in some cases, necessity, of hurricanes.

Galveston and the Early-Twentieth Century

The early years of the twentieth century witnessed the culmination of what historians of American Religion, such as David Moberg and George Marsden, have called the 'great reversal.' They argue that, in the decades after the Civil War, evangelicalism underwent a marked retreat from the social reform movements that had characterized it earlier in the nineteenth century.[10] One of the significant factors for this change was the growing affiliation of the Social Gospel movement with advanced liberal Christian social thought, culminating in the work of prominent theologians like Walter Rauschenbusch (1861–1918). Rauschenbusch rejected the standard evangelical view of the atonement—that Christ's death was a punishment for the sins of humanity—and instead emphasized Jesus's suffering love as a demonstration of the power that could be restorative and transforming in the believer's life

[10] David O. Moberg quoted in Garth Rossell *The Surprising Work of God: Harold Jay Ockenga, Billy Graham and the Rebirth of Evangelicalism* (Michigan: Baker Academic, 2008), 164. George Marsden *Fundamentalism and American Culture: The Shaping of twentieth Century Evangelicalism, 1870–1925*. (New York: Oxford University Press, 1980), 90.

when following his example. His view of theodicy followed from this belief. He argued that the 'innocent suffering' caused by natural disasters was not to be seen as primarily punitive but is, on the one hand, 'the price we pay for living on a lovely planet,' and at a deeper level, evidence of the fact that, 'social suffering serves social healing.'[11] For Rauschenbusch, 'solidarity…acts as a theodicy.' In its simplest form, Rauschenbusch argued that the reason for suffering in the world is that it causes people to love one another.[12] The question of God's role in human suffering was implicit in the Social Gospel movement and the subsequent infamous fundamentalist-modernist controversy—behind the more prominent issues of biblical criticism, miracles, and evolution. Drawing lessons from what he termed Christ's 'social conscience,' Henry Sloane Coffin (1877–1954), Presbyterian minister in New York and a leading theological liberal, believed that 'There is no needless suffering… in all the world that does not in a very genuine sense come home to you and me as something for which we are personally blameworthy.'[13] In this belief, Coffin laid the responsibility for the suffering in the aftermath of natural disasters such as hurricanes squarely at the feet of society rather than God. Harry Emerson Fosdick (1878–1869), a prominent New York minister and famous theological liberal opposed to fundamentalist literalism, agreed and rejected the idea of personal punishment by God for specific sins. Instead, he advanced the idea that societies face their own punishment in the consequences of their failures to act socially to alleviate suffering.[14] The view put forward by these theologians, on one hand, highlights the human role in human suffering, implying that social action in the future can reduce vulnerability to the effects of hurricanes, yet on the other hand, rests on the idea that natural disasters are to be expected, and actively serve to develop humanities social awareness.

Amid the development of the theological liberalism held by these figures, as well as the emergence of the Social Gospel movement, stands the Galveston Hurricane of 1900 with a death toll of over 6000. At the dawn of the twentieth century, as a growing nation full of optimism and self-confidence was embracing the new and powering ahead on the road of progress, something old with a power from another era smashed into the coast. It brought unprecedented water surges and the inevitable wake of death and destruction. The hurricane made landfall on September 6 near Galveston, Texas, the town that lent its name to the storm. The scale of the destruction gripped the nation as newspaper inches were filled with stories of loss and heroism. The city had thirty-seven protestant churches and at least six Catholic, of which

[11] Paul Minus, *Walter Rauschenbusch: American Reformer* (New York: Macmillan, 1988), 43,47.

[12] Walter Rauschenbusch *A Theology for The Social Gospel* (New York: Macmillan, 1917), 191–193.

[13] Henry Sloane Coffin *Social Aspects of the Cross* (New York: Hodder & Stoughton, 1911), 16.

[14] Harry Emerson Fosdick *A Guide to Understanding the Bible* (New York: Harper and Brothers, 1938), 157.

twenty-five were destroyed entirely, and twelve were heavily damaged. Only three Catholic churches remained once the flood waters receded and the winds died down.[15]

The following Sunday, September 16, the country began to speculate about the lessons to be learned from this great catastrophe. Where was God? What was his intention? Why had he allowed this to happen? Had he caused it? Rev. Dr. Conwell of Temple of Grace Baptist Church in Philadelphia reminded his congregation of St. Paul's teaching in his epistle to the Romans that 'all things work together for good to them that love God.' He acknowledged that it was hard to understand but reassured his hearers that Galveston was an event under God's guidance and good would come out of the suffering. Not content to simply make the assertion, Rev. Conwell supplied his hearers with some ways in which good might be realized. In words remarkably similar to Rauschenbusch's later writings on social suffering and social healing, he opined, "The people who now live in Galveston will be better all their lives. This experience has deepened their natures, enriched their sympathies, enlarged the boundaries of their feelings, and the people of that city will be blessed by that awful experience." He thought that the residents of Galveston would be more loving toward others, more affectionate, and in effect they would be "different men." He also believed "the people of Galveston have been taught that there is something more than dollars in this world, the rich will now feel what it is to be poor. It does man good to feel the depths of life. Many of the survivors will thank God they have to begin life over again." Conwell also argued that the whole country would experience the good that would come from the bad of the hurricane, saying "when it arouses the sympathies of many tens of thousands, it must be a gigantic force to work out an ultimate good." He predicted that "They will order lumber from the North, where the suffering people are waiting for the order." He left his congregation with the thought that "when you consider how that disaster locally is going to bless so many people outwardly, then the measure of its good may be far greater than the measure of its evil."[16]

Other preachers also highlighted the connection between the nation and those who suffer. Rev. Frank Dewitt Talmage, preaching in Chicago, reassured his congregation with notions of the "brotherhood of mankind." Talmage admitted that neither he nor others could know why the hurricane happened but stated that lessons could be learned, primarily "God has made of one blood all nations. The misfortunes of mankind make us one."[17] The Rev. W. H. Carwardine echoed those sentiments and explained, "The highest light that comes to victims of the most appalling disaster of the century is the unfolding of the world's friendship." These sentiments are embedded in the turn-of-the-century theological liberalism that dominated northern pulpits and would be the foil against which Fundamentalism would react. Little to no

[15] Clarence Ousley Ed. *Galveston in Nineteen Hundred* (Atlanta: William Chase, 1900), 78. Nathan Green, *The Story of the 1900 Galveston Hurricane* (Louisiana: Pelican, 2000), 3. Kristine Brennan, *The Galveston Hurricane* (Philadelphia: Chelsea House, 2002), 12.

[16] Paul Lester *The Great Galveston Disaster* (Philadelphia: Globe Bible Publishing, 1900), 311. 8 Lester The Great Galveston Disaster, 314.

[17] Lester *The Great Galveston Disaster*, 314.

mention of judgment is to be found; in fact, the congregants of Oxford Presbyterian church in Philadelphia were told that "Christ declares that disasters are not be interpreted as judgment" but rather "the object of every disaster is to bring us fast to repentance." since "we don't know what tomorrow will bring."[18] A Reformed Episcopal minister declared to his Chicago congregation, "The lesson of self-help which this calamity teaches will not be lost. God intended man to conquer nature, to bind its forces, to ride triumphantly on its seemingly resistless energies." He concluded that "another lesson of American pluck and energy will thus be learned by mankind."[19] This little sample of sermons in response to Galveston shows the concerns of the time, the brotherhood of man, the progressive providence of God, and encouraging American pluck and a can-do attitude. Each sermon also focused on positive effects of the hurricane that outweighed the suffering, the good that will come from the bad. In doing so, they both accept the presence of hurricanes and minimize the need for future avoidance of the same.

In the decades following Galveston, God's sovereignty over natural disasters such as hurricanes became a critical point of contention in the fundamentalist-modernist debate. Responding to Modernist ministers such as Harry J. Fosdick, the conservative theologian, and leader of the fundamentalist wing of the Presbyterian Church in the United States of America, J. Gresham Machen (1881–1937) was concerned with maintaining the link between God and everything that comes to pass, writing that "Human suffering may be unpleasant, but it is real, and God must have something to do with it."[20] The English-born minister and fundamentalist A.W. Pink (1886–1952), while pastor in Spartanburg, South Carolina, defended the classic Calvinistic doctrine of God's intimate control over physical events, which he felt was too little believed in his day, in his influential book *The Sovereignty of God* (1918). He advocated the view that God is in minute and detailed control over each facet of the weather and climate.[21] Yet, these fundamentalist writers, primarily concerned with what they saw as liberalism's theological errors, were not especially likely to call on God's judgment as a reason for catastrophe. Seeing social issues and, to some extent, human suffering as the remit of the Social Gospel and theological liberalism, they were rather more concerned to exalt a high view of the sovereignty of God and the otherness of his ways than offering social critiques of the society around them, as Cornelius van Til (1895–1987) eventually concluded in

[18] Ibid. 313.

[19] Ibid. 314.

[20] J. Gresham Machen *Christianity and Liberalism* (New York: Macmillan, 1923), 133. For Machen and the Fundamentalist-Modernist Controversy, see Ned Stonehouse *J. Gresham Machen: A Biographical Memoir* (Grand Rapids, MI: Eerdmans, 1954), and Daryl G. Hart, *Defending the Faith: J. Gresham Machen and the Crisis of Conservative Protestantism in Modern America* (Baltimore, MD: John Hopkins University Press, 1994).

[21] A. W. Pink *The Sovereignty of God* (London: I.C. Henderson, 1928), 14, 32. For the continued influence of Pink in Conservative Reformed Evangelicalism, see Ian Murray's *The Life of Arthur W. Pink* (Edinburgh: Banner of Truth, 2004).

1923, 'God is his own theodicy.'[22] In this view, God is the author of natural disasters, regardless of attempts at risk reduction, and thus such measures become a moot point.

Hazel and the Mid-Twentieth Century

In the middle decades of the century, evangelical figures such as the minister Harold Ockenga (1905–1985) and popular preacher and evangelist Billy Graham (1918–2008) became mainstream fixtures in American public life. In a step away from Fundamentalism, the evangelicalism of this period embraced culture and society. Carl Henry (1913–2003), a leading Evangelical theologian, criticized what he saw as Fundamentalism's disengagement with society and culture and indicated a new way of integration for evangelicalism.[23] In recent years, historians of American evangelicalism have uncovered the ways in which mainstream evangelicalism was funded and supported by big business and was thoroughly entwined with the dominant political and economic status quo of postwar America. A significant cause of this situation was the socially cohesive effects of widespread anti-communist fears and the unifying experience of a perceived external threat to 'American values.'[24] Therefore, evangelicals were more than happy to court political involvement in the mid-twentieth century and exuded a consequent self-confidence.[25] In-depth views of the role of God in natural catastrophes and suffering were not pressing issues for a movement that was at home and at ease in the world. For example, reporting on Hurricanes by the Baptist press in the mid-century evinces a lack of questioning God's purposes or grappling with big questions. Avoiding speculations about God's judgment and calls to repentance fit neatly with an American dream-affirming mode of Christianity developed in the third quarter of the century by popular and influential ministers such as Norman Vincent Peale (1898–1993) and Robert Schuyler

[22] Cornelius Van Til, *Evil and Theodicy* Handwritten, two bound notebooks. This paper dealing with the problem of evil, dated March 31st, 1923, won the middle-year contest for the A. A. Hodge prize in systematic theology and was presented to C. W. Hodge, Jr. https://www.monergism.com/thethreshold/sdg/vantil/Evil%20And%20Theodicy%20-%20Cornelius%20Van%20Til.pdf, accessed 06/06/2022

[23] Carl Henry wrote, 'Fundamentalism is the modern priest and Levite, bypassing suffering Humanity' in Carl Henry *The Uneasy Conscience of Modern Fundamentalism,* (Grand Rapids: Eerdmans, 1947), 2.

[24] Kevin M. Kruse *One Nation Under God: How Corporate America Invented Christian America* (New York: Basic Books, 2015), Darren E. Grem *The Blessings of Business: How Corporations Shaped Conservative Christianity,* (Oxford: Oxford University Press, 2016), Angela M. Lahar *Millennial Dreams and Apocalyptic Nightmares: The Cold War Origins of Political Evangelicalism* (Oxford: OUP, 2007), and Matthew Sutton *American Apocalypse: A History of Modern Evangelicalism* (Cambridge, MA: Harvard University Press, 2014).

[25] Benjamin T. Lynerd *Republican Theology: The Civil Religion of American Evangelicals* (Oxford: OUP, 2014), 160

(1926–2015), for whom suffering and brokenness had little place in the Christian life in the face of the power of positive thinking.[26]

At the midpoint of the twentieth century sits Hurricane Hazel. It is particularly remembered in Canada and caused widespread damage in parts of the Caribbean, but its presence was also felt in the Carolinas. It made landfall in the United States on October 14, 1954, registering 150 mph winds; it made such an impact on North Carolina that people in that region marked time by simply referring to before or after 'the storm.' As with Galveston and others, reportage focused on the destruction of churches in a way that highlighted the interrelatedness of religion and community.[27] For the Southern Baptists of North Carolina, reportage of Hurricane Hazel focused mainly on the resilience of its members and their fortitude in rebuilding and keeping on with the business of church life. The congregants of Carolina Beach Baptist Church, whose building had been severely damaged, were lauded for coming out in large numbers at the first possibility following the storm. It was observed that some were "in second-hand clothing" and were praised for voting to purchase a house for the church's use despite many of them now being homeless.[28] For the people of these Baptist churches, continuing God's work, despite hardships, was a sanctifying experience that displayed true discipleship and faithfulness to their most deeply held beliefs. So, too, was the case for the North Carolina Baptist State Association, whose buildings were damaged but repaired in time for the assembly meeting to proceed, a sign of blessing from God, a cause for thanksgiving, but also an indication of the attitude of the members who repaired in time.[29] This attitude continued and formed a theme in the commentary on further hurricanes. Keeping on despite warnings or inclement weather became a mark of true courage and commitment to church activities and the community's life, the elevation of bravery and the refusal to acknowledge vulnerability play into this attitude.[30]

The theme of God bringing good from evil also emerged, as in the case of Shiloh Baptist Church in Wilmington. The city, being close to the coast, experienced hurricane-force winds that leveled several houses. However, at the same time, many trees were also uprooted, which enabled the members of Shiloh Baptist to rebuild their pastor's house. For these Christians, The hurricane was an occasion to illustrate the lesson of God's provision and wisdom on display in the midst of

[26] See works such as Norman Vincent Peale's *Power of Positive Thinking* (New Jersey: Prentice Hall, 1954) and Robert Schuller's *God's Way to the Good Life* (New York: Eerdmans, 1963)

[27] Jay Barnes, *North Carolina's Hurricane History* (Chapel Hill: University of North Carolina Press, 2013), 12. and Jay Barnes, *Hurricane Hazel in the Carolinas* (Charleston: Arcadia Publishing, 2010) and Jim Gifford, *Hurricane Hazel: Canada's Storm of the Century* (Toronto: Dundurn, 2004).

[28] "Editorial Brevities," *Biblical Recorder*, November 13th, 1954, 4.

[29] "Damage at Caswell assembly," *Biblical Recorder*, November 06th, 1954, 9.

[30] See "Liberty Association meets," *Biblical Recorder*, October 24th, 1959, 9. and "Rev. W. Howard Goes to South Carolina" *Biblical Recorder*, April 06th, 1957, 4.

destruction.[31] Yet, not all in the community showed such resilience and strength of faith; indeed, one reader of the North Carolina Baptist newspaper, *The Biblical Recorder*, wrote to the editor encouraging people to 'Thank God' rather than grumbling or complaining, significantly neither option—thankfulness or grumbling—was the traditional diagnoses of societal sins and God's judgment.[32] The trend of ignoring serious reflection on God and suffering that this correspondent was observing was commented upon by Carl Henry, founding editor of *Christianity Today*, when he argued that contemporary society boiled down the traditional theodicy of Leibnitz—the attempt to reconcile God's omniscience, omnipotence and goodness—to the question 'Why has *This* happened to *me*?'[33]

The response to Hurricane Hazel also displayed a tendency within the Baptist community toward missionary efforts. Haiti, in particular, had borne the brunt of Hazel's wind, and the country was short of animal livestock. The Southern Baptist Convention started an effort to meet this need, and even those in North Carolina who were themselves hit by the hurricane were encouraged to contribute and were keen to do so.[34] The focus on others despite the experience of personal hardship was displayed as a sign of spiritual maturity and humility. For the Baptist community of North Carolina, distress afforded not just an opportunity for self-reflection but selflessness during loss and suffering. However, this was not just about serving others' physical needs but also their spiritual needs by spreading the Gospel. The coupling of mission with disaster relief is a clear indication of evangelicalism's attempt to join religious orthodoxy with humanitarian concern following the example of Carl Henry and in distinction to Fundamentalism's rejection of social concern. Notably, in 1966, when the Southern Baptist Convention first voted the amount of $50,000 to be devoted to use in disaster relief, it was to be administered by the Home Mission Board.[35]

Hugo and the Late-Twentieth Century

In the third quarter of the twentieth century, while evangelicals were ignoring theodicy in favor of a culture-affirming combination of religious orthodoxy and social concern, others in mainline and liberal Protestantism were developing their own responses to the problem of evil and suffering. British Born theologian and Professor John Hick (1922–2012) attempted to formulate a solution to the problem of evil and

---

[31] "Hurricane Hazel was an Ill wind that blew some good for Shiloh congregation," *Biblical Recorder,* March 20th, 1965, 18.

[32] "letter to the Editor," *Biblical Recorder,* October 30th, 1954, 20.

[33] Carl Henry *God Revelation and Authority Volume IV: God who Stands and Stays Part II* (Waco: Word Books Publishing, 1983), 269.

[34] "Livestock Shipped to Haiti," *Biblical Recorder,* December 04th, 1954, 19.

[35] *Involving Southern Baptists in Disaster Relief: Serving Christ in Crisis.* Southern Baptist Historical Library and Archives, Nashville, TN.

suffering in his book *Evil and the God of Love* (1966). He proposed a Soul-Making Theodicy, which he also termed an Irenaean Theodicy, after the second-century Bishop and theologian Ireneus. Hick distinguished this from standard Augustinian theodicies that emphasize humanity's free will as the cause of evil. His Soul-Making Theodicy view rejected sin or punishment as the cause of suffering. Instead, it stressed the didactic nature of the universe in which mankind is taught to develop its moral and spiritual capabilities. Thus, natural disasters are to be seen as part of the created order intended to aid mankind in its growth in attributes such as compassion, self-sacrifice, and solidarity.[36] Even more radical in its move away from classic theism, theologians and philosophers of religion such as David Ray Griffin (1939–2022) advocated for a Process Theodicy. In this view, God is not omnipotent and not in control of creation or nature, but rather is in a process of development in and with creation. This means that mankind cannot blame God for suffering but must work with God in order to overcome evil and realize a greater potentiality for humanity and each individual.[37] Alongside these debates in the fields of philosophy and theodicy, developments in theology more broadly recaptured a focus on the imminence of God. This trend was seen especially in the theology of Jürgan Moltman (1926–), which he developed, along with all subsequent theologians and philosophers of theodicy, in the light of the enormity of human evil displayed in the holocaust. His book The *Crucified God* (1972) powerfully emphasized the fact that God has experienced suffering by becoming human and has understood the sense of loss, defeat, and betrayal evidenced in the ultimate suffering act, the cross. Therefore, suffering is given meaning because, in the truest and most profound sense, God is not impassible but suffers with us.[38]

A generation later, these academic discussions had had time to filter through seminaries into the pulpit and finally enter the religious consciousness of American Protestants. In their distinct ways, each of these theodicy options focused on human development, growth, and personal spiritual appreciation of God's involvement in suffering and humanity's response to it. When this was combined at a popular level with the influence of individualist consumer culture and market forces on American Christianity, there was a perceptible move from hurricanes being seen as opportunities to develop social solidarity to being seen as part of an internalized narrative of personal or communal growth.[39] In this move, any discussion of the alleviation of

---

[36] John Hick *Evil and the God of Love* (New York: Harper Collins, 1966)

[37] See David Ray Griffin God, *Power and Evil: A Process Theodicy* (Philadelphia: Westminster Press, 1976).

[38] D. K. Chester "The Theodicy of Natural Disasters." *Scottish Journal of Theology* 1, no. 4 (1998): 485–506, Jürgan Moltman, T*he Crucified God*, trans. R. A. Wilson and John Bowden (London: SCM Press, 1974).

[39] For the individualization and consumer market-driven nature of American Christianity and especially evangelicalism, see Kristin Kobes Du Mez *Jesus and John Wayne: How White Evangelicals Corrupted A Faith and Fractured a Nation* (New York: Liveright, 2020) and Daniel Vaca *Evangelicals Incorporated: Books and the Business of Religion in America* (Cambridge, CT: Harvard University Press, 2019). The therapeutic aspect of Christianity continues into the twentieth century, and authors such as S.N. Arel and Rambo have developed a theology of trauma that is

suffering that the social gospel movement had entertained was replaced by a utilization of suffering in aid of personal development, and thus vulnerability and risk were sidelined and overlooked as categories of theological analysis. The theologian A. J. Conyers (1946–2004) was alarmed at this trend, which he observed in the wake of Hurricane Hugo of 1989. He wrote *The Eclipse of Heaven* (1999) as a result, portions of which were published in *Christianity Today*. He wondered why people were not asking big questions in the wake of disaster but rather focusing on their own experiences, writing, "I became aware, then, of just how godless our world has become." Conyers was not referring to sins that he thought may or may not have been being judged in the hurricane but rather the fact that after the hurricane, Americans "no longer ask how this illuminates the mystery over us, but instead attend to the enigma within us."[40] However, regardless of the accuracy of Conyers' observation, religion remained an integral part of many individuals' attempts to comprehend and process Hurricane Hugo. At the popular level, this is seen in reports of a photograph captured during the storm showing the shape of Jesus in the clouds presiding over the event.[41]

Hurricane Hugo rampaged across the Carolinas on September 21, 1989, causing an estimated $10 billion of damage, claiming the lives of over 100 people and leaving 100,000 homeless.[42] For those who lived through it, the storm became a defining moment in their personal stories as well as the narrative of cities such as Charleston, whose physical landscape was forever changed. This was also true of churches, which began to use natural disasters as identity-forming experiences that contributed to the life and history of the congregation. For example, the history of Second Presbyterian Charleston retells that while the hurricane did significant damage to the sanctuary, which could not be used for a year, "The Education Building accommodated worship services during the year the sanctuary was out of service. In turn, the Education building was improved by repairing and upgrading the Fellowship Hall, kitchen, classrooms, offices and storage areas."[43] Similarly, Sunrise

in direct continuity with Christian counseling trends. A. Groenewald "Trauma is Suffering that Remains: The contribution of Trauma Studies to Prophetic Studies," *Acta Theologica* Vol 26, 2018, 88–102 and see Shelly Rambo, *Spirit, and Trauma: A Theology of Remaining*, (Louisville, KY: Westminster John Know Press, 2010) Stating 'Disasters remind humans of the need to remain spiritual and retain faith in God.' reminiscent of the formula first proposed by Christian Smith and Melinda Denton Of 'moralistic therapeutic deism.' Christian Smith and Melinda Lundquist Denton *Soul Searching: The Religious Lives of American Teenagers* (Oxford: OUP, 2005).

[40] A. J. Conyers *The Eclipse of Heaven: Rediscovering the Hope of a World Beyond* (Illinois: Intervarsity Press, 1992) and "After the Hurricane" in *Christianity Today*, November ninth, 1992.

[41] "Believers say Photo Shows Jesus During Hugo" UPI Archives, April 12th, 1990. Accessed September 15th, 2023. https://www.upi.com/Archives/1990/04/12/Believers-say-photo-shows-Jesus-during-Hugo/8690639892800/

[42] Mickey Noah "Hugo, 20 years ago was a disaster relief catalyst," Baptist Press, August 26th, 2009. Accessed September 15th, 2023. https://www.baptistpress.com/resource-library/news/hugo-20-years-ago-was-disaster-relief-catalyst/

[43] "Our living history," Second Presbyterian Church of Charleston, accessed May 30th, 2022. https://www.2ndpc.org/our-living-history

Presbyterian church on Sullivan's Island, South Carolina, includes Hurricane Hugo in what they call 'our story.' It retells how the church "underwent a rebirth and became Sunrise Church. In 1989, Hurricane Hugo reduced the church's buildings to rubble. With faith and confidence, the congregation rebuilt."[44] For these and many other similarly affected churches, the destruction of the hurricane was not so much a reason to question God or wonder at his purposes, much less cast around blame and judgment, but rather a time of rebirth. The biblical principle, 'Except a corn of wheat fall into the ground and die' (Jn, 12:24. KJV), was lived out in the life of these faith communities. The death of their buildings and old patterns of ministry gave way to growth, development, renewal, and new life. For these churches, and many others, Hugo became another page in the story of God's faithfulness to them as congregations, or even as individuals within those congregations, that could then form part of their shared history and identity as churches, woven into a narrative of a congregation's life. The unstated assumption of these narratives is that any future hurricanes or other natural disasters would work in the same way.

Alongside these positive narratives of the experience of Hugo, other voices turned back to older interpretations of the purpose of natural disasters. As the evangelical consensus of the mid-century fractured and evangelicals began to perceive their loss of grip on American politics and culture, the religious right properly emerged as a force in American society along with the infamous moral majority.[45] In this movement, the theme of weather and God's sovereignty emerged once again; for example, one of its leaders, Pat Robinson, claimed to have averted a hurricane away from the coast of Virginia in 1985 by prayer.[46] Leaders of this movement were more than happy to ascribe any number of natural disasters to God's judgment for the country's failure to live up to their view of biblical morality, an inevitable force that would not be stopped by attempts to reduce risk and vulnerability. Consequently, there were those who looked for reasons why Hugo had hit America in the usual explanation of judgment for sins. However, the establishment mainline evangelicals distanced themselves from this rhetoric without downplaying the role of God as the ultimate cause of weather and natural disasters. The area of Myrtle Beach, South Carolina, received a visit from the great statesman of American evangelicalism, Billy Graham, then nearing his 71st year of age. He quickly made clear that he did not believe the storm was divine judgment and was cautious and indeed ambiguous in his response to questions, saying, "Maybe God is saying something to us about

---

[44] "Our story," Sunrise Presbyterian Church, accessed May 30th, 2022. https://www.oursunrise.org/about/our-story/

[45] The literature on the history of the religious right and the moral majority is extensive, and helpful overviews about the origin and development of the movement include Clyde Wilcox's *Onward Christian Soldiers? The Religious Right in American Politics* (Boulder, CO: Westview Press, 1996), Ruth Murray Brown *For A "Christian America" A History of the Religious Right,* (Amherst, NY: Prometheus Books, 2002), and Matthew Avery Sutton *Jerry Falwell and the Rise of the Religious Right* (Boston: Bedford, 2013).

[46] Daniel K. Williams *God's Own Party: The Making of the Christian Right* (Oxford: OUP, 2010), 217.

changing our lives, changing our lifestyles." He offered the crowd that gathered the opinion that it was possible "God was sparing you, and he spared you to serve him and to serve others.''Service was ostensibly the reason for his visit. His charity, overseen by his son Billy Graham Jr., Samaritans Purse, was doing large-scale aid relief in the area.[47]

The development of modern Christian disaster relief forms another theme of the Christian response to Hurricane Hugo. Mickey Noah of the Southern Baptist Convention claimed that Hugo was the catalyst for developing the multibillion-dollar enterprise that the SBC disaster relief fund has become. Other states had their own, but South Carolina did not when Hugo hit, and they were reliant upon other states for aid. Following this experience, developments were made to enact a fund for the state convention of South Carolina.[48] The hurricane also worked as a personal incentive for volunteering and feeling something of the joy of serving those in need. As one man stated, "it's like drinking coke; you try it once, and you hooked. I just enjoy helping people." When Hurricane Andrew hit just five years later, helpers from areas afflicted by Hugo, motivated by a shared sense of endured hardship, volunteered and offered others the kinds of help they had received a short while before. Indeed, the staff of the SBC believed that the giving of Christians in the affected areas would fall, but they found it actually increased.[49]

Katrina and the Twenty-First Century

As America moved into the twenty-first century, the country experienced increased exposure to natural disasters worldwide and at home. Subsequently, the issue has returned to the forefront of the question of God and evil after being overshadowed in the mid-twentieth century by theodicies more concerned with the role of human evil in suffering.[50] As the classic construction of the religious right declined in influence, not least because of George Bush's mishandling of Katrina, a section of conservative Christians began to disentangle themselves from such overt political involvement while still maintaining continuity with parts of earlier Fundamentalism.[51] For one subsection of this group, dubbed the Young, Restless, and Reformed by the journalist Colin Hansen, God's sovereignty is a central tenant of their belief. But as sociologist Brad Vermeulen has shown, the covenantal nature of the relation between

47 Zane Wilson, "How Rev. Billy Graham lifted spirits during his grand strand visit after Hurricane Hugo," The Sun News, Updated February 21, 2018. https://www.myrtlebeachonline.com/news/local/article201282614.html

48 Noah, "Hugo"

49 James Dotson "SBC joins the 'First wave of Christ's arms,' September 19th, 2003. https://www.baptistpress.com/resource-library/news/sbc-joins-the-first-wave-of-christs-arms/

50 Chester "The Theodicy of Natural Disasters," 496.

51 Williams *God's Own Party,* 271.

God and America has been jettisoned.[52] Multiple significant figures in the movement have written books on questions of suffering. These works do not shy away from God's direct role in events but do not focus on the judgment of sins or even present that as a biblically viable option. The Presbyterian theologian and pastor R.C. Sproul (1939–2017) wrote *Surprised by Suffering* in 1989, and John Piper and Justin Taylor, spearheads of the influential conservative evangelical group The Gospel Coalition, as well as the evangelist Jerry Bridges authored books in the year after Katrina on the issue of God and suffering. These books dealt directly with natural disasters.[53] For these authors, the crucial goal is the personal conversion of individuals and the apologetic equipping of the church, not the prophetic condemnation of society. There is no notion of a shared moral or religious consensus that would allow for a message of judgment to make sense; instead, these authors acknowledge that Christianity is one voice among many that must explain itself to a public square that does not know or understand the moral and theological framework that undergirds Christian theodicy.

The rise of the Internet as a network tool for Christian organizations and denominations increased the Christian response to natural disasters. In the wake of Hurricane Katrina (2005), the Internet was a buzz with commentary, opinions, and assertions about God and the hurricane. It was a hurricane of intense controversy, and still has the power to evoke strong emotions: Billy Graham called it the worst disaster to hit the United States since the Civil War.[54] Christians may have gotten involved in the debates over the government's responsibility and culpability in the handling of the hurricane and subsequent floods, but what is most apparent in its wake was the outpouring of apologetic material. There were, of course, responses including all the themes we have seen so far, secular sources reporting on the sterling work churches were doing in relief efforts, as well not a few churches showing their own work among refugees and survivors. There were also Christian-run charity concerts, such as the one hosted by the megachurch pastor T. D. Jakes, and the expected personal stories and narratives of renewal, growth, and reconstruction after the event.[55] Yet, the sheer quantity of think pieces about God and suffering was unprecedented and set the tone for all subsequent natural disaster events down to today. All traditions of Christianity reached into their theological repositories to

---

[52] Colin Hansen, *Young, Restless, Reformed: A Journalist's Journey with the New Calvinists* (Wheaton: Crossway, 2008) and Brad Vermurlen *Reformed Resurgence: The New Calvinist Movement and the Battle Over American Evangelicalism* (Oxford: OUP, 2020), 52–53.

[53] R.C. Sproul, *Surprised by Suffering* (Orlando: Ligonier Publishing, 1989) reissued in 2010, John Piper and Justin Taylor *Suffering and the Sovereignty of God* (Wheaton: Crossway, 2006) and Jerry Bridges *Is God really in Control: Trusting God in a World of Hurt* (Colorado Springs: NavPress, 2006)

[54] "Billy Graham: A Voice of Hope in Crisis—Hurricane Katrina, 2205," Billy Graham Library, August 20th, 2020. https://billygrahamlibrary.org/blog-billy-graham-a-voice-of-hope-in-crisis-hurricane-katrina-2005/

[55] Justin Camacho "T.D. Jakes and Kirk Franklin Team up for Katrina Benefit Concert" The Christian Post, September 16th, 2005, Accessed September 15th, 2023. https://www.christianpost.com/news/t-d-jakes-and-kirk-franklin-team-up-for-katrina-benefit-concert.html

bring out the question of God and evil. The 9/11 attacks had given this theme impetus. However, events such as the 2004 Indian Ocean Tsunami and then 2005's Hurricane Katrina raised the more specific question of purely natural disasters. The mainline Presbyterian Church USA issued a statement that 'God is present in the midst of pain and panic,' a clear derivative of a Moltman-inspired theodicy emphasizing the fact that God experiences our suffering with us.[56] This pastorally sensitive message sidesteps the issue of judgment or even sovereignty in suffering and focuses simply on God's presence.

More conservative authors such as the influential Minneapolis Baptist preacher, John Piper, took a standard conservative message using the event as a call for self-reflection and repentance. The hurricane was not a punishment for specific sins but a reminder that we are all sinners who will face death and are, therefore, called to repent. He wrote, "Whatever judgment has fallen, it is we who deserve it—all of us. And whatever mercy is mingled with judgment in New Orleans neither we nor they deserve."[57] This is, in effect, an evangelistic theodicy and a self-conscious attempt to be 'Gospel-centered.' The standard approach these figures take is that of Jesus in Luke 13:7–8. In this passage, Jesus was asked about the tower of Siloam, which had fallen over and killed eighteen people; his questioners desired to know who had sinned in order for this disaster to have occurred. He says, "Do you think they were more guilty than all the others living in Jerusalem? I tell you, no! But unless you repent, you too will all perish." The message is clear: do not worry about the sin of others, but use every reminder of the presence of death to reflect on your own status before God. Disaster is preached, therefore, as a call to repentance in the face of death, a classic evangelistic technique that maintains the sovereignty of God over natural disasters but avoids having to be precise about the reasons why they may have happened. This is in keeping with the oft-cited verse, Isaiah 55:9 (KJV): "For as the heavens are higher than the earth, so are my ways higher than your ways and my thoughts than your thoughts."

This message was echoed by Al Mohler, the conservative chairman of the Southern Baptist Convention and prominent Christian public thinker, who took a similar line and wanted to defend the idea of God's sovereignty: "First, the Bible reveals that God is omnipotent and omniscient. These are unconditional and categorical attributes. The sovereignty of God is the bedrock affirmation of biblical theism. The Creator rules over all creation."[58] However, Billy Graham took a more widely appealing approach and offered this advice to those questioning: "The Bible

[56] Rick Ufford-Chase, Cliff Kirkpatrick and John Detterick. "A message on Hurricane Katrina from Top Leaders of PC(USA)," *Christianity Today*, September 06th, 2005. Accessed September 15th, 2023. https://www.christiantoday.com/article/a.message.on.hurricane.katrina.from.top.leaders.of.pcusa/3871.htm

[57] John Piper, "Was Katrina Intelligent Design?," Desiring God, September 02nd, 2005. https://www.desiringgod.org/articles/was-katrina-intelligent-design

[58] Albert Mohler "Hurricane Katrina: Why Does God Allow Disaster?" Crosswalk, August 31st, 2005. https://www.crosswalk.com/faith/spiritual-life/hurricane-katrina-why-does-god-allow-disaster-1348449.html

says evil is a mystery. Someday we will understand, but not now. I do know this, however: God knows what we are going through, and He still loves us and cares about us. In the midst of suffering and tragedy we can turn to Him for the comfort and help we need."[59] One musician in the conservative denomination, the Presbyterian Church in America, penned a hymn days after the hurricane, which aimed to sum up some of the prayers and thoughts people would have after the disaster. It contained the uneasy juxtaposition of knowledge and mystery implicit in this evangelical preaching in lines such as,

> Yours the power of devastation, yours to gather, help and heal:
> We know not your ways of wisdom: let your light our paths reveal.

The song also called for repentance without explicitly naming the hurricane as an act of judgment or specifying the sins for which repentance was needed. Ultimately, the hymn again returns to an evangelistic note in the following.

> Call our nation to repentance; gather us that we may see
> You're the God of love and mercy; rescue us and set us free.
> Let our deeds announce redemption, saving grace for all our days.
> Let your song go forth to cheer us, lifting heart and hands to praise.[60]

Evangelical Protestant responses to Katrina were both evangelistic and apologetic, and each Christian group or network was quick to offer its own explanation for why such an event such as Katrina happened via the Internet. Such writings, as well as podcasts and other media, display an attempt to reassure the faithful and offer answers to those interested: seeking or purely skeptical. There is an assumption that everyone is asking the question, "Why did God let this happen?" Even Answers in Genesis, the young earth creationist think tank, waded into the debate with an article arguing that the church cannot give satisfying answers on the existence of evil because "most have either given up or ignored the foundational importance of the history in Genesis chapters 1–11."[61] This apologetic and evangelistic tendency must be placed in the context of the rise of the New Atheism, a phenomenon that would flourish briefly in the years following Katrina. Books such as Richard Dawkins's *The God Delusion* (2006) specifically use the existence of suffering following natural disasters to discredit faith in a divinity. Dawkins even cited conservative Christian use of judgment theodicies to explain Katrina as a particularly egregious offense of religion against society and norms of rational morality.[62] The presence of figures such as these heightened the felt need to respond to natural disasters in pastorally sensitive ways that also warned of the neglect of faith and

[59] "Billy Graham: A Voice of Hope in Crisis"

[60] Stephen D. Lawton 'God of Nature, At Whose Voice' quoted in "PCA Congregations and Ministers Affected by Hurricane Katrina" PCA history, September 08th, 2005. Accessed September 15th, 2023. https://www.pcahistory.org/churches/katrina.html

[61] Ken Ham "They're Not Connecting it! Hurricanes Katrina and Rita ... and a loving God," Answers in Genesis, September 23rd, 2005. https://answersingenesis.org/suffering/natural-disasters/theyre-not-connecting-it/

[62] Richard Dawkins, *The God Delusion* (London: Black Swan, 2007), 270.

reinforced the possibility of belief in the face of natural disasters. What is conspicuously absent in these apologetic uses of theodicy in the wake of Katrina is any mention of any responsibility for reducing risk and vulnerability with regard to future natural disasters, but rather an implicit expectation that regardless of man's efforts, this would be a perpetual question to be addressed by much the same arguments in the event of the next hurricane.

Conclusion

American Protestant Christianity has undergone many changes throughout the twentieth and twenty-first centuries. Alongside those developments, splits, divergences, mergers, and movements, theodicy has evolved and receded into the background or emerged to the fore, depending on the year. The call of judgment for sins, as those employed by the religious right, still occurs, and, predictably, at each stage, religious conservatives have discussed the relationship between America and God's pleasure or displeasure.[63] This is too deeply embedded in the psyche of American Christianity not to emerge and too sensational not to be commented upon by the broader culture. But the reality of American Christianity more broadly is that theodicy plays a far more diverse set of roles within the life of the church and believers.[64] It allows disasters and suffering to be turned into calls to social action and justice, as with the liberals and modernists of the early twentieth century. It can be a way to experience, assess, and realize one's personal spiritual growth or form an integral part of a church's self-identity and be used as an apologetic tool to convert nonbelievers or skeptical outsiders. The result is that by examining theodicy along with natural disasters, a picture of American Christianity emerges that is complex and diverse, not open to generalization or simple characterization. Movements and trends within it are historically contingent and have their own emphases, preoccupations, and influences that change how theodicy is deployed. The Internet has proliferated this process and increased the plurality of ideas and theological influences. Consequently, there exist far more options for Christians, even conservative evangelicals, when constructing a theodicy in the wake of natural disasters as we move into the twenty-first century than ever before. Yet, the majority of these views, in their own ways, assume, reinforce, or accept the presence of natural disasters,

[63] Matthew Yglesias "Hagee on Katrina" in *The Atlantic,* March 05th, 2008 https://www.theatlantic.com/politics/archive/2008/03/hagee-on-katrina/45516/ and for conservative conjecture on the relation of America to God see John MacArthur "What's Wrong with America" *Grace to You,* February 21st, 1993. https://www.gty.org/library/sermons-library/80-112/whats-wrong-with-america

[64] As Charlene P. E. Burns has commented, the calls of judgment from the likes of Robinson, Falwell, and others are deeply unhistorical in terms of the entirety of Christian history. Charlene P.E. Burns, *Christian Understandings of Evil: The Historical Trajectory* (Minneapolis: Fortress Press, 2016), 6.

caused by God or otherwise, and attempt to utilize them for their own ends and purposes. As such protestant theodocies can be seen as contributing to a limiting of the conceptual frameworks through which to view the question of vulnerability and risk from a religious perspective.

References

Barnes, Jay. 2013. *North Carolina's Hurricane History*. Chapel Hill: University of North Carolina Press.

Barnes, Jay. 2010. *Hurricane Hazel in the Carolinas*. Charleston: Arcadia Publishing.

"Believers say Photo Shows Jesus During Hugo" UPI Archives, April 12, 1990. Accessed September 15, 2023. https://www.upi.com/Archives/1990/04/12/Believers-say-photo-shows-Jesus-during-Hugo/8690639892800/

Bellah, Robert N. Civil Religion in America, Dædalus, *Journal of the American Academy of Arts and Sciences* 96, No. 1 (Winter 1967): 1–21.

Billy Graham Library. *Billy Graham: A Voice of Hope in Crisis – Hurricane Katrina*, 2205, August 20, 2020. https://billygrahamlibrary.org/blog-billy-graham-a-voice-of-hope-in-crisis-hurricane-Katrina-2005/.

Bridges, Jerry. 2006. *Is God really in Control: Trusting God in a World of Hurt*. Colorado Springs: Navpress.

Burns, Charlene P. E. 2016. *Christian Understandings of Evil: The Historical Trajectory*. Minneapolis: Fortress Press.

Brennan, Kristine. 2002. *The Galveston Hurricane*. Philadelphia: Chelsea House.

Brown, Ruth Murray. 2002. *For A "Christian America" A History of the Religious Right*. Amherst, NY: Prometheus Bookds.

CBS News. *Pat Robinson Controversies*. Last modified January 13, 2010. https://www.cbsnews.com/pictures/pat-robertson-controversies/

Chester, D. K. 1998. The Theodicy of Natural Disasters. *Scottish Journal of Theology* 1 (4): 485–506.

Conyers, A. J. 1992a. *The Eclipse of Heaven: Rediscovering the Hope of a World Beyond*. Illinois: Intervarsity Press.

Conyers, A.J. "After the Hurricane" in *Christianity Today,* November 9, 1992b.

Coffin, Henry Sloane. 1911. *Social Aspects of the Cross*. New York: Hodder & Stoughton.

Dawkins, Richard. 2007. *The God Delusion*. London: Black Swan.

Dranseika, V., & Gordijn, B. (eds.), *Disasters: Core Concepts and Ethical Theories. Advancing Global Bioethic*s, vol 11. 2018.

Fosdick, Harry Emerson. 1938. *A Guide to Understanding the Bible*. New York: Harper and Brothers.

Friedman, Meghan. *God's Wrath Caused Katrina*, TIME, September 16, 2011. http://content.time.com/time/specials/packages/article/0,28804,1953778_1953776_1953771,00.

Gardella, Peter. 2014. *American Civil Religion: What America Holds Sacred*. Oxford: Oxford University Press.

Gifford, Jim. 2004. *Hurricane Hazel: Canada's Storm of the Century*. Toronto: Dundern.

Green, Nathan. 2000. *The Story of the 1900 Galveston Hurricane*. Louisiana: Pelican.

Grem, Darren E. 2016. *The Blessings of Business: How Corporations Shaped Conservative Christianity*. Oxford: Oxford University Press.

Griffin, David Ray. 1976. *God, Power and Evil: A Process Theodicy*. Philadelphia: Westminster Press.

Groenewald, A. 2018. Trauma is Suffering that Remains: The contribution of Trauma Studies to Prophetic Studies. *Acta Theologica* 26:88–102.

Gorski, Philip. 2017. *American Covenant: A History of Civil Religion from the Puritans to the Present*. New Jersey: Princeton University Press.

Ham, Ken. *They're Not Connecting it! Hurricanes Katrina and Rita ... and a loving God*. Answers in Genesis, September 23, 2005. https://answersingenesis.org/suffering/natural-disasters/theyre-not-connecting-it/

Hansen, Colin. 2008. *Young, Restless, Reformed: A Journalist's Journey with the New Calvinists*. Crossway: Wheaton.

Hart, Daryl G. 1994. *Defending the Faith: J. Gresham Machen and the Crisis of Conservative Protestantism in Modern America*. Baltimore, MD: John Hopkins University Press.

Henry, Carl F. H. 1983. *God Revelation and Authority Volume IV: God who Stands and Stays Part II*. Waco: Word Books Publishing.

Henry, Carl F. H. 1947. *The Uneasy Conscience of Modern Fundamentalism*. Grand Rapids: Erdmans.

Hick, John. 1966. *Evil and the God of Love*. New York: Harper Collins.

Involving Southern Baptists in Disaster Relief: Serving Christ in Crisis. Southernn Baptist Historical Library and Archives, Nashville TN.

Kruse, Kevin M. 2015. *One Nation Under God: How Corporate America Invented Christian America*. New York: Basic Books.

Lahar, Angela M. 2007. *Millennial Dreams and Apocalyptic Nightmares: The Cold War Origins of Political Evangelicalism*. Oxford: Oxford university Press.

Lester, Paul. 1900. *The Great Galveston Disaster*. Philadelphia: Globe Bible Publishing.

Lynerd, Benjamin T. 2014. *Republican Theology: The Civil Religion of American Evangelicals*. Oxford: OUP.

MacArthur, John. Whats Wrong with America. *Grace to You,* Feb 21st, 1993. https://www.gty.org/library/sermons-library/80-112/whats-wrong-with-america

Machen, J. Gresham. 1923. *Christianity and Liberalism*. New York: Macmillan.

Marsden, George. 1980. *Fundamentalism and American Culture: The Shaping of 20th Century Evangelicalism, 1870–1925*. New York: Oxford University Press.

Miller, Perry. 1939. *The New England Mind: The Seventeenth Century*. New York: The Macmillan Company.

Du Mez, Kristin Kobes. 2020. *Jesus and John Wayne: How White Evangelicals Corrupted A Faith and Fractured a Nation*. New York: Liveright.

Minus, Paul. 1988. *Walter RauschenBusch: American Reformer*. New York: Macmillan.

Mohler, Albert. *Hurricane Katrina: Why Does God Allow Disaster?* Croswalk, August 31, 2005. https://www.crosswalk.com/faith/spiritual-life/hurricane-katrina-why-does-god-allow-disaster-1348449.html.

Moltman, Jürgan. 1974. *The Crucified God.* trans. R. A. Wilson and John Bowden. London: SCM Press.

Moore, Russell. *2 Chronicles 7:14 isn't about American Politics*. RussellMoore.com, January 14, 2016. https://www.russellmoore.com/2016/01/14/2-chronicles-714-isnt-about-american-politics/

Morrissey, Mary. 2000. Elecy Nations and Prophetic Preaching: Types and Examples in the Pauls Cross Jerimiad. In *The English Sermon Revised: Religion, Literature and History*, ed. Laurie Ferrell and Peter McCulloch, 1600–1750. Manchester: Manchester University Press.

Murphey, Andrew R. 2008. *Prodigal Nation: Moral Decline and Divine Punishment from New England to 9/11*. Oxford: OUP.

Murray, Ian. 2004. *The Life of Arthur W. Pink*. Edinburgh: Banner of Truth.

Noah, Mickey. *Hugo, 20 years ago was a disaster relief catalyst*. Baptist Press, August 26, 2009. https://www.baptistpress.com/resource-library/news/hugo-20-years-ago-was-disaster-relief-catalyst/.

Ousley, Laurence, ed. 1900. *Galveston in Nineteen Hundred*. William Chase: Atlanta.

O'Mathúna, D.P. 2018. Christian Theology and Disasters: Where is God in All This? In: O'Mathúna, D., Dranseika, V., Gordijn, B. (eds) *Disasters: Core Concepts and Ethical Theories. Advancing Global Bioethics*, vol 11.

Pink, A. W. 1928. *The Sovereignty of God*. London: I.C. Henderson.
Piper, John, and Justin Taylor, eds. 2006. *Suffering and the Sovereignty of God*. Crossway: Wheaton.
Piper, John. *Was Katrina Intelligent Design?* Desiring God, September 02, 2005. https://www.desiringgod.org/articles/was-katrina-intelligent-design.
Rauschenbusch, Walter. 1917. *A Theology for The Social Gospel*. New York: Macmillan.
Rambo, Shelley. 2010. *Spirit and Trauma: A Theology of Remaining*. Louisville, KY: Westminster John Know Press.
Rossell, Garth. 2008. *The Surprising Work of God: Harold Jay Ockenga, Billy Graham and the rebirth of Evangelicalism*. Michigan: Baker Academic.
Reichenbach, Bruce R. 1982. *Evil and A Good God*. New York: Fordham University Press.
Schuller, Robert. 1963. *God's way to the Good Life*. New York: Erdemans.
Scott, Mark S. M. 2015. *Pathways in Theodicy: An Introduction to the Problem of Evil*. Minneapolis: Fortress Press.
Second Presbyterian Church of Charleston. *Our living history*, accessed May 30, 2022. https://www.2ndpc.org/ourliving-history.
Smith, Christian, and Melinda Ludquist Denton. 2005. *Soul Searching: The Religious lives of American Teenagers*. Oxford: OUP.
Sproul, R. C. 1989. *Surprised by Suffering*. Orlando: Ligonier Publishing.
Stonehouse, Ned. 1954. *J. Greshcm Machen: A Biographical Memoir*. Grand Rapids, MI: Erdemans.
Sunrise Presbyterian Church, *Our story*, accessed May 30, 2022. https://www.oursunrise.org/about/our-story/.
Sutton, Matthew. 2014. *American Apocalypse: A History of Modern Evangelicalism*. Cambridge, MA: Harvard University Press.
Sutton, Matthew. 2013. *Jerry Falwell and the Rise of the Religious Right*. Boston: Bedford.
Thuesen, Peter J. 2020. *Tornado God: American Religion and Violent Weather*. Oxford: Oxford University Press.
Ufford-Chase, Rick, Cliff Kirkpatrick and John Detterick. *A Message on Hurricane Katrina from Top Leaders of PC(USA)*, Christianity Today. September 06, 2005. www.christiantoday.com/article/a.message.on.hurricane.katrina.from.top.leaders.of.pcusa/3871.
Vaca, Daniel. 2019. *Evangelicals Incorporated: Books and the Business of Religion in America*. Cambridge, CT: Harvard University Press.
Van-Til, Cornelius. *Evil and Theodicy* Handwritten, two bound notebooks. This paper dealing with the problem of evil, dated March 31, 1923, won the middler year contest for the A. A. Hodge prize in systematic theology and was presented to C. W. Hodge, Jr. https://www.monergism.com/thethreshold/sdg/vantil/Evil%20And%20Theodicy%20-%20Cornelius%20Van%20Til.pdf, accessed 06/06/2022.
Vermurlen, Brad. 2020. *Reformed Resurgence: The New Calvinist Movement and the Battle Over American Evangelicalism*. Oxford: OUP.
Vincent Peale, Norman. 1954. *Power of Positive Thinking*. New Jersey: Prentice Hall.
Wilcox, Clyde. 1996. *Onward Christian Soldiers? The Religious Right in American Politics*. Boulder, CO: Westview Press.
Williams, Daniel K. 2010. *God's Own Party: The Making of the Christian Right*. Oxford: OUP.
Yglesias, Matthew. "Hagee on Katrina" in *The Atlantic*, March 5th, 2008. https://www.theatlantic.com/politics/archive/2008/03/hagee-on-katrina/45516/.

Through the Eyes of the White, Innocent Child: Whiteness, Vulnerability, and (Environmental) Crisis in Lauren Tarshis's *I Survived* Series

Lea Espinoza Garrido

Abstract This chapter explores the potential of children's literature to promote environmental justice. Drawing on (ecocritical) affect studies and (eco)narratology, it demonstrates that Lauren Tarshis's *I Survived* series, on the one hand, uses natural catastrophes as an access point to children's environmental education and thus offers the potential to educate them as "ecocitizens" (Massey and Bradford): The affective strategies employed in the series can not only help readers to develop (narrative) empathy for the human protagonists and their struggle for survival amidst natural catastrophes but also provide a starting point for discussing the relevance of environmental protection and more-than-human companionship in the Anthropocene. On the other hand, this chapter shows that many of the texts in the series perpetuate stereotypical depictions of non-white Americans and center the emotions of white protagonists through their narrative and affective strategies, their invocation of historically specific American myths, and their erasure of Black and Indigenous populations. In particular, this chapter highlights that the foregrounding of natural catastrophes as outstanding *events* manageable through personal action obscures underlying histories of marginalization and oppression which have rendered some bodies and some peoples more vulnerable to such catastrophes than others. More generally, this chapter makes the case that centering whiteness and white suffering in narratives of the climate crisis potentially obscures the unequal distribution of risk and vulnerability in our current ecological moment and thus contributes to the settler-colonial projects that have produced the now unfolding crisis in the first place.

L. Espinoza Garrido (✉)
John-F.-Kennedy Institute, Free University Berlin, Berlin, Germany
e-mail: lea.espinoza.garrido@fu-berlin.de

N. Rauscher, W. Werner (eds.), *Natural Disasters in the United States*, Risk, Governance and Society 22, https://doi.org/10.1007/978-3-031-96436-7_8

Introduction

In the past decade, the increasingly visible consequences of the unfolding climate crisis in the form of so-called natural catastrophes or disasters[1] that proliferate around the globe have sparked heightened scholarly interest in the role that literature can play in communicating environmental concerns. "While few scholars are so bold as to state in print that a 'literary genre' could 'save the planet,'" as Matthew Schneider-Mayerson argues, "the ecopolitical value of environmental literature has been a key subtext for the growing interest in climate fiction in (liberal) popular discourse and the academic fields of ecocriticism and environmental humanities" (2018, 474). As scholars such as Kathryn Yusoff, Jennifer Gabrys, or Antonia Mehnert suggest, literature can explore the "cautionary offerings of the disaster" (Yusoff and Gabrys 2011, 520) and offer "insight into the ethical and social ramifications of this unparalleled environmental crisis" (Mehnert 2016, 4).[2] Besides mass extinction events, natural catastrophes are perhaps the most imminent manifestations of this accelerating and unprecedented crisis. In this context, survival is not only the most pressing issue for many humans and non-humans around the globe but has also become a dominant trope in literary and cultural imaginations of natural catastrophes and the climate crisis more broadly.

Whereas Yusoff and Gabrys observe that "science-fiction imaginings of *possible* catastrophes" (Yusoff and Gabrys, 2011, 519, my emphasis) have become a globally dominant mode of representing natural catastrophes, the *New York Times* bestselling book series *I Survived* turns toward *true* natural disasters of the past to negotiate the possibility for survival. In the series, author Lauren Tarshis not so much engages with the apocalyptic futures that may lie ahead for humanity but re-imagines catastrophic historical events, always told through the eyes of fictional survivors. The books' protagonists are children whose "resilience and strength in the midst of unimaginable disasters" (Scholastic) are depicted in accessible and engaging prose

---

[1] I am using the terms "natural disaster" and "natural catastrophe" interchangeably in this chapter for several reasons: While I find a distinction between the two based on their differences "in scale and level of sustained support required to get to recovery" (Gulliver-Garcia 2019, no pag.; see also Quarantelli 2011) helpful if we want to discuss the damage they cause and the need to mobilize resources for recovery efforts at a different speed, I find them less helpful for my chapter, which focuses primarily on fictionali(ized) literary representations of these events. I find it particularly hard to measure the 'impact' on affected communities solely based on statistics, death tolls, and financial damages when what is at stake for these communities is their social and emotional vulnerability and protectedness more broadly. In other words, who gets to decide what is "catastrophic" and for whom? And how can we measure the damage done by such events without committing to exercises of abstraction? Finally, the literature that I find most fruitful for addressing these issues, is not consistent in its use of these terms.

[2] For helpful discussions of literature's potential to represent the climate crisis and the unfolding catastrophes in and of the Anthropocene, see also Lawrence Buell's *Writing for an Endangered World: Literature, Culture, and Environment in the U.S. and Beyond* (2001); Erin James's *Narrative in the Anthropocene* (2022), Adam Trexler's *Anthropocene Fictions* (2015); or Amitav Ghosh's, *The Great Derangement* (2017).

suitable for readers between the ages of 8 and 12. Many of the stories focus on natural catastrophes such as *The California Wildfires, 2018*; *Hurricane Katrina, 2005*; *The Japanese Tsunami, 2011*; *The Eruption of Mount St. Helens, 1980*; or *The Joplin Tornado, 2011*, while some also cover other moments of crisis such as *The Sinking of the Titanic, 1912*; or *The Attacks of September 11, 2001*.[3]

In this chapter, I use this series as a starting point and draw on (ecocritical) affect studies and (eco)narratology to examine the potential of children's literature to further environmental justice.[4] I will argue that the *I Survived* series, on the one hand, can make natural disasters tangible and create empathy not only for their victims and survivors but also for non-human others like animals affected by such catastrophes. On the other hand, I will show that many of the texts in the series perpetuate stereotypical depictions of non-white Americans and center the emotions of white protagonists through their narrative and affective strategies, their invocation of historically specific American myths, and their erasure of Indigenous populations. In particular, I will argue that the foregrounding of natural catastrophes as outstanding *events* in the series obscures underlying histories of marginalization and oppression which have rendered some bodies and some peoples more vulnerable to such catastrophes than others. More generally, I make the case that centering whiteness and white suffering in narratives of the climate crisis potentially obscures the unequal distribution of risk and vulnerability in our current ecological moment and thus contributes to the settler-colonial projects that have produced the current climate crisis in the first place.[5]

---

[3] Primarily geared toward the US American book market, most books in the series are also set within the boundaries of the United States, with some set, for instance, in Hawai'i or Alaska. The few exceptions to this are the titles that deal with global historical events—*The Japanese Tsunami, 2011*; *The Nazi Invasion, 1944*; *The Battle of D-Day, 1944*; or *The Destruction of Pompeii, 79 A.D.*—all of which are either well-known in the United States or relate directly to US (military) history.

[4] As Kyle Powys Whyte suggests, environmental justice is typically concerned with "how the distribution of certain environmental nuisances [...] burden populations who already face multiple forms of oppression, from structural racism to systemic poverty" (2016, no pag.). In addition to that, Whyte highlights further dimensions of environmental justice that are frequently neglected: He argues that it must also address how settler colonialism "wrongfully interferes with and erases the socioecological contexts required for indigenous populations to experience the world as a place infused with responsibilities to humans, nonhumans and ecosystems" (ibid). In this chapter, I follow this broader perspective on environmental justice as it aligns not only with the concept of "relational responsibility" (2021, 10) that I have proposed with Ina Batzke and Linda M. Hess to make sense of the challenges of the climate crisis but also with a central aim of environmental literature geared towards children and young adults; "to socialize young people into becoming the *responsible* and empathetic adults of tomorrow by positioning readers as ecocitizens" (Massey and Bradford 2011, 109, my emphasis). For a more detailed discussion of Massey and Bradford's concept of the "ecocitizen," see further below.

[5] Following the works of J. Kēhaulani Kauanui (2016), Patrick Wolfe (2006), and others, I understand settler colonialism as "a land-centered project entailing permanent settlement" (Kauanui 2016, no pag.) that follows the "logic of eliminiation" and needs to be understood as "a structure not an event" (Wolfe 2006, 388). For helpful overviews of the influence of the concept see also Jean O'Brien's "Tracing Settler Colonialism's Eliminatory Logic in *Traces of History*" (2017);

Affect, Empathy, and the Ecocritical Potential of Children's Literature

The growing interest in the potential of literary texts to communicate central environmental concerns related to anthropogenic climate change and the devastating consequences of the climate crisis is closely tied to questions of representation, narrative form, and literature's ability to engage readers emotionally. Silvia Schultermandl, Jana Aresin, Dijana Simić, and Si Sophie Pages Whybrew suggest that literary texts do so through forms of "affective worldmaking" as their "affective narrative strategies form bases for identification and belonging amongst protagonists and […] offer potential points of recognition to audiences" (2022, 15). In the context of literature for children, such questions are also linked to children's cognitive development. As David Comer Kidd and Emanuele Castano have shown, reading literary fiction enables children to "detect and understand others' emotions" (1996, 377) and, thereby, also to increase empathy (1996, 377, 380).[6] According to Suzanne Keen, however, readers' emotional responsiveness is dependent on several different factors both inside and outside of the narrative text itself. In her widely cited work on the connection between empathy and the novel, Keen lists "character identification" and "narrative situation" as the two aspects most commonly associated with the creation of empathy in fictional literary texts (Keen, 2006, 216). While character identification is not a narrative strategy itself but is located within the reader as a result of their reading experience, according to Keen, it "may be precipitated by the use of particular techniques of characterization" (2006, 216).[7] In turn, the narrative situation as the result of a text's composition is a characteristic of the narrative text itself whose emotional effect(iveness), however, may be influenced by various factors inside the reader as well. In Tarshis's *I Survived* series, both character identification and narrative situation play a crucial role in the series' affective worldmaking, in which "real" historical events are presented as fictional(ized) narratives that are told through the eyes of their child protagonists who are approximately the same age as the readers in the series' target audience.

Jennifer McLerran's "Theorizing Settler Colonialism: Alternative Indigenous Methodologies" (2019); Lorenzo Veracini's "'Settler Colonialism': Career of a Concept" (2013). For an insightful discussion of how this connects to the climate crisis, see Daniel R. Wildcat's *Red Alert!: Saving the Planet with Indigenous Knowledge* (2009), or Kyle Powys Whyte's "Indigenous Experience, Environmental Justice and Settler Colonialism" (2016).

[6] See also, Matthijs Bal and Martijn Veltkamp's "How Does Fiction Reading Influence Empathy?" (2013).

[7] For a more detailed discussion of the connection between narrative perspective and the creation of readerly empathy, see also Sylvia Adamson's "The Rise and Fall of Empathetic Narrative: A Historical Perspective on Perspective" (2001); Wayne C. Booth's *The Rhetoric of Fiction* (1983); Dorrit Cohn's *Transparent Minds: Narrative Modes for Presenting Consciousness in Fiction* (1978); Carolina Fernandez-Quintanilla's "Textual and Reader Factors in Narrative Empathy" (2020); or Suzanne Keen's *Empathy and the Novel* (2007).

While the texts in the series introduce readers to new lifeworlds that will most often lie outside of their own immediate lived reality, they take their readers' age and their presumably similar range of experience as an entry point to depict the multiple challenges that the child protagonists have to navigate. The protagonists, then, are not presented as superhuman but as relatable characters who are struggling, who are scared, who suffer, but who are also brave and strong and who eventually overcome their fears to survive. In these depictions, the texts also draw on the ambivalent aesthetic potentiality of the child as both victim and survivor. As Katherine Capshaw Smith argues, "[b]ecause children are imagined as innocent, they are figured almost iconographically as the ultimate victims of trauma" (Capshaw Smith 2005, 116). At the same time, Capshaw Smith contends, they are often "also figured as the survivors of trauma, those who can offer adults spiritual advice in how to triumph over pain through simple, honest, essential values like love, trust, hope, and perseverance" (2005, 116). Tarshis's series oscillates between these dualistic figurations of the innocent child as a victim and survivor of trauma, both of which are central to the texts' affective dimensions. Told "through the eyes of a kid who lived to tell the tale" (Scholastic), the texts' predominantly fixed internal focalization thus invites readers to empathize with the relatable protagonists and their fight for survival amidst natural catastrophes.

In its investment in cultivating empathy for the loss experienced in such catastrophes, *I Survived* also makes use of the affordances of historical fiction. As Alexa Weik von Mossner suggests, young adult climate change fiction texts "develop fictional risk scenarios that allow their young readers to deal on the imaginative level with the fragility of their current lifestyles and the vulnerability of the social orders in which they live" (Weik von Mossner 2017, 554). Weik von Mossner focuses explicitly on dystopian cli-fi texts geared toward young adult readers, but I would argue that the semi-fictional/fictionalized narratives for younger children in the series fulfill similar functions: They can visualize catastrophes that are not only imaginable but that have already occurred beyond the realm of fiction. Since the series is keen on highlighting this aspect in the books' paratext and marketing, the *I Survived* texts are potentially even more "realistic" explorations of risk scenarios, fragility, and vulnerability than completely fictional texts—regardless of how these scenarios are mediated. While the colorful attention-grabbing covers of the individual books lure young readers in with the promise of an exciting story—they usually depict children looking at or running away from raging fires, erupting volcanoes, or blustering waves—the texts serve an educational purpose as well: They are supplemented with lists of reading recommendations, Q&A sections, and additional material offering background knowledge about the risks associated with natural disasters and/or the historical context in which they occurred. These include information on how to behave, and thus survive, in a natural disaster, e.g., in *I Survived: I Galveston Hurricane 1900*, where readers receive four tips on "How to stay safe in a hurricane": (1) Know your risk, (2) Help your family make a plan, (3) Don't go outside during a hurricane, and (4) Help those in need. These tips are framed by a note by the author who explains that "Being prepared can actually make us feel less scared—and safer if a storm actually hits." These sections frame natural catastrophes as a condition that can be managed. Risk mitigation, then, is something that can be taught, even to young children. Whereas "[c]limate futures arguably require approaches that are not only characterized by calculability and risk, but also mobilize imaginative acts that open new spaces and practices for dealing with the effects of living with uncertain futures" (Yusoff and Gabrys 2011, 518), Tarshis's exploration of a catastrophic past thus seems to offer a sense of control amidst the uncertainty of the climate crisis.

The series' focus on a catastrophic, yet manageable, past is complemented by explicit references to the relevance of the catastrophes presented in the books for US American history in the texts' marketing: On the publisher's website, *The Eruption of Mount St. Helens, 1980*, for example, is described as "one of the most environmentally devastating events in recent U.S. history," *The Children's Blizzard, 1888* as "one of the deadliest blizzards in American history," *The California Wildfires, 2018* as "the deadliest in 100 years and the most destructive in history," and *The Galveston Hurricane, 1900* as "still America's deadliest disaster" (Scholastic). Such contextual and paratextual cues, teeming with superlatives, predefine the reading experience and are crucial for readers' affective engagement with the natural catastrophes depicted in the texts' diegesis. After all, as Gulsin Ciftci reminds us, "[a]ffects are at work even before the act of reading takes place. A text's materiality constitutes one of our initial encounters with a work of literature" (2022, 14). As such, the series' affective potential—its call upon readers to emotionally engage with natural catastrophes—not only sets the stage for but already begins to unfold before the readers' immersion into the narrative itself.

Yet, the narrative strategies employed in the series crucially determine how (and how much) readers delve into the stories and emotionally respond to them. Unlike documentaries or other forms of factual narratives about historical events and natural disasters, fictional narratives allow their readers to "try on" what Marie-Laure Ryan (2014) calls "possible worlds" in her eponymous study. In the opening chapter of *I Survived: The Joplin Tornado, 2011*, readers are instantly immersed in the world of disaster as the narrator details how the text's protagonist, 11-year-old Dexter James, is exposed to the full force of the EF-5 tornado that destroyed the city of Joplin in Missouri:

> Dex was in an SUV when the tornado hit, and now he was trapped. The ferocious winds roared and sent tree limbs and rocks smashing against the SUV.
>
> And then,
>
> *Smash!*
>
> A window shattered. The tornado's fury blasted into the SUV. Dex was attacked by swirling winds filled with bits of wood and metal and glass. The pain was like being stung by thousands of scorpions, over and over again.
>
> And then the wind grabbed hold of Dex. It wrapped around him like invisible tentacles, pulling him toward the open window.
>
> Dex had always wanted to see a tornado for real.
>
> And now here it was—the evil, swirling darkness.
>
> Dex was being sucked into the tornado, and he knew there could be no escape. (9, original emphasis)

The sensational and accessible language used in this passage is typical for the series. It makes the events tangible and vividly describes the power of natural forces, immediately translating them into binary oppositions ("evil") and human categories ("fury"), to attach them to young readers' presumed registers of knowledge. The text also acknowledges a certain fascination with natural catastrophes mediated via Dex's child-like curiosity about tornadoes—a curiosity presumably shared with the readers who are investing time to read a story *about* a tornado after all. This serves as a point of contact between reader and protagonist within the series' politics of identification, underwritten by the text's narrative perspective. It is particularly the series' dominant fixed internal focalization, then, that enables its readers to witness the risk that these events pose for the safety of human beings, more specifically other children, first-hand: Through Dex's perspective, readers learn about the physical and emotional consequences of surviving a natural disaster. In its fictional re-imagination, *I Survived: The Joplin Tornado* thus enables its readers to "try on" the fears and feelings of the survivors and to relate to—without appropriating—the loss and suffering that they have likely experienced. In short, the relatable perspective of these texts makes it possible to identify and empathize with the child. The *I Survived* series thus combines the emotional affordances of fictional narrative texts with the verisimilitude of factual narratives.

While the natural forces in the texts are primarily depicted as evil and ruthless, threatening human existence, the series also emphasizes relations between humans and the non-human world. In many of the stories, animals function as protectors and companions to the child protagonists. They particularly encourage empathy for domestic animals like cats and dogs; for instance, for Nya, the cat in *I Survived: The Japanese Tsunami, 2011*, Dexter's dog Zeke in *I Survived: The Joplin Tornado, 2011*, or Barry's dog Cruz in *I Survived: Hurricane Katrina, 2005*. Moreover, the latter includes a section in the paratext of the book that informs readers about the consequences of the hurricane for animals and pets. While these examples do not exactly "decenter the human," as many scholars in the broader field of posthumanism call for,[8] they do shift the focus toward human–nonhuman relations and foreground the relevance of looking beyond human perspectives, even amidst natural catastrophes. These perspectives are also explored in other works in the series: In *I Survived: The California Wildfires, 2018*, Holly's explanation that her family operates an animal shelter, which presently houses a Savannah monitor lizard and a Burmese python, serves as a starting point for a discussion about the appropriateness of reptiles as domestic animals, which is taken up in the paratext of the title as well: Here, Tarshis encourages her readers: "If you love reptiles, show your love by learning all you can about them and letting them live in the wild, where they belong."

At the same time, in titles like *I Survived: The Shark Attacks of 1916* and *I Survived: The Attack of the Grizzlies, 1967*, the series taps into imaginations of sharks and bears as evil and ruthless monsters popularized through films such as *Jaws* (2015) or, more recently, *The Revenant* (2015). The shark that attacks

[8] See for example Cynthia Huff (2017), or Rosi Braidotti (2016).

protagonist Chet in the former text, for instance, is described as a "monster shark" with "black killer eyes" and "bloody teeth glistening," whereas the latter text also describes bears as "monsters" with "dripping claws" and "dagger claws." Yet, in stark tension with these representations, Tarshis uses the books' supplementary material to contextualize the attacks she depicts in her books. In *I Survived: The Attack of the Grizzlies, 1967*, she writes: "I hope this book has not made you afraid of grizzlies. My goal was to spark your curiosity so you'll want to learn more. I also hope this book inspires you to want to help protect our wilderness and the animals that live there." In addressing these aims explicitly, Tarshis not only echoes the findings of many empirical studies suggesting that fictional texts can change readers' behavior toward animal welfare and environmental issues (cf. Małecki et al. 2016, 9),[9] but also uses specific instructions to guide her audience through this process.

Other titles in the series extend these concerns about animals to plants and the environment more broadly, by raising awareness about environmental issues such as deforestation: In *I Survived: The Eruption of Mount St. Helens, 1980*, we are told that "[l]umberjacking was dangerous. [...] But it wasn't the danger that got to Clive. It was watching those big trees disappear" (18). While the text thus stresses the often-devastating impact of human behavior on the environment, they also foreground nature's ability to recover: "'The forest will grow back one day,' Mom had said, in her usual hopeful way. 'Nature has an amazing power to heal itself.' Dr. Morales had told them that. And it was true. The trees *would* grow back. The animals *would* return" (*I Survived: I Eruption of Mount St. Helens, 1980*, 66, original emphasis). Although these issues are mostly marginal for the plot, these examples raise awareness about the entanglement of humans and their "natural" environment, and, in the progression of the story, the Volcano's eruption is predated by extractive practices. Although not a consequence of deforestation, the natural disaster is nevertheless placed in narrative proximity to it, evoking a moral connection between the story's initial lumberjacking and its eventual projection of regrowth. As such, the novel here illustrates "the ways in which literary forms can encode environmental meaning" (James 2015, 29). By including the negotiation of environmental issues within the texts' adventure narratives, the series uses its ecocritical potential to educate children as what Geraldine Massey and Clare Bradford have termed "ecocitizens" (109). According to Massey and Bradford,

> [c]hildren's environmental texts—that is, texts which thematize contemporary ecological issues—reflect shifting global agendas and predict future possibilities. One of their primary functions is to socialize young people into becoming the responsible and empathetic adults of tomorrow by positioning readers as ecocitizens, dedicated both to sustainable development in the local sphere and also to global responsibility. (Massey and Bradford 2011, 109)

[9] For other works that suggest the potential of literary texts to communicate concerns about animal rights and the environment more broadly, see Mobley, Vagias, and DeWard, "Exploring Additional Determinants of Environmentally Responsible Behavior" (2010); Hormes et al., "Reading a Book Can Change Your Mind" (2013); or Lawrence Buell, *Writing for an Endangered World: Literature, Culture, and Environment in the U.S. and Beyond* (2001).

Overall, then, the texts in the *I Survived* series commit to this function not only by creating spaces to address contemporary ecological issues but also through their narrative and affective strategies. In particular, they use animal companionship to create what Keen has termed "narrative empathy" (2011, 136) for non-human others and thematize ecological issues in their diegesis but also in their paratext as a way of educating children as ecocitizens. In this sense, the specific disasters in these texts present potential access points to more general ecological questions and the fight for ecological justice.

Myths of Wilderness and Whiteness

At the same time, the series' negotiation of environmental issues via its representation of natural catastrophes is underwritten by racial politics that reproduce myths and racial hierarchies which have significantly shaped the unequal distribution of risk and vulnerability in our current ecological moment. These racial politics become visible, for instance, in the series' reiteration of "wilderness" as a spatial marker through which the consequences of natural catastrophes for the environment are supposed to be understood: In the aforementioned *I Survived: The Eruption of Mount St. Helens, 1980*, readers learn that the text's eponymous disaster resulted in devastating transformations of what is referred to as "wilderness": "Two hundred and thirty miles of wilderness had been burned, ripped apart, or buried under a hundred feet of mud and debris" (ch. 16). This term is repeated in the Q&A section supplementing the story ("The St. Helens eruption destroyed 230 miles of wilderness"), highlighting that it is not only the perspective of one of the fictional characters, but an assumption shared by the author as well. The wilderness discourse—and the nature/culture divide that accompanies it—is often at the forefront of environmental activism, i.e., the idea that "untouched nature" has to be protected. This idea is promoted throughout the *I Survived* series and its attempt to educate children as ecocitizens through their engagement with natural catastrophes in historical fiction.

Yet, this notion has come under scrutiny from various disciplines[10]—not because "nature" is not worth protecting, but because the very notion of "wilderness" as a pristine, untouched space separate from human influence is steeped in settler-colonial ideologies and practices. As Malcom Ferdinand argues, drawing on Dorceta Taylor's work, "the creation of wilderness areas in the United States involved the removal of Native Americans, the creation of a myth that they never inhabited there, a racist and prejudiced culture tainted with eugenics rhetoric, the exclusion of ethnic minorities from these areas, and the later segregation of these spaces and their management" (2022, 182–183). Importantly, Ferdinand asks us to consider: "Under

[10] See for example, Ursula Heise's "The Hitchhiker's Guide to Ecocriticism" (2006); William Cronon's "The Trouble with Wilderness; or, Getting Back to the Wrong Nature" (1996); Roderick Neumann's *Imposing Wilderness* (1998); or Yrjö Haila's "Beyond the Nature-Culture Dualism" (2000).

which silences can 'wilderness' be defended?" (2022, 184). Given that the wilderness discourse is central to the environmental messages promoted in the *I Survived* books, it is also worth examining the silences that emerge in their depiction of natural disasters.

The notion of wilderness that Ferdinand discusses has historically shaped policies and practices related to land use, disaster management, and risk mitigation strategies, leaving some communities more vulnerable to the impacts of natural disasters than others. This holds for the United States, but also on a planetary scale, considering how the structures of colonial and capitalist exploitation continue to shape not only the distribution of capital but also of risk and vulnerability on our planet. As Jeremy Williams demonstrates, comparing the global distribution of carbon emissions and climate vulnerability, "there is a stark disconnect between the causes of climate change and its consequences. Those who are most responsible for damaging the atmosphere face much lower risks, while the greatest dangers fall on those who are least responsible. This is the injustice of climate change" (2021, no pag.). Williams' argument aligns with the perspectives of scholars such as Dipesh Chakrabarty (2009), Kathryn Yusoff (2018), Stacia Ryder et al. (2021), Ina Batzke et al. (2021), and Rob Nixon (2011) who highlight the Anthropocene's profound implications for the occurrence and impact of natural disasters and who collectively challenge the conceptualization of the Anthropocene and the climate crisis as a universal crisis affecting all humans equally. This injustice, Williams and these scholars suggest, is particularly detrimental to countries in the so-called Global South.[11] In light of this disparity, the exclusive focus of the *I Survived* series on events in the so-called Global North, narrated through the perspectives of predominantly white, US American children, is particularly noticeable.

Moreover, as Andy Horowitz and Jacob A.C. Remes propose in their recent "Introduction to Critical Disaster Studies," "[s]cholars have come to accept the once controversial maxim that there is no such thing as a natural disaster. The causes and consequences of disaster are not defined by an autonomous natural order, nor are they inevitable. Rather, they are bound up in human history, shaped by human action and inaction" (2021, 1). As such, vulnerability to such disasters is not an intrinsic characteristic of a human body but is discursively constructed and thus shaped by systemic power structures and social relations (cf. Butler, *Precarious Life* 2004, 32; see also Mad'arová et al. 2020, 12). "We are," Judith Butler suggests, "never simply vulnerable, but always vulnerable to a situation, a person, a social structure, something upon which we rely and in relation to which we are exposed" (*The Force of Nonviolence* 2020, 46). Analyzing fictional re-imaginations of natural disasters thus also calls for reading them in the context of the human histories and social structures with which they are entangled.

Against this backdrop, I will examine how the emphasis on white suffering and white heroism in the *I Survived* series obscures the underlying social structures *of* and the lived realities caused *by* the historically differential allocation of

[11] See also Richard Marcantonio et al. (2021).

vulnerability and risk. With very few exceptions, all texts in the *I Survived* series focus on white US American children through whose eyes the catastrophic events are mediated. While I have outlined the ecocritical potential that the emotional investment in the perspective of the innocent child affords, the trope of the—specifically white—innocent child also invokes hegemonic narratives of national victimhood closely intertwined with a national imagination that is constructed around whiteness as a norm. Particularly in the dominant literary and medial trauma narratives of the post-9/11 era, this trope was frequently used to negotiate grief, loss, and a lack of understanding, but also to promote notions of American exceptionalism and the loss of innocence. These notions, in turn, were crucial for legitimizing racial and Islamophobic profiling, the curtailment of citizenship rights for non-white immigrants and citizens, and multiple state invasions during the War on Terror.[12] At the same time, they are, as Susan Faludi's influential study *The Terror Dream: Fear and Fantasy in Post-9/11 America* (2007) has demonstrated, closely linked to a retreat into traditional gender roles and a resurgence of certain US American mythmaking mechanisms after 9/11, for instance, in superhero comics, in the glorification of NYC firefighters, or in the stylization of George W. Bush as what she calls "the president of the wild frontier" (2007, 146). Faludi thus particularly highlights the relevance of the frontier myth and its connection to whiteness and (cis) male heroism, which seems to be at the core of the *I Survived* series and its representational strategies as well.

In this context, it is worth noting that Lauren Tarshis generally seems perceptive to questioning the politics of representations in her books. In an epilogue to *I Survived: I Eruption of Mount St. Helens, 1980*, Tarshis explicitly addresses the gender imbalance in her choice of protagonists for her books. Delineating that many of her fans had reached out to her to inquire about the lack of female representation in her books ("Mrs. Tarshis, why are there no girls on your covers?"), the author explains that her books

> are not about boys or girls, are they? They're about young people who go through difficult and frightening experiences and discover their inner strength. They're about how people can find inspiration and comfort from their family and friends and their faith. The I Survived books are not boy books or girl books. They are, I hope, human books.

While the author does not venture past the perimeters of binary constructions of gender identity, with these explanations and the introduction of Jess as the series' first female protagonist in *I Survived: The Eruption of Mount St. Helens, 1980*, the *I Survived* books have at least tentatively addressed the need to create a more balanced representation of and for children in the context of natural catastrophes.

Yet, a similar coming to terms with the series' racial politics does not seem in sight. Through its storytelling and marketing, particularly its motto "when disaster strikes, heroes are made," the *I Survived* series frequently taps into the mythical

[12] For a more detailed discussion of whiteness, citizenship and American exceptionalism post-9/11, see for instance Carol Fadda-Conrey's *Contemporary Arab-American Literature* (2014); and Evelyn Alsultany's *Arabs and Muslims in the Media* (2012).

imaginations of heroism and wilderness as well; of a nature that cannot be tamed but by select humans who persevere due to their willpower and exceptional courage. As the works of Ferdinand and others have shown, however, such myths are steeped in settler-colonial ideologies that cannot be separated from the racial politics that they underwrite. Moreover, there is an inextricable link between race and US American nationalism which is fostered through myth-making: As Mahshid Mayar reminds us, myths "have lent themselves to various historical moments and functioned as the context for various grand narratives about the 'new world'; and [...] have made it at times reductively and discriminatorily possible for the white founders and inhabitants of this 'new world' to imagine it as one exceptional, racially homogeneous nation" (2022, ch. 4, FN 11). As such, myths are central in America's self-fashioning as an "imagined community" (cf. Benedict Anderson) that centers whiteness and peripheralizes people of color. Thus, what is at stake if a series that depicts fictional child protagonists surviving different kinds of disasters invokes these myths? I argue that the functionalization of both, American myths and the trope of the white, innocent child, is symptomatic of the dominance of white perspectives in the series and ties in with larger debates about the ethics of representation, the modalities of storytelling,[13] and the overbearing "racial blindness" (Yusoff 2018) in the Anthropocene.

Against this backdrop, it is necessary to examine the racial politics of representation in the *I Survived* series as well. Its prioritization of white protagonists becomes particularly egregious when we consider the series' imagination of Native American history (or the lack thereof), its location vis-à-vis white settler colonialism, and its imagination of the nineteenth century. In this imagination, Indigenous perspectives are often merely presented as "afterthoughts," best exemplified in *I Survived: The Children's Blizzard, 1888*. The main story in this book foregrounds the experiences of white settlers (called "pioneers" in the book) during this natural disaster. Yet, shortly before the blizzard strikes, John, the protagonist of the text, makes a striking observation:

> John found arrowheads all the time around here. They belonged to Sioux Indian people who used to live on this land. Probably a Sioux hunter shot this arrow at a buffalo [...] John had never met any Sioux people. That's because the American government had made them all leave here. The buffalo were gone, too. Every last one. [...] John held the arrowhead in his hand. He wondered who it had belonged to. Maybe it was a Sioux boy, a kid like him, hunting with his father like John hunted with Pa. Maybe that kid had a sister like Franny. He could have sat right here, in this exact spot. Where was that boy now? John could have wondered about this all day. But just then he heard voices. (22)

The narrative immediately interrupts John's thoughts as the natural catastrophe draws closer. However, like a belated response to John's realization, the final question in the Q&A section at the end of the book is dedicated to the absence of Indigenous peoples: "What happened to the Native American people who had been living on the prairie?" (86), the text provokingly asks. And the answer follows

[13] See also Nixon "The Great Acceleration" (2011); Amitav Ghosh's *The Great Derangement* (2017); or Marco Caracciolo's *Narrating the Mesh* trilogy (Caracciolo, 2021, 2022a, 2022b).

promptly: "The story of what happened to America's native peoples is one of the most shameful in *our* history" (86, my emphasis). While the inclusion of this perspective openly addresses the nefarious treatment of the Indigenous population during the period of Westward expansion, I want to draw attention to the narrative details of problematic nature of the text's representation of Indigenous peoples:

Firstly, John's imagination of the arrowhead belonging to "a kid like him" illustrates that, in the text, Indigenous peoples can only be made legible through the lens of whiteness. They can only be understood and related to with reference to a normalized white American citizen. This normalized, and ultimately hegemonic, whiteness reinforces the invisibility of Indigenous voices and experiences and perpetuates their erasure in literary and cultural representations. According to Laura M. Jiménez, one of the most pressing representational issues in children's literature is "that the White experience is so pervasive that it is seen as an unquestioned norm" (2016, 67).[14] The fact that children's literature is "produced against a backdrop of white supremacy" (2016, 380), as Brynn F. Welch calls it, not only "contributes to the cultivation of racial biases and stereotypes" but also affects children's general ability to foster empathy and compassion for others (Welch 2016 369).[15] As such, we have to also critically interrogate how far the "pervasive whiteness of children's literature" (Welch 2016, 369) that is traceable in the *I Survived* series also undermines the potentially positive outcomes of the series' affective strategies outlined above. More importantly, even if these affective strategies succeed in educating children as "ecocitizens" (Massey and Bradford 2011, 109), they concurrently promote a version of ecocitizenship that is intertwined with hegemonic whiteness similar to US citizenship in general. Hence, in reproducing this form of hegemonic whiteness in children's literature, the series contributes to the continuation of a settler-colonial project that now extends into the sphere of environmental criticism as well.

Secondly, the text presents the white innocent child as the primary subject affected by natural catastrophes in the present. In the broader context of environmental justice, Indigenous scholars like Kyle Whyte (2016) or Grace Dillon (2021) have emphasized that such representations "can erase Indigenous peoples' perspectives on the connections between climate change and colonial violence" (Whyte 225) and obscure the fact that many "Indigenous persons see our current situation as already having been through a crisis that is ongoing" (Whyte 2016, 227). As Dillon and Pedro Neves Marques (2021) argue in their critique of the appropriation of (post-)apocalyptic scenarios in contemporary representations of climate disaster in speculative fiction, "[w]hile Indigenous peoples have lived through apocalypses

[14] See also Rebecca Aanerud's "Fictions of Whiteness" (1997).

[15] Welch also discusses why the purchase and consumption even of individual literary works is relevant despite the fact that this might not shift the whole book market in one particular direction: "A single purchase likely makes no difference to whether there are in general racial biases and stereotypes, or whether there are in general obstacles to cultivating compassion across racial lines, or whether there are in general differences in access to valuable goods between white children and children of color. However, that purchase might very well make a difference to whether a particular child experiences those outcomes" (Welch 2016 386).

for centuries, white people now steal even that space, or airtime if you will, of both Native trauma and survivance, only to push their own anguish, seemingly erasing any colonial differences with statements like, 'We're all together in this climatic mess'" (2021, 4). If we accept that any contemporary fictional(ized) representation of natural catastrophes in the past is influenced by our historical situatedness in the *current* climate crisis, the prioritization of white perspectives as the specter through which these catastrophes are to be read and understood contributes to the silencing of such ongoing and accumulated crises for Indigenous populations.[16]

Thirdly, John's observations in *I Survived: The Children's Blizzard, 1888* as well as Tarshis's editorial comment implicitly relegate Indigenous peoples to a paradoxical position both outside of "our" history as well as outside of "our" present. They are not given a voice in the narrative and are only presented in the form of remnants and traces. Even in 1888, Indigenous presences are relegated so far into the past that John can only find arrowheads, with the presumption that, much like archaeological traces, the wooden arrows have been dissolved via a "natural" degradation of the material. US American history on the whole, then, is stylized as white American history, confining Indigenous populations to a presumably static past, barely traceable even in the nineteenth century. This confinement and the focus on Indigenous absence obscures how Indigenous populations not only form an integral part of US American society today but implicitly also render the entire history of the twentieth century devoid of Native American presences and agency.

It is significant in this context that the series continues these racial politics in other titles, which do not deal with natural catastrophes, as well: While *I Survived: The Battle of Gettysburg* indeed focuses on the perspective of a formerly enslaved Black boy, Thomas, the text chooses his role in supporting white Unionist troops at Gettysburg as "his" defining catastrophe that he is narratively tasked with navigating—not his former enslavement. It is from this battle and only through his allegiance to the "true" American troops, that he, like all children in the series, derives a sense of identity as a survivor. In such a version of American history, as Nathan Irvin Huggins writes, "racial slavery and oppression [are treated] as curious abnormalities— aberrations—historical accidents to be corrected in the progressive

[16] It is worth highlighting that white imaginations of apocalypse appropriate not only experiences that the brutal 'conquest' of the Americas brought upon its Indigenous populations, but that the contemporary popularization of apocalypse-narratives also structurally negates transatlantic Black experiences: "While the apocalypse is generally conceived as a dystopic possible futurity, the African diaspora has already undergone brutalities so vile and degrading and so historically unprecedented in scope and scale, that only Armageddon can accurately describe the advent of modernity on our collective past, and only the post-apocalypse can define our present" (Maynard 2018, 30–31). For a discussion of how these experiences also encompass a quintessentially modern(ist) alienation and prefigure twentieth-century genres, see Toni Morrison's interview with Paul Gilroy, (1994), and Kodwo Eshun's "Further considerations of Afrofuturism" (2003).

upward reach of the nation's destiny" (1990, xii).[17] The narrative absence of the institution of slavery—in conjunction with the erasure of Indigenous apocalypses—as the defining catastrophes on which the United States was built thus echoes what Huggins has criticized as a "deforming mirror of truth" (xi).

Within this hegemonic approach to children's suffering, there are only two titles in the series that deal with natural disasters whose protagonists are not white: *I Survived: Hurricane Katrina, 2005*, narrated through the eyes of a Black American boy, Barry, and *I Survived: The Japanese Tsunami, 2011*, narrated through the eyes of a Japanese American boy, Ben. This shift in perspective prompts numerous questions: Is the inclusion of non-white voices in these few texts a productive step toward representational diversity, or is it mere tokenism? Do these titles complicate the series' overall dominance of white emotional centering (cf. Cargle 2018) by foregrounding non-white experiences? Or does this apparent diversification perpetuate stereotypical depictions of non-white Americans? With these questions in mind, it is striking that the inclusion of Black and Asian American voices only happens in very specific geographical and historical contexts—as (formerly) enslaved people (*The Battle of Gettysburg, 1863*), located in Japan (*The Japanese Tsunami, 2011*), or confined to the Lower Ninth Ward in New Orleans (*Hurricane Katrina, 2005*). As such, non-white perspectives are only ever included concerning events that, in the public imagination, are already linked to the racializing dynamics of American politics (the Civil War, the Katrina fallout), or spaces imagined as racially Other on the whole (Japan). This contingent inclusion, therefore, does not constitute a real break with the hegemonic whiteness of the overall series. On the contrary, it relegates non-white perspectives outside or at the margins of the United States and thus perpetuates imaginations of the US American nation as white. In short, even if the focalizers and protagonists are not white themselves, their experiences are still mediated through the specter of whiteness.

This racialized mediation provides the frame for the representation of the natural disasters in the *I Survived* series. In turn, these disasters—or more precisely, how human bodies are exposed and made vulnerable to them—are also entangled in and shaped by racial hierarchies. As Jeremy Williams maintains in *Climate Change Is Racist*, "[c]ommunities in poorer neighbourhoods consistently face greater environmental risk. When disaster inevitably strikes, there is less support for recovery. The most notorious example of this was Hurricane Katrina in 2005. The poorest communities were already without protection, on the wrong side of the levees and facing the full wrath of the storm" (2021, no pag.). These systematic and individual failures in the emergency management and disaster mitigation of local and state authorities in the context of Hurricane Katrina are well documented. *I Survived: Hurricane Katrina, 2005* addresses these issues through the eyes of its Black protagonist Barry, by contrasting his expectations for help with the lack thereof that he experiences first-hand:

[17] For more detailed discussions about the erasure of racial slavery in US American history and the United States as a settler-colonial project, see for example Saidiya Hartman's (1997) or Ajamu Baraka's work (2019).

> Of course there would be. Police and firemen. Doctors and nurses. Barry knew what happened in disasters. He'd seen it on the news. He imagined a big tent on the bridge, with cots set up in neat rows. He would get fresh clothes, more water, good food. The people there would know how to track down Mom and Dad and Cleo. […] Finally the bridge appeared in the distance. And even from far away, Barry could see that there were no tents. No flashing lights or police cars or fire engines or ambulances. (96–97)

Barry's expectations invoke familiar images of disaster response which are shaped by the media and what he had "seen on the news" (96), from tents with ample supplies to families reunited in blinking lights. In equating the absence of these generic markers of emergency management with Barry's loss of newfound hope, security, and comfort that comes with them, *I Survived: Hurricane Katrina, 2005* invites its readers to relate to the emotional toll that the failures in the emergency management in New Orleans have taken on the city's marginalized communities most strongly affected by the disaster. This critical stance is valuable and necessary. Yet, the racial undertones of these failures and their connection to systemic issues are not acknowledged in the text—and it seems at odds with the otherwise overwhelmingly positive and uncritical depiction of the military, the police, and other state institutions in the series.[18] As such, the lack of care that Barry experienced during Hurricane Katrina, as a marker of the ways in which these institutions have failed him and his community, is presented as exceptional rather than symptomatic and thereby continues the framing of systemic anti-Black violence as transhistorical "aberrations."

Conclusion: The Slow Violence of Natural Disasters

What is the function of natural catastrophes/disasters in my reading then? On the one hand, natural disasters in the *I Survived* series are presented as a connecting element, a threat that is intelligible for "everyone." They are framed as a universal signifier that cuts across different cultural and linguistic communities: "Ben didn't speak Japanese. But he understood one word. Tsunami!" (*I Survived: The Japanese Tsunami, 2011*, ch. 1). They function as an access point to children's environmental education and thus offer the potential to educate them as "ecocitizens" (Massey and Bradford 2011, 109). The affective strategies employed in the series not only help children to develop (narrative) empathy for the human protagonists and their plight

[18] The series employs various visual and narrative strategies to present these institutions in a positive light. Helmets of soldiers and firefighters, as well as canons are used as visual cues at the beginning of chapters in multiple titles in the series, and numerous (male) relatives of the protagonists (particularly fathers and brothers) are firefighters, policemen or soldiers. In *I Survived: The Japanese Tsunami, 2011*, for instance, Ben's admiration for his father is closely linked to his military past particularly in the War on Terror: "Ben had to be tough, like Dad. During Dad's last tour in Afghanistan, when Ben was a baby, the engine of Dad's F-16 exploded. He had to eject from the plane over enemy territory. He broke his ankle when he parachuted down. But he still managed to escape into the mountains before enemy fighters found him. For six days, he'd hidden in a cave, until he was finally rescued by a helicopter filled with US Marines" (*The Japanese Tsunami, 2011*).

for survival amidst natural catastrophes but also serve as an access point to a broader discussion about the relevance of environmental protection and more-than-human companionship in the Anthropocene.

On the other hand, this environmental education only happens via the erasure of Black and Indigenous apocalypses and the reproduction of the pervasive whiteness within the genre of children's literature. Against this backdrop, the foregrounding of the natural catastrophe as an *event* obscures the underlying histories of marginalization and oppression, of making certain bodies more vulnerable to these catastrophes. With Rob Nixon (2011), I would argue that the series thus adds to the invisibility of the *slow violence* of our current ecological moment. Slow violence, Nixon suggests, is

> a violence that occurs gradually and out of sight, a violence of delayed destruction that is dispersed across time and space, an attritional violence that is typically not viewed as violence at all. Violence is customarily conceived as an event or action that is immediate in time, explosive and spectacular in space, and as erupting into instant sensational visibility. (2011, 2)

The spectacular *events* that the natural catastrophes present in the *I Survived* series upstage the slow violence operating in the background of these disasters.

As Jeremy Williams argues in the broader context of the climate crisis, "minority ethnic communities often find themselves carrying a greater burden of risk and inconvenience, sometimes fatally so. This disadvantage springs from existing patterns of inequality and privilege, making environmental justice impossible to tease apart from other injustices—but no less significant for it" (2021, no pag.). Therefore, according to Williams, it is crucial to think of environmental justice *as* social justice and to critically examine the historically and locally specific, but always racialized, patterns that have created social and environmental injustices in the first place. As Oronde Drakes and Eric Tate's recent review of research on the interplay between social vulnerability and multiple hazards has shown:

> There continues to be limited attention paid to [...] how these interactions develop to produce the unique social and structural circumstances shaping the perilous conditions in which populations exist. Overall, there is need for additional research on how relationships between the determinants of social vulnerability evolve across disaster stages and across events. The paucity of such research is striking, particularly given the stark impacts of multi-hazard events in the 21st century. (2022, 13)

In light of the now widely accepted causal relation between anthropogenic climate change and an increased frequency and severity of natural disasters, we are thus also tasked with thinking about natural disasters *in* and *on* these terms. Natural catastrophes are not only embedded in patterns of disadvantage but, as exceptional moments of crisis, they also function as points of condensation in which these patterns converge, accumulate, and potentially become apparent.[19] Hence, if we study the ways

[19] For a helpful discussion of the histories of oppression underlying the seemingly exceptional catastrophe(s) of the present, see also the conversation between Sylvia Wynter and Katherine McKittrick in "Unparalleled Catastrophe for Our Species?" (2015).

in which they are represented in fictional texts, they can function as potential access points to America's complex histories of marginalization in our equally complex ecological, social, and political climate as well. Whereas contemporary science fiction has developed "a distinct fascination with disaster at moments in history when it became easier to imagine the end of the world than alternative futures" (Yusoff and Gabrys 2011, 520), the *I Survived* series turns to natural catastrophes in the past to make sense of our current climate crisis. The "manageable catastrophes" depicted in the series—through the focus on past catastrophes with calculable outcomes, and ultimately, also with happy endings for all protagonists—and the "hands-on" advice in the books' supplementary material, then, speak not only to the series' young target audience but also to its racial politics and oversights. Risk mitigation and the fight for survival, it seems, are presented as a *personal* responsibility, not a *structural* one. In so doing, the series eschews viewing natural catastrophes as embedded within structures of inequality. Despite the transformative potential of catastrophes, (the emplotment of) crisis in the series must thus ultimately be understood "as a predominantly conservative modality which seeks to stabilize an existing structure within a radically contingent world" (Espinoza Garrido 2024, 74). While the series offers a distinct ecocritical potential, its complicity in the settler-colonial projects that have created today's climate crisis in the first place stands in the way of truly imagining alternative futures and furthering climate justice.

References

Aanerud, Rebecca. 1997. Fictions of Whiteness: Speaking the Names of Whiteness in US Literature. *Displacing Whiteness: Essays in Social and Cultural Criticism*, 35–59.

Adamson, Sylvia. 2001. The Rise and Fall of Empathetic Narrative: A Historical Perspective on Perspective. *New Perspectives on Narrative Perspective*, ed. Willie van Peer and Seymour Chatman. State University of New York Press, 83–99.

Alsultany, Evelyn. 2012. *Arabs and Muslims in the Media: Race and Representation after 9/11*. New York University Press.

Bal, P. Matthijs, and Martijn Veltkamp. 2013. How Does Fiction Reading Influence Empathy? An Experimental Investigation on the Role of Emotional Transportation. *PLoS ONE*, ed. Liane Young, 8 (1): e55341. https://doi.org/10.1371/journal.pone.0055341.

Baraka, Ajamu. 2019. Foreword: Countering the Violence of Imposed Forgetting. In *American Exceptionalism and American Innocence: A People's History of Fake News—From the Revolutionary War to the War on Terror*, ed. Danny Haiphong and Roberto Sirvent. Skyhorse.

Batzke, Ina, et al. 2021. Introduction: Life Writing in the Posthuman Anthropocene. *Life Writing in the Posthuman Anthropocene*, ed. Ina Batzke, Lea Espinoza Garrido, and Linda M. Hess, Palgrave Macmillan, 1–19.

Booth, Wayne C. 1983. *The Rhetoric of Fiction*. University of Chicago Press.

Braidotti, Rosi. 2016. The Critical Posthumanities; or, Is Medianatures to Naturecultures as Zoe is to Bios? *CulturalPolitics*, 12 (3): 380-390.

Buell, Lawrence. 2001. *Writing for an Endangered World: Literature, Culture, and Environment in the US and Beyond*. Harvard University Press.

Butler, Judith. 2004. *Precarious Life: The Powers of Mourning and Violence*. Verso.

Butler, Judith. 2020. *The Force of Nonviolence: An Ethico-Political Bind*. Verso.

Capshaw Smith, Katharine. 2005. Trauma and Children's Literature. *Children's Literature* 33 (1): 115–119.
Cargle, Rachel. 2018. When Feminism is White Supremacy in Heels. *Harper's Bazaar*, August 16. https://www.harpersbazaar.com/culture/politics/a22717725/what-is-toxic-white-feminism/.
Caracciolo, Marco. 2021. *Narrating the Mesh: Form and Story in the Anthropocene*. University of Virginia Press.
Caracciolo, Marco. 2022a. *Slow Narrative and Nonhuman Materialities*. University of Nebraska Press.
Caracciolo, Marco. 2022b. *Contemporary Fiction and Climate Uncertainty: Narrating Unstable Futures*, Bloomsbury Academic.
Chakrabarty, Dipesh. 2009. The Climate of History: Four Theses. *Critical Inquiry*, 35 (2): 197–222.
Ciftci, Gulsin. 2022. The Affects of Reading. *Current Objectives of Postgraduate American Studies* 23 (2): 10–29.
Cohn, Dorrit. 1978. *Transparent Minds: Narrative Modes for Presenting Consciousness in Fiction*. Princeton University Press.
Cronon, William. 1996. The Trouble with Wilderness: Or, Getting Back to the Wrong Nature. *Environmental History* 1 (1): 7–28.
Dillon, Grace L., and Pedro Neves Marques. 2021. Taking the Fiction out of Science Fiction: A Conversation about Indigenous Futurisms. *E-Flux Journal*.
Drakes, Oronde, and Eric Tate. 2022. Social Vulnerability in a Multi-Hazard Context: A Systematic Review. *Environmental Research Letters* 17 (3): 1–17.
Eshun, Kodwo. 2003. Further Considerations of Afrofuturism. *CR: The New Centennial Review* 3 (2): 287–302. https://doi.org/10.1353/ncr.2003.0021.
Espinoza Garrido, Lea. 2024. Affect and the (Un)Doing of Trust in 20th and 21st Century Black and Asian American Fiction and Nonfiction. *KWZ: Kulturwissenschaftliche Zeitschrift* 1: 71–94. https://doi.org/10.28937/9783787348770_6.
Fadda-Conrey, Carol. 2014. *Contemporary Arab-American Literature: Transnational Reconfigurations of Citizenship and Belonging*. New York University Press.
Faludi, Susan. 2007. *The Terror Dream: Fear and Fantasy in Post-9/11 America*. Macmillan.
Ferdinand, Malcom. 2022. Behind the Colonial Silence of Wilderness. *Environmental Humanities* 14 (1): 182–201. https://doi.org/10.1215/22011919-9481506.
Fernandez-Quintanilla, Carolina. 2020. Textual and Reader Factors in Narrative Empathy: An Empirical Reader Response Study Using Focus Groups. *Language and Literature* 29 (2): 124–146.
Gaard, Greta. 2009. Children's Environmental Literature: From Ecocriticism to Ecopedagogy. *Neohelicon* 36 (2): 321–334. https://doi.org/10.1007/s11059-009-0003-7.
Ghosh, Amitav. 2017. *The Great Derangement: Climate Change and the Unthinkable*. University of Chicago Press.
Gilroy, Paul. 1994. Living Memory: A Meeting with Toni Morrison. In *Small Acts: Thoughts on the Politics of Black Cultures*, ed. Paul Gilroy, 175–182. Serpent's Tail.
Gulliver-Garcia, Tanya. 2019. Disasters versus Catastrophes: The Difference Matters. *Center for Disaster Philantrophy*. https://disasterphilanthropy.org/blog/disasters-versus-catastrophes-the-difference-matters/.
Haila, Yrjö. 2000. Beyond the Nature-Culture Dualism. *Biology & Philosophy* 15 (2): 155–175. https://doi.org/10.1023/A:1006625830102.
Hartman, Saidiya. 1997. *Scenes of Subjection: Terror, Slavery, and Self-Making in Nineteenth-Century America*.Oxford University Press.
Heise, Ursula K. 2006. The Hitchhiker's Guide to Ecocriticism. *PMLA* 121 (2): 503–516.
Hormes, Julia M., et al. 2013. Reading a Book Can Change Your Mind, but Only Some Changes Last for a Year: Food Attitude Changes in Readers of The Omnivore's Dilemma. *Frontiers in Psychology* 4: Art. 778, 1-8.
Huff, Cynthia. 2017. After Auto, after Bio: Posthumanism and Life Writing. *a/b: Auto/Biography Studies*, 32 (2): 279-282.

Horne, Gerald. 2018. *The Apocalypse of Settler Colonialism: The Roots of Slavery, White Supremacy, and Capitalism in Seventeenth-Century North America and the Caribbean.* Monthly Review Press.
Huggins, Nathan Irvin. 1990. *Black Odyssey: The African-American Ordeal in Slavery.* Vintage Books.
James, Erin. 2015. *The Storyworld Accord.* NE, USA: University of Nebraska Press Lincoln.
James, Erin. 2022. *Narrative in the Anthropocene.* Ohio State University Press.
Kauanui, J. Kēhaulani. 2016. 'A Structure, Not an Event': Settler Colonialism and Enduring Indigeneity. *Lateral: Journal of the Cultural Studies Association* 5 (1).
Keen, Suzanne. 2006. A Theory of Narrative Empathy. *Narrative* 14 (3): 207–236.
Keen, Suzanne. 2007. *Empathy and the Novel.* Oxford University Press.
Keen, Suzanne. 2011. Fast Tracks to Narrative Empathy: Anthropomorphism and Dehumanization in Graphic Narratives. *SubStance* 40 (1): 135–155. https://doi.org/10.1353/sub.2011.0003.
Kidd, David Comer, and Emanuele Castano. 2013. Reading Literary Fiction Improves Theory of Mind. *Science* 342 (6156): 377–380. https://doi.org/10.1126/science.1239918.
Maďarová, Zuzana, et al. 2020. What Makes Life Grievable? Discursive Distribution of Vulnerability in the Pandemic. *Mezinárodní Vztahy* 55 (4): 11–30.
Małecki, Wojciech, et al. 2016. Literary Fiction Influences Attitudes Toward Animal Welfare. *PLOS ONE*, edited by Edna Hillmann, 11(12): e0168695. https://doi.org/10.1371/journal.pone.0168695.
Marcantonio, Richard, et al. 2021. Global Distribution and Coincidence of Pollution, Climate Impacts, and HealthRisk in the Anthropocene. *PLoS One*, 16 (7): e0254060.
Massey, Geraldine, and Clare Bradford. 2011. Children as Ecocitizens: Ecocriticism and Environmental Texts. In *Contemporary Children's Literature and Film: Engaging with Theory*, ed. Kerry Mallan and Clare Bradford, 109–126. Palgrave Macmillan.
Mayar, Mahshid. 2022. *Citizens and Rulers of the World: The American Child and the Cartographic Pedagogies of Empire.* University of North Carolina Press.
Maynard, Robyn. 2018. Reading Black Resistance Through Afrofuturism: Notes on Post-Apocalyptic Blackness and Black Rebel Cyborgs in Canada. *TOPIA: Canadian Journal of Cultural Studies* 39: 29–47. https://doi.org/10.3138/topia.39.04.
McKittrick, Katherine, and Sylvia Wynter. 2015. Unparalleled Catastrophe for Our Species? Or, to Give Humanness a Different Future: Conversations. In *Sylvia Wynter: On Being Human as Praxis*, 9–89. Duke University Press.
McLerran, Jennifer. 2019. Theorizing Settler Colonialism: Alternative Indigenous Methodologies. *Feminist Studies*, 45 (2): 455–466.
Mehnert, Antonia. 2016. *Climate Change Fictions: Representations of Global Warming in American Literature.* Springer.
Mobley, Catherine, et al. 2010. Exploring Additional Determinants of Environmentally Responsible Behavior: The Influence of Environmental Literature and Environmental Attitudes. *Environment and Behavior* 42 (4): 420–447.
Weik von Mossner, Alexa. 2017. Vulnerable Lives: The Affective Dimensions of Risk in Young Adult Cli-Fi. *Textual Practice* 31 (3): 553–566.
Neumann, Roderick P. 1998. *Imposing Wilderness: Struggles over Livelihood and Nature Preservation in Africa.* University of California Press.
Nixon, Rob. 2011. *Slow Violence and the Environmentalism of the Poor.* Harvard University Press.
Nixon, Rob. 2014. The Great Acceleration and the Great Divergence: Vulnerability in the Anthropocene. *Profession* 14, no pag.
O'Brien, Jean M. 2017. Tracing Settler Colonialism's Eliminatory Logic in *Traces of History*. *American Quarterly* 69 (2): 249–255.
Pugh, Tison. 2010. *Innocence, Heterosexuality, and the Queerness of Children's Literature.* Routledge.
Quarantelli, E. L. Catastrophes Are Different from Disasters: Some Implications for Crisis Planning and Managing Drawn from Katrina. *Items: Insights from the Social Sciences*, 2011 June 2006.

https://items.ssrc.org/understanding-katrina/catastrophes-are-different-from-disasters-some-implications-for-crisis-planning-and-managing-drawn-from-katrina/.
Reddy, Maureen T. 1998. Invisibility/Hypervisibility: The Paradox of Normative Whiteness. *Transformations: The Journal of Inclusive Scholarship and Pedagogy* 9 (2): 55–64.
Remes, Jacob A. C., and Andy Horowitz. 2021. *Critical Disaster Studies*. University of Pennsylvania Press.
Ryan, Marie-Laure. 2014. Possible Worlds. *Handbook of Narratology*, ed. Peter Hühn et al., 726–742. DeGruyter.
Ryder, Stacia, et al., eds. 2021. *Environmental Justice in the Anthropocene: From (Un)Just Presents to Just Futures*. Routledge.
Schneider-Mayerson, Matthew. 2018. The Influence of Climate Fiction. *Environmental Humanities* 10 (2): 473–500. https://doi.org/10.1215/22011919-7156848.
Schultermandl, Silvia, et al. 2022. Introduction: Affective Worldmaking: Narrative Counterpublics of Gender and Sexuality. In *Affective Worldmaking: Narrative Counterpublics of Gender and Sexuality*, ed. Schultermandl, Silvia et al., 13–44. Transcript.
Taylor, Dorceta E. 2016. *The Rise of the American Conservation Movement: Power, Privilege, and Environmental Protection*. Duke University Press.
Trexler, Adam. 2015. *Anthropocene Fictions: The Novel in a Time of Climate Change*. University of Virginia Press.
Veracini, Lorenzo. 2013. 'Settler Colonialism': Career of a Concept. *The Journal of Imperial and Commonwealth History*, 41 (2): 313–333.
Welch, Brynn F. 2016. The Pervasive Whiteness of Children's Literature: Collective Harms and Consumer Obligations. *Social Theory and Practice* 42 (2): 367–388.
Whyte, Kyle Powys. 2016. Indigenous Experience, Environmental Justice and Settler Colonialism. *SSRN Electronic Journal*. https://doi.org/10.2139/ssrn.2770058.
Whyte, Kyle Powys. 2018. Indigenous Science (Fiction) for the Anthropocene: Ancestral Dystopias and Fantasies of Climate Change Crises. *Environment and Planning E: Nature and Space* 1 (1–2): 224–242. https://doi.org/10.1177/2514848618777621.
Wildcat, Daniel R. 2009. *Red Alert!: Saving the Planet with Indigenous Knowledge*. Fulcrum Publishing.
Williams, Jeremy. 2021. *Climate Change Is Racist: Race, Privilege and the Struggle for Climate Justice*. Icon Books.
Wolfe, Patrick. 2006. Settler Colonialism and the Elimination of the Native. *Journal of Genocide Research* 8 (4): 387–409.
Yusoff, Kathryn. 2018. *A Billion Black Anthropocenes or None*. University of Minnesota Press.
Yusoff, Kathryn, and Jennifer Gabrys. 2011. Climate Change and the Imagination. *WIREs Climate Change* 2 (4): 516–534.

Reporting Catastrophes: Mainstream Media and Vulnerability to Disasters in the United States

Natalie Rauscher

Abstract As a country that experiences natural disasters on a regular basis, the United States today is highly vulnerable to these events—more so than other countries. What could be the reasons behind this high vulnerability and lack of resilience in the United States today? This chapter explores the role of mainstream media in disaster situations such as hurricanes by arguing that through the structure and focus of mainstream media in a politically polarized context, there is little reflection of underlying reasons for high risk and vulnerabilities to disaster including a lack of discussion of government policy in this area. By investigating a textual corpus of media data around natural disasters based on transcripts from the New York Times, ABC News, and Fox News between 2017 and 2020 using quantitative and qualitative methods from corpus linguistics, this chapter contributes to previous research illustrating that there is little reflection in US mainstream media of underlying causes of high vulnerability to disasters. This likely contributes to a lack of awareness of the origins of disaster situations as well as possibly an unwillingness to support political reform in this area. Media discourses thus play a crucial role in the cycle of failed disaster risk reduction in the United States.

Introduction

The last few decades have seen an enormous rise in costs and damages from natural catastrophes around the globe. In the United States, natural catastrophes are frequent and often more damaging compared to other developed regions, for example in Western Europe. Several factors, such as climate change play a role. Moreover, policy failure on various levels of government and population movement and density in disaster-prone areas are adding to the problem. Extreme weather and disaster events are affecting everyone in a given place. Yet, in relative terms, there are

N. Rauscher (✉)
Würzburg University, Würzburg, Germany
e-mail: Natalie.rauscher@uni-wuerzburg.de

N. Rauscher, W. Werner (eds.), *Natural Disasters in the United States*, Risk, Governance and Society 22, https://doi.org/10.1007/978-3-031-96436-7_9

members of a community that are much more affected by these events than others. They are thus more vulnerable to disaster events and natural hazards. Social vulnerability to disasters and the (lack of) resilience are thus studied to arrive at a more nuanced picture of the impact of natural disasters in the United States (Tierney 2019; Cutter 2016a; Cutter and Finch 2008). In this equation, not only institutions and governments play a role, but public discourses do as well.

Different disciplines, also in the social sciences, including in Economics (e.g., Robert Shiller) or Political Science (e.g., Taylan Yildiz et al. 2018), have shown that discourses and narratives play a role in reflecting events and attitudes by different stakeholders, but they can also reinforce the status quo or bring about change. Narratives usually include elements of truth but can also be based on myths, misinformation, or even falsehoods (Shiller 2017, 968). Through their dispersion and popularity, narratives form and influence public and political perception about a given topic and thus drive the respective response (or the lack thereof). Discourses shape people's attitudes, knowledge, and awareness of certain events and issues and ultimately also influence how they will act on these perceptions.

A crucial lens for the perception of certain events is mass media outlets. In a democratic society, media assumes the role of the "fourth branch" of government and thus has the purpose of informing and critically reflecting upon issues, actors, and institutions (Patterson 2000, 248). Most citizens rely on the media to become informed about politics and public affairs (Cushion 2019, 303). However, the mass media landscape in the United States, just like many institutions in the country, has become polarized in recent years (Pew-Research-Center 2020) and is much more driven by market logic than public media elsewhere (Patterson 2000). Several studies found that a focus on entertainment and market logic of media outlets as "profit-seeking entities" (Patterson 2000, 244) leads to less informed citizens (Cushion 2019), which is in itself a critical issue for a democracy that relies on an informed citizenry to hold politicians accountable (Curran et al. 2009). More than that, the media sphere is also contributing to the further polarization of the country by offering explicitly partisan views through outlets like Fox or MSNBC (Levendusky 2014). The effect is that groups of people seek out media that is in line with their preconceived political beliefs, while others avoid political news altogether (Levendusky 2014).

For the field of disaster research, and the US context in particular, these findings are highly relevant. As Timothy Recuber argues, the American public is "consuming catastrophe" (Recuber 2016) extensively through mass media whenever a disaster occurs, while deeper knowledge about possible prevention or reaction remains insufficient among many citizens. In disaster events, mass media takes on different roles, including issuing warnings, providing information, describing events, or "even contributing to individual and community recovery and to community resilience" (Houston et al. 2012, 607). Yet, as several studies have shown, news media's depictions of disasters often focus little on any causal explanations of disaster events but rather emphasize dramatic descriptions for a short period of time with a focus on the affected region only (Houston et al. 2012; Rauscher 2024).

Media coverage has the power to draw attention to and "normalize dominant cultural assumptions" as well as "influence social attitudes (…)" (Cox et al. 2008, 470). Thus, news and print media influence the cultural understanding of a crisis event and the role people have in it (Davis and French 2008, 243). People also evaluate the response to the disaster (Cox et al. 2008, 470), and how they ultimately recover from it. The implication is that the structure of America's mass media and its focus is ill-suited to reflect on reasons for the US's vulnerability and lack of resilience to disaster. This chapter thus argues that media discourse plays a role in the cycle of failed disaster risk reduction in the United States.

After briefly introducing the most relevant concepts of (social) vulnerability, resilience, and uncertainty, this chapter reviews the literature on mass media's role in society and disaster situations. Additionally, it presents more research in line with previous findings, showcasing through quantitative and qualitative discourse analysis that media discourses on natural disasters do not reflect on underlying causes of high vulnerability, specifically on government policy, such as flood insurance, which is crucial in this area. This one-sided approach in turn contributes to a lack of awareness about the origins of (recurring) disaster situations among the American population and possibly an unwillingness to support direly needed policy reform in this area.[1]

Natural Hazard Vulnerability, Resilience, and Uncertainty

Although an extreme event affects the whole population in a given region in absolute terms, certain groups in a community are much more affected in relative terms and are even more (socially) vulnerable to natural hazards. The concept of vulnerability is used across many fields including disaster management, development, economics, sociology, anthropology, geography, health, global change, and environmental studies, which leads to a fuzzy definition of the term (cf. Bergstrand et al. 2015, 392).

In the sociological study of disasters, for example through Kathleen Tierney, vulnerability refers to "the potential for experiencing adverse impacts from

[1] This chapter will present a case study based on textual sources. Different textual corpora based on media outlets were compiled. The corpora include articles from The New York Times, news transcripts from ABC News as well as Fox News Network between 2017 and 2020. The data was collected over the Nexis Uni platform using the search terms 'hurricane' and 'Louisiana.'

Hurricanes the consequent flooding are one of the most frequent and common natural disaster events to occur in the USA every year. Floods continue to be one of the costliest natural disaster event in the USA. Louisiana is frequently hit by these events and some of the biggest disaster events of U.S. history devastated New Orleans (Hurricane Betsy, Hurricane Katrina) which triggered national discussions over vulnerability and resilience in the region and in the USA more widely. Louisiana also continues to be one of the poorest states and thus displays high levels of socially vulnerable groups (low socio-economic status, dense urban areas, mixed ethnic and racial make-up of the population).

disasters and having poorer outcomes after the occurrence of disasters" (Tierney 2019, 7). The use of vulnerability in risk management literature, for example by researchers like Susan Cutter, focuses on conditions that make communities vulnerable to natural hazards, such as social systems, development level, infrastructure, demographic structure, and more generally the socioeconomic make-up of a community (Bergstrand et al. 2015, 392). Additionally, one must recognize that all severe events becoming a disaster are "the consequence of human decisions" (Drabek 2010, 235). Only by the "human factor" do events develop into a disaster with severe consequences for the civilization living in its vicinity. Thus, events are usually measured in the magnitude of their social impact on (vulnerable) communities, not merely by the physical force involved (Tierney 2019, 4). With disasters originating in human agency, the role of people in disasters and social support after disasters is especially crucial (Zakour and Gillespie 2013, 12).

Experts like Susan Cutter have intensively studied (the changes in) social vulnerability to natural hazards in different regions of the United States. She concludes that the most vulnerable counties are not necessarily located where natural disasters are most frequent, which reflects the interdependence and complexity of social issues more broadly (Drabek 2010, 211). Generally, the most vulnerable populations are located in dense urban areas, are racially and ethnically diverse, and have a low socioeconomic status (Cutter et al. 2008). Overall, however, there are many factors that need to be taken into account when social vulnerability is evaluated including social, economic, and environmental factors, community capital, institutional factors as well as housing and infrastructure (Cutter 2016b). Additionally, measuring social vulnerability takes into account a wide array of additional factors such as age disparities, socioeconomic status, employment, or special needs as well as variability and capacity of preparedness, response, and recovery at the county level (Disasters 2012, 98). Although lower socioeconomic status plays a major role in how vulnerable communities are to disasters, the rising damages from these events are increasingly also hitting middle-class households hard, illustrating that disasters "play a growing role in further dividing the rich and the poor and also in hollowing out the middle class, propelling some households toward greater wealth and others toward downward mobility" (Rhodes and Besbris 2022, 147).

Compared to vulnerability, resilience "refers to coping with and recovering from a hazard that has already occurred" (Bergstrand et al. 2015, 392). Just like vulnerability, the idea of resilience is studied in many fields such as economics, ecology, psychology, sociology, anthropology, public health, geography, or disaster management and generally means the capacity to absorb disaster shocks, cope with disaster impact, and successfully adapt or improve after a disaster (Tierney 2019, 7). When talking about resilience, one often hears phrases such as "bounce back" or "build back better" (UNDRR 2017), which can be understood as a focus on preparing and building capacity to prevent disasters from leading to a long-term downturn in a given community. Resilience capacity includes, for example, technological capacity, skill and education levels, high economic status, quality of the natural environment and resource management institutions, stable political structures, adequate infrastructure, or flows of knowledge and information between political levels and inside communities (UNDRR 2017). Thus vulnerability to natural hazards can be higher if any of these factors is not adequately managed. Additionally, studies of risk in society have shown that underlying cultural ideas can influence the perception of possible risk in a given society (Wakeham 2015, 720). Thus, cultural and ideological factors also play a role in vulnerability to natural hazards and disasters.

Vulnerability to natural disasters in the United States has changed over the last decades, and certain communities are more affected by frequent and recurring natural catastrophe events than others. As mentioned, the most vulnerable communities are those that are located in dense (urban) areas, are racially and ethnically diverse, and have a lower socioeconomic status than the average population (Cutter and Finch 2008). Thus, resilience to natural disaster events is not easily achieved for these groups. Although there are efforts by local groups to achieve "community resilience" more effectively, the sheer magnitude of some disaster events demands the support and responsibility of local, state, and federal governments. Additionally, even if some members of a community are better prepared than others, discussions "of community resilience often note that the 'whole is more than the sum of its parts,' meaning that a collection of resilient individuals does not guarantee a resilient community (…)" (Norris et al. 2008). In general, both vulnerability and resilience are often discussed as linked concepts (Cutter et al. 2008): "Vulnerability speaks to the inherent social system that exists before events like disasters occur that contribute to the amount of risk of exposure as well as the degree of harm, while resilience is the conditions that help social systems to absorb, cope with, and adapt to hazards and disasters" (Bergstrand et al. 2015).

In their study, Bergstrand et al. have found that indeed, communities in the United States that display high levels of vulnerability also show low levels of resilience. Geographically, the West of the United States is especially susceptible to hazards (high vulnerability), while the South-East is low on its ability to recover (low resilience) (Bergstrand et al. 2015, 394). In the case study of this chapter, the focus will also lie on the low-resilience region of the southern United States where severe weather events are frequent but where the response to disasters is often ineffective.

Underlying the issues of vulnerability and (a lack of) resilience is the factor of uncertainty in the context of natural hazards and disasters. The sheer complexity of certain problems—such as natural hazards or climate change—leads to uncertainty (Wakeham 2015, 716). For example, uncertainty about how to make sense of hazards or prepare for them. This can be true for both decision-makers and the population. Risk and uncertainty are often combined as concepts. While risks are generally understood as *known* potential negative outcomes of a course of action, the perception of large risks is more likely to produce uncertainty about possible action (Wakeham 2015, 717). If it is unclear in what way a situation will develop, it is difficult to decide a course of action which could potentially be costly. In the face of very complex risk situations, the level of uncertainty is therefore high.

While it does seem persuasive that risks can be modeled and anticipated, in the real world, uncertainties remain and not all negative outcomes can truly be known. Yet, in the face of (growing) natural hazards action still needs to be taken. However, here, cultural ideology plays a strong role in the perception of risks. Often, a certain culture can prevent "people from anticipating the worst possible outcomes, skewing their perceptions to be overly optimistic" (Wakeham 2015, 720). In this context, uncertainty continues to loom in the background, and culture "shapes what areas of uncertainty receive more attention with a given social context, and thus, what areas of uncertainty people may fail to see coming" (Wakeham 2015, 720). Cultural ideals of individualism, self-reliance, and limited government—underlying societal and cultural norms that are strongly intertwined with American culture and society—are factors in risk reduction and mitigation measures in the United States. Therefore, it seems likely that these ideals can skew the perception of looming risks toward being "overly optimistic."

In this context, it is crucial to communicate risks in an effective way. But what is effective communication of risks and uncertainties and how can citizens be reached and skewed perceptions of risks overcome? One aspect is the reflection of risk in the public—for example, through media portrayals and reporting. Yet, as this chapter argues, mass media in the United States is unlikely to play a constructive role in communicating effectively in the context of natural disasters.

The American Media System and Its Role in Disaster Situations

It is clear that in our "network society" (Castells 2010), we rely on a myriad of information streams and are constantly connected online. The media, once dominated by print and TV, has also adapted, and many (print) outlets have a strong presence online. Today, most Americans receive their information through digital devices, with TV being the second most common news source after radio and print. News websites or apps lead the way in news access among Americans (Pew-Research-Center 2022). Thus, our society is thoroughly "mediatized" (Cushion 2019, 304) with media outlets having a systematic impact on shaping different facets of our society, penetrating institutions, cultures, and shaping how these "are understood and practiced" (Cushion 2019, 305).

Yet, unlike many other Western countries, the United States does not have a publicly funded media system but a market-driven system. A system like this tends to be less regulated and does not have to produce strictly impartial and objective news—or news at all. Although this does not mean that US media does not produce any news, commercial interests play a much bigger role and partisan reporting has been on the rise for a long time, which was accelerated through the abandonment of the Fairness Doctrine in the 1980s (Cushion 2019, 311). This has led to the development and a growing influence of partisan media in the United States "with new content and social media platform exacerbating public concerns about disinformation and so-called 'fake news'" (Cushion 2019, 304).

From this follows that the US's hyper-commercialized media sees audiences as consumers first, and citizens second (Cushion 2019, 308). This has very real consequences: As several studies have shown, public service media rather than commercial television and broadsheet newspapers, produce news that has a higher democratic and informative value for its citizens (Cushion 2019, 310). This leads to greater knowledge and understanding of public affairs among citizens, which could also be proven in comparative studies (Cushion 2019, 312).

Over the last few decades and especially since the development of commercial media in the United States, especially Cable TV, news consumption has declined overall. Additionally, TV news changed to include more "human-interest stories" and dramatic appeals; TV segments and articles also became shorter, more focused on conflict, and used more story-telling, which was believed to appeal more to audiences (Patterson 2000, 247–8).

But declining news consumption is not the only concerning development here:

> [E]nhancing knowledge in the US it [sic] not only about supplying more information because its increasingly partisan news environment is undermining the accuracy and credibility of coverage. This has led to an audience behavior known as 'partisan selective exposure,' which characterizes how people turn to news they trust and ideologically agree with (Stroud 2011). (Cushion 2019, 312)

So-called echo chambers and filter bubbles emerge from the consumption patterns of news in the United States, which makes it even less likely that people are exposed to different viewpoints and opinions. While this is not a US-only phenomenon, it is strongly amplified through the structure of the media system in the United States.

Inside the echo chambers of (both sides of) partisan media, "interpretative reporting" (Patterson 2000, 249) is used to present news. While it is not a new form of news coverage, it has become the norm in the United States in recent decades: "Whereas descriptive reporting is driven by the facts, the interpretative form is driven by the theme around which the story is built" (Patterson 2000, 250). What this form of reporting is trying to achieve is to attract and hold viewers for advertising space, not primarily to foster "a free marketplace of ideas" (Patterson 2000, 253).

Overall, we primarily think of the media as a source of information, yet one has to acknowledge that news "is a construct. It is not, as some journalists like to say, a mirror held up to reality (Mickelson 1972)" (Patterson 2000, 241). On the contrary:

> News is the product of conventions (…). In the United States, these conventions have developed in the context of a press that is decidedly commercial and adversarial in its orientation. This orientation has intensified in recent decades, accentuating what was already a distinctive construction of the news. The values at stake in political conflict and society's underlying problems, for example, are not conspicuously aired in the news. (Patterson 2000, 241)

Instead, the press is dealing with "charges and countercharges, rarely digging into the details of political positions or social conditions underlying policy problems" (Patterson 2000, 255). As mentioned before, the result of this reporting is very limited issue awareness among the American public (Patterson 2000, 257). In a rather gloomy fashion, Patterson concludes that the "media provide a narrow and biased version of political reality, one that focuses on the political game and political shortcomings and that weakens rather than strengthens the relationship between the American people and their political leaders and institutions" (Patterson 2000, 264). Partisan media choice, as it has developed in the United States, is likely accelerating the issue: "the growth of media choice strengthens the extremes while hollowing out the center, making the electorate more divided" (Levendusky 2014). Other audiences increasingly succeed in avoiding news altogether in favor of entertainment programs (Levendusky 2014).

But what role does the American media environment play in the context of disasters? As we have seen so far, the American media is structured to attract consumers and retain them for as long as possible. In order to do that the news is often framed in a dramatic way and includes so-called human-interest stories rather than featuring more informative but less entertaining reports. Disaster events are prime examples of when such reporting happens as the events are both dramatic and filled with individual tragedy. Additionally, it seems that recent years are filled with enormous disaster events. In his book *Consuming Catastrophe: mass culture in America's decade of disaster*, Timothy Recuber describes how the early twenty-first century has been shaped by a myriad of disasters:

> In the first decade of the twenty-first century, Americans watched as mass media, consumer culture, big business, and national politics both shaped and were shaped by the September 11 terrorist attacks in 2001, Hurricane Katrina in 2005, the Virginia Tech shootings in 2007, and the financial crisis in 2008. Although each of these disasters was the focus of extensive news coverage, political debate, and popular culture, efforts to rebuild have been slow, at best, and often ineffective, particularly in instituting the safeguards necessary to prevent similar calamities in the future. (Recuber 2016, 3)

Here, Recuber points to the main issue of disaster perceived in the public eye through media consumption: Even though coverage is extensive, dramatic, and reoccurring event after event, we do not seem to see any improvement in preparedness for the next disaster: "This shock at the repetition of disasters, this disappointment at the lack of lessons learned, is a common feature of disaster media" (Recuber 2016, 8). We might assume that "the harrowing spectacle of mass-mediated disasters provides [a] guarantee that institutions will take the appropriate measures to safeguard us from future calamities. Far from it. Instead, the assumption that consuming catastrophe allows us to learn from these tragedies and improve ourselves and our society as a result is often shattered when the next disaster strikes" (Recuber 2016, 8).

Why is this the case? The above-mentioned market-oriented media system is certainly one aspect of it. Disasters tend to attract a large number of viewers. In fact, media, especially TV, now reports in a "crisis mode" constantly, even on trivial matters which "fatigues" audiences and does not contribute to the understanding and knowledge about actual crises and disaster situations (McRobbie and Thornton 1995; Moeller 1999; Recuber 2016, 18). Additionally, with technological changes more recently, "(…) the speed, frequency, and intimacy, of our exposure to others' suffering has increased exponentially with the growth of consumer culture and the rapid development of a host of mass-media technologies" (Recuber 2016, 9).

Thus, even though disasters are reflected in the media in various ways, the knowledge and preparedness toward possible future disasters does not seem to increase: "Rather than being informed of potential disasters and then acting preventatively, governments, media, and consumers often seem to only really appreciate a threat that has already materialized as a spectacular catastrophe" (Recuber 2016, 17). Instead of reflecting disasters holistically, a "consumption of disasters" occurs including "a heavily mediated experience of disaster in a consumer society (...) using up the raw material of human tragedy, devouring the spectacular, tragic, and even the mundane aspects of catastrophes" (Recuber 2016, 12). Soon, the novelty of the disaster event fades and although "the suffering they caused may persist, and the work of rebuilding may be undone, once disasters lose their novelty or their resonance, one can expect to see fewer reminders of them on the evening news, in TV or magazine advertisements, and on the shelves of stores" (Recuber 2016, 12). Overall, Recuber argues that the consumption of "of disaster thus encourages a kind of 'political anesthesia' (Szasz 2007) that reduces one's ability to recognize the collective solutions to problems, as well as one's willingness to work towards them" (Recuber 2016, 19).

As Recuber has pointed out, the media plays a crucial role in how we "consume catastrophes" but the representation of disasters in the media is deeply flawed. Media representations of disasters are lacking any true reflection of why disasters happen in the first place and how they could effectively be prevented (e.g., through government policy). This is consistent with other research looking into disaster news coverage. J. Brian Houstan et al. found that mass media depicts major American natural disasters only for a short time span, shorter than the coverage of other issues. Additionally, there is a focus on what is happening at that moment in time and on the region the disaster occurs. While this is of course helpful during the occurrence of the disaster, this coverage does not help in "preparedness, recovery, and resilience" (Houston et al. 2012, 619) of the regions or the nation as a whole. Temporal and geographically limited media coverage of disasters in mass media can hardly be beneficial for communities located outside of these areas that are not currently experiencing such events. Moreover, it is questionable if wider political conversations about disaster risk reduction policy, climate change policy, or disaster aid on the national level are in any way supported by this form of media coverage (Houston et al. 2012, 620).

It would be especially necessary to illuminate and critically assess existing and future government policy in this area. Yet, this is mostly not the case. As Logan Strother found, for example, the National Flood Insurance Program (NFIP), a key measure in responding to floods across the United States, which occur frequently and are extremely costly (FEMA 2010), is barely ever discussed in the media. Although the program has many flaws and is working ineffectively in many ways, flood insurance per se is a crucial measure of disaster mitigation. In their study of a Texan community after Hurricane Harvey, Anne Rhodes and Max Besbris found that those households who held flood insurance were more financially resilient a few years later, while other households struggled significantly. Even though the community investigated in this study was overall middle class, insurance coverage was meager. The researchers argue that "the absence of an insurance mandate leaves homeowners with relatively little information about their vulnerability, and what information they do have is often difficult to translate into a definitive conception of risk" (Rhodes and Besbris 2022, 147). Given this circumstance, it would be paramount for citizens to gain more information about concrete disaster mitigation principles (and reforms) through the media they consume.

Yet, this is largely not the case, which has real consequences for citizens but also for political actors. Strother discusses reasons why necessary reform of the NFIP has not materialized, also taking into account a lack of knowledge of this highly specialized policy field. He concludes that the lack of media attention is most likely contributing to the lack of badly needed and often rejected reform of the NFIP (Strother 2018, 470). The NFIP was established in the late 1960s in order to address the continuously rising costs from natural disasters, especially severe hurricanes and consequent flooding. Almost from the beginning, it was clear that the program was not capable of working effectively to prevent flooding costs and damages from rising (Knowles and Kunreuther 2014), especially because it is not a mandatory program. Additionally, insurance premiums are subsidized (and therefore too low to reflect actual risk), which leads to a flawed perception of risk in many cases (Rhodes and Besbris 2022, 150). Because of these low insurance premiums and extremely costly events such as Hurricane Katrina or Harvey, the program has not been able to cover losses and has accumulated enormous amounts of debt (FEMA 2022). Nowadays, many researchers are arguing that the program and other recovery measures after the disaster might have actually contributed to even more movement to and continued construction in disaster-prone areas (Rhodes and Besbris 2022, 149) instead of preventing this by a true reflection of the risks of living in a flood-prone area through risk-adjusted insurance premiums (Knowles and Kunreuther 2014). This and other factors such as climate change, wealth increase in disaster-prone areas, and other government policy failures have led to an increase in costs and damages from disasters overall (Rauscher and Werner 2022). In response, the NFIP needs reform to make it more effective. Yet, time and again, reform of the program has been rejected—most likely due to a lack of knowledge about the program and its purpose. This shows that media attention to underlying reasons for vulnerability to disaster in the United States is needed to improve knowledge among citizens and political actors, to help reform and improve resilience.

We thus see that the mass media sphere plays a role in the multi-faceted malaise in the United States when it comes to disaster risk reduction. Climate change is going to exacerbate the already existing issues in effectively addressing natural disasters, which include population movement and density in disaster-prone areas and government failures to increase resilience in these areas more broadly through policy. Political gridlock and extreme polarization also make it questionable if any reform to address these issues is possible nationwide. On top of that comes the fact that the media is also not reflecting on (natural) disasters in a way that enhances knowledge about why the United States is so vulnerable to disaster events and how these could be addressed effectively.

In line with findings from other researchers, the following discourse analysis will show that specific policy failures and programs do not play an important role in the representation and reporting about disasters in the United States, which ultimately exacerbates issues surrounding disaster risk reduction.

Disaster Discourse in American Mainstream Media

Media discourses reporting on natural disasters are taking part in portraying vulnerability and resilience to disasters in affected regions. How natural disasters are portrayed and how specific issues and policies are reflected upon, shapes the way the population is informed about disasters and their vulnerability to them. This could ultimately shape their behavior in a disaster situation and their attitudes toward disaster risk reduction policy.

In the case study of this chapter, the analysis of discourse will focus on the United States and more narrowly on the issue of hurricanes that hit the South-Eastern coast of the United States frequently.

Methodology

The following case study uses several different news outlets (the New York Times, ABC News Transcripts, Fox News Network) between the years 2017 and 2020 as source material to investigate media discourses around disasters and the discussion of specific policies of disaster risk reduction, in this case, flood insurance. As was explained previously, flood insurance is a key element in disaster mitigation efforts, but the NFIP has not been working effectively for several decades. The case study will thus illuminate if and how this crucial policy is reflected in media discourses which ultimately shows whether Americans have access to this information through the media sphere. Using the three different outlets, more mainstream liberal but also more conservative viewpoints are included in the investigation. By using national news outlets and print media, the case study investigates a bird's eye or even external perspective on local communities affected by disasters. These news outlets reference a broader perspective on natural catastrophe events in the United States and influence notions and perspectives on disaster events across the United States.

In Table 1, the different corpora used for this discourse study are displayed. The sizes of the different corpora rank from 250,822 tokens[2] (ABC) to 1,606,677 tokens

[2] Token refers roughly to 'word.'

Table 1 Corpora sizes

Outlet	Number of articles/transcripts	Corpus size
NYT	829	1,606,677 tokens
ABC News	266	250,822 tokens
Fox News Network	317	2,594,117 tokens

(NYT) and 2,594,117 tokens (Fox).[3] As a reference corpus, a sample Corpus of the COCA (Corpus of Contemporary English) was used (approximately nine million words). COCA stands for what can be considered regular use of the English language.

In the first step, the corpora are analyzed quantitatively using the corpus linguistic software LancsBox (Brezina et al. 2018). The quantitative study of discourses includes keyword analysis through frequency, as well as collocation analysis. In the second step, the corpora are investigated qualitatively looking at text samples. Keywords are used to examine specific contexts, in this case, flood insurance. By looking more closely at the texts themselves, certain recurring narratives around the topic *flood insurance* can then be identified.

Quantitative Discourse Analysis: Frequency and Collocation Analysis

The corpora were first examined using the keyword list (cf. Table 2). The keyword list was informed by close reading of text samples from the corpora, as well as through engagement with natural hazard literature. Using the COCA sample corpus as a reference point, it can be determined whether a certain keyword occurs more or less often in the investigated discourse than in "regular discourse" which can then inform about significant points in the discourse.

The keyword list was mainly used to examine whether certain contexts were discussed in the material and to what extent. The list was divided into different keyword categories. The relative frequency of these terms was then computed with LancsBox and is displayed below.

In the keyword category *Natural Catastrophes* (Fig. 1), we can see that beyond *hurricane*, *flood** is the most used other keyword in this list, although it does not occur as often in the Fox material as in the other outlets. Flooding is the most common and one of the costliest disasters to occur in the United States every year, especially after hurricanes and other storms. Surprisingly, *climate change* as a term

[3] Although the corpora have different sizes, measures like word frequency can be displayed in relative terms which makes it more comparable. We also see a different amount of articles and transcripts. This is due to the fact that written articles (NYT) are shorter than TV segments. Fox segments are also longer than ABC News segments and therefore the total size of the Fox corpus is larger than the others, although the number of transcripts is comparable to ABC.

Table 2 Keyword list[a]

Natural catastrophes	Impact/Effects	Local communities
flood*	mitigat*	communit*
hurricane*	failure*	Communal*
climate change	respon*	local*
catastroph*	risk*	neighbor*
natural	resistan*	actor*
disaster*	vulnerab*	activis*
hazard*	resilien*	group*
	marginal*	minorit*
	strength*	woman
	policy	women
	policies	immigrant*
	responsibility	immigration
	flood insurance	poor
	managed retreat	poverty
		working-class
		middle class
		wealth

[a]* refers to so-called wildcard terms, taking into account different forms of the same word e.g. flood, floods, flooding, etc.

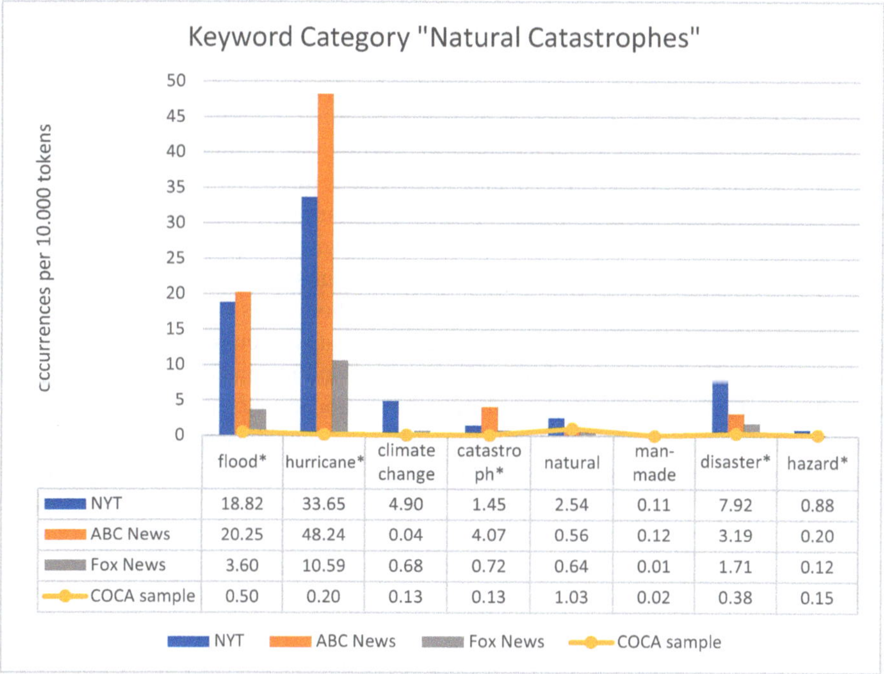

Fig. 1 Keyword category Natural Catastrophes

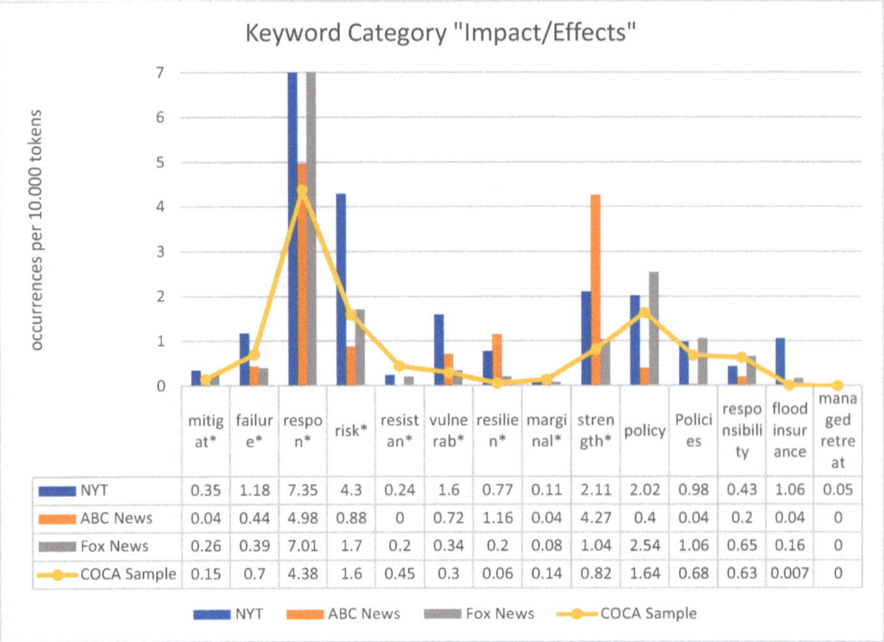

	mitigat*	failure*	respon*	risk*	resistan*	vulnerab*	resilien*	marginal*	strength*	policy	Policies	responsibility	flood insurance	managed retreat
NYT	0.35	1.18	7.35	4.3	0.24	1.6	0.77	0.11	2.11	2.02	0.98	0.43	1.06	0.05
ABC News	0.04	0.44	4.98	0.88	0	0.72	1.16	0.04	4.27	0.4	0.04	0.2	0.04	0
Fox News	0.26	0.39	7.01	1.7	0.2	0.34	0.2	0.08	1.04	2.54	1.06	0.65	0.16	0
COCA Sample	0.15	0.7	4.38	1.6	0.45	0.3	0.06	0.14	0.82	1.64	0.68	0.63	0.007	0

Fig. 2 Keyword category Impact/Effects

does not occur as often, in most cases not more often than in "regular" discourse (compare to yellow line), much like *hazard**. While this might not be so surprising for the Fox material, which, as a conservative outlet might be questioning the relevance of climate change altogether, it is a bit more surprising for the NYT and ABC material. Both outlets do not seem to discuss the changing climate much in relation to extreme weather. *Disaster** and *catastrophe** occur somewhat more frequently in the material investigated, especially *disaster**. Although, as we heard in the previous part of this chapter, the media in the United States often reports on events in a "crisis mode" regardless of the issue, hurricanes, flooding, and other extreme events actually constitute disasters, and it is the most common term to talk about such events.

In the keyword category *Impact/Effects* (Fig. 2), we can see that the terms vulnerability and resilience play a significant role—the two key concepts when we talk about disaster risk reduction. We also see a quite large frequency of *strength** which could refer to the strength of the weather or other contexts, like the strength of communities hit by disaster. *Respon** is also a strong term, but it is very broad as a wildcard term because it can refer to a wide array of contexts such as response or responsibility. What is of interest to this chapter, however, is the frequency of specific policies in disaster risk reduction, especially flood insurance. One can see that *flood insurance* does not occur frequently (or almost at all) in the ABC material, and very rarely in the Fox News Material, which is consistent with the notion that mainstream media does not talk enough about crucial policies around disaster risk reduction. Only the NYT material displays a higher frequency of the term than in "regular" discourse and a much larger one than in the ABC or Fox News material. Outlets based on written content are certainly more detailed in their analysis than TV outlets, so it is only natural that the topic of flood insurance would be discussed in more detail in the NYT than on TV. On the other hand, one has to say that the total amount of occurrences of the term *flood insurance* in the NYT (170 times) can be pinpointed to only a few articles over the years (some of these articles mention the term up to 30 times). The frequency is thus higher than average in the NYT, which indicates somewhat of a discussion of the issue of flood insurance, yet the overall absolute number and number of articles that actually discuss the topic remains rather low.

In an additional step of the quantitative analysis, one can investigate certain keywords in more context than just frequency. In what context are the keywords occurring? Collocation analysis can help with this question. Collocates are words that are likely to co-occur with the "node"—the keyword to be investigated (Vaclav Brezina 2018, 67). Using LancsBox, collocates can be displayed in network graphs.[4]

Graphs 1 and 2 display the collocates of the term *flood insurance* in the NYT and Fox material.[5]

As one can see, some of the strongest collocates in the NYT material are *broke* and *broken*, which stems from the fact that an article appeared in the NYT called *A Broke, and Broken, Flood Insurance Program* (2017), which talks both about the

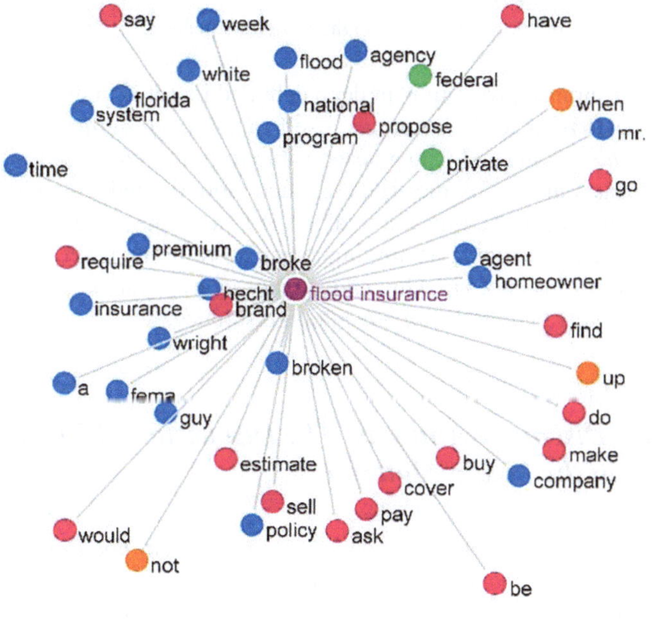

Graph 1 Collocates of "flood insurance" in the NYT

[4] The network graphs also display certain factors about the keyword and its collocate. Closeness to the node indicates strength of connection; right or left indicates where in the sentence the collocates occur; color of dots indicates the word class of the collocate.

[5] Because the keyword *flood insurance* only occurs one time in the ABC material, there are no collocates displayed for the ABC material.

Graph 2 Collocates of "flood insurance" in the Fox News material

fact that the insurance program is in deep debt and that it is in dire need of reform. Other collocates refer to the insurance itself (*national, program, premium, homeowner, company, cover*) and who is responsible for it (*federal, agency, fema*).

The collocates in the Fox material are much more limited which is unsurprising as the term only occurs 42 times in total in the Fox corpus, whereas it appears 170 times in the NYT. The contexts are thus much more limited. Yet, the graph shows clearly what most instances in which *flood insurance* appears in the corpus actually discuss: That *many people do not have* flood insurance.

Qualitative Discourse Analysis: The Discussion of Flood Insurance in the Media Material

Having looked at the quantitative occurrence of the term *flood insurance* in more detail and concluding that overall, the policy topic does not occur frequently in the investigated material compared to many other keywords, we nevertheless want to know how this policy measure is discussed in the instances it does occur. There are certain narratives and elements that can be identified when flood insurance is discussed in the three outlets.

The first narrative surrounding flood insurance in the discourse, is, in fact, the lack of flood insurance and thus the lack of adequate federal assistance despite severe and devastating disaster events. This is portrayed as an enormous issue for households struggling after disaster events and a failure on the side of the government. The single instant in which flood insurance is discussed in the ABC material shows this:

> The hardworking mom said the hardest hit came after Hurricane Harvey destroyed her house. (...) It's really clear that the road to recovery is just beginning for this family. (...) 550,000 victims have now applied for financial assistance from FEMA (...). She was approved for a one-time $500 payment, but still hasn't received it. (...) So, no assistance so far in any of this? (ABC)

The concept of flood insurance and why it is important for people's recovery is not discussed here. Instead, there is a focus on the immediate consequences of disaster and the personal stories of victims. Also, blame is put on government agencies by stressing that assistance is not coming easily and fast enough, again without discussing the larger contexts of this assistance and the work and failures of FEMA policy vis-à-vis different regions. There is also a lack of discussion and understanding of what FEMA actually does, which is mainly the provision of loan assistance and other similar measures to rebuild after disaster events.

As was indicated above in the collocate analysis, the Fox material also mainly talks about flood insurance in the context of people not having it, even though their homes were destroyed. There is also the allusion in many of these instances that flooding to the degree it happened (for example after Hurricane Harvey) is indeed very unusual and exceptional—not taking into account the frequency of flooding in the United States every year.

> many people don't have flood insurance. Many people, Mara, who have—who have never flooded before who are now on their second flood or rooftop because the first floor is flooded. There will be hundreds and hundreds of thousands homes impacted. (...) (Fox)

The exceptional circumstances are discussed as well as the fact that the federal government (and here Congress is blamed) is not acting fast enough even though so many people are affected:

> (...) the majority of homes in this area never see water like this. So many homeowners say they do not have flood insurance, and tonight amidst all this disaster and across the country for the third time today, the House failed to sync with the Senate and pass a $19 billion disaster aid bill (...). (Fox)

In another instance, a Fox segment actually informs audiences about what happens to people when they do not have flood insurance, having the director of the NFIP live on TV. He stresses that FEMA can only provide some disaster assistance but that it is not comparable to an insurance policy:

> Roy Wright is director of the National Flood Insurance Program over at FEMA. It's a program that needs an act of Congress to keep running before the fund runs out. (...) we have had statistics to the effect that four out of five homeowners didn't have any flood insurance. (...) What happens with them? So if you don't have flood insurance, you are going to be registering for disaster assistance [from the federal government]. (...) It provides some initial kind of support related to housing and the like. But it's not an insurance policy. (Fox)

There is also mention of the fact that people living along the coasts do not have flood insurance—again without the context of why this is the case: high risk.

> most of those homes or a big portion of the homes actually don't even have flood insurance because they're right there on the coast. (…) And that is what FEMA is designed to do. FEMA is not the operational arm to come in. That is the state and locals, but FEMA comes in and provides the assistance needed after the event. (Fox)

Here, we can once again see that high-risk areas, which are prone to flooding and are therefore uninsurable, are not adequately addressed in these discourses. Yet, in this segment, there is a bit more discussion of the role of FEMA.

In some instances, there are some details on the flood insurance program, but these details are mainly presented to blame the malfunctioning and inadequacy of the system without giving more information:

> And just to update you on that aid and what could be coming, many people who even have flood insurance in this neck of woods are coming to find (…) that it's usually topped out at around $250,000. (Fox)

Overall, the most common context in which flood insurance is mentioned in the Fox material are the political struggles to further fund disaster aid and the NFIP. Here, partisan "powerplay" in Congress from different sides is blamed for the bad functioning of the program.

> [Gov Chris Christie on Fox:] a lot of them didn't even vote for the funding for the national flood insurance plan. There was nothing else involved in that, except for funding flooding—for flood insurance that people had already paid for. And so a lot of that is disingenuous and political rewriting. (Fox)

In one segment in the Fox material, a roundtable is hosted on the flood insurance funding problems, among other things relating to the debt ceiling standoff in Congress. This roundtable that invites experts from several areas details a few key factors around the issue of flood insurance:

> [Jason Riley, Manhattan Institute Senior Fellow:] I think that the flood insurance program will be allowed to increase its debt (…). It would be hard to come up with a program more in need of reform. Around 50 years, we have had the federal government sending taxpayer backed insurance to people who build in flood prone areas. It makes no sense whatsoever. All the wrong incentives are in place. (…) So the goal here should be to get the federal government out of the flood insurance business and let the insurance price this to market. (Fox)

The segment continues with a discussion of the debt ceiling where the NFIP also plays a role, but there is consensus among the guests that it will be increased, and especially in emergency situations, Congress will send disaster aid. The crucial points raised by expert guest Riley are not further discussed.

Overall, as we could see in the quantitative analysis of discourse, the discussions around the policy of flood insurance are very limited and the benefits of a well-working flood insurance program are not further discussed. There are references to the fact that the program is not working properly and that many people affected by disasters or living in high-risk areas do not have insurance, which is often not further problematized. Emblematically, some Fox anchors talk about flood insurance in the following way:

> You know what a really boring segment would be but in fact, one that people have to know about is flood insurance (…). The whole controversy behind flood insurance is absolutely astounding. The types of repeats floods and what happened and how much is paid out. It would bore you to tears but it is probably the most important stories out there. (Fox)

Even though flood insurance is identified as one of the "most important stories out there," it is not discussed or further explained. The segment actively turns away from this subject because it is deemed too boring for audiences—an emblematic picture of the disaster coverage by mainstream media in the United States.

As was discussed as part of the quantitative analysis, *flood insurance* as a term appears most frequently in the material of the NYT, which indicates a somewhat more detailed discussion. Yet, even though the frequency of the term is much higher than in the other outlets, flood insurance as a topic is essentially only discussed in a handful of articles over the years. Apart from mentions of people in disaster situations (like Hurricanes) that do not have flood insurance, which is a common occurrence across all the outlets, there are articles going into more detail on the political decisions around the NFIP and how it lacks the funds to continue unless Congress acts to renew it. For example, in the article *White House Asks Congress for $29 Billion in Hurricane Aid* (2017), the struggles over immediate disaster aid are presented, where flood insurance and the NFIP are only a part. In 2017, Puerto Rico was hit severely by Hurricane Irma and needed federal emergency funds.

> The White House asked Congress on Wednesday to approve $29 billion in additional hurricane relief and debt forgiveness, seeking to help Puerto Rico and other battered areas and to shore up the troubled federal program that provides flood insurance to homes and small businesses. (…) (NYT)

In the same article, the difficult financial situation of the NFIP is in fact mentioned, as well as struggles between different states (which are not equally affected by repeated flooding) to finance the program and raise premiums, which is highly unpopular among lawmakers. The article also mentions the Trump Administration advising some changes to the program in order to make it more financially viable—something that has been suggested by other administrations before but mostly failed to be implemented:

> Fiscal conservatives in Congress, especially those from states without much flooding, may object to the debt forgiveness because it means the taxpayers they represent will not be repaid for helping to pay flood claims in Louisiana, Texas, New York and New Jersey. To make taxpayer losses more palatable, the White House also proposed to raise flood insurance premiums (…) and to make changes that would allow insurance companies to underwrite flood insurance themselves. Currently, insurance companies can sell flood insurance, but they do not pay the claims with their own money. Instead, they (…) pass the risk on to the federal government. (…) lawmakers have been unwilling to impose higher premiums on constituents in flood zones. (NYT)

Again, there is also reference to political struggles over issues of disaster aid, with both partisan sides regularly blaming each other when disasters occur and both hindering most of the much-needed reform discussions over disaster mitigation policy:

> [Senator Schumer:] Republicans should not put much-needed aid at risk by insisting on the proposed ideological changes to the flood insurance program and forest management policy (…) Those changes should be debated separately from this aid package. (NYT)

Another article in the NYT titled *A Broke, and Broken, Flood Insurance Program* (2017) discusses the NFIP as the only source of flood insurance in the United States and that it can and will not be able to cope with rising damages from flooding. This article alone mentions the term *flood insurance* about 30 times. As the article before, it is mentioned that the Trump Administration, together with an "unusual coalition of insurers, environmentalists and fiscal conservatives," is calling for reforms of the NFIP, especially through the private insurance market. Issues that this coalition agrees on include the burden for taxpayers after each disaster event and when the NFIP needs to be financially bolstered again as well as the fact that the NFIP might have been an "enabler of construction on flood-prone coastlines, by charging premiums too low to reflect the true cost." The article even mentions a few key historical facts about the founding of the NFIP and the most commonly leveled criticisms of the program, for instance, the lack of updated flood-plain mapping. The article also mentions other risk mitigation policies such as government buy-outs of repeated flood properties (which is still a very rare policy) and the fact that reform was attempted by Congress in the preceding years. This article is a rather detailed account of the issues and problems of the NFIP, which is overall a rare occurrence among all the articles in the corpus.

Another article in the NYT called *Ignoring Climate Science, FEMA Is Mired in Cycle of Repairs* (2018) details some other fundamental problems of disaster risk reduction policy through the federal government, namely that FEMA is often releasing funds for repetitive loss properties, paying over and over for properties that are flooded regularly without the possibility to break this cycle and spend the funds on prevention and adaption rather than as a reaction to disaster—a vicious cycle indeed.

Other articles such as *When the Levees Break Again* (2019), an opinion piece, talk about local struggles in Louisiana and how the current flood insurance system does not help them as the system remains ineffective. As Congress struggles to once again increase the debt limit of the NFIP, this will not fundamentally change the outlook and issues of the affected regions:

> The plan is to do the bare minimum. (…) But even if the funding is forthcoming, piecemeal fixes like that are what got us to this point, and they won't get us much further. (…) Louisiana's levee system stands, for now, as a sinking monument to America's dangerously shortsighted climate policies. The corps acknowledged as much from the start. While it had called its pre-Katrina levees a "hurricane protection system," it described the post-Katrina project as only a "risk reduction system." (NYT)

Another article shows once more that the flood insurance program as it exists today might in fact encourage building in high-risk areas (*Homes Are Being Built the Fastest in Many Flood-Prone Areas, Study Finds,* 2019) and that people are drawn to these areas, often ignoring glaring risks:

> people's memories of Katrina have ebbed, replaced by the enduring appeal of being by the water. (NYT)

Apart from these articles, there are many more alluding to the aforementioned debate in Congress to increase the debt limit, which includes funds for the NFIP.

One article in the NYT stands out, however, in its detailed analysis of insuring coastal regions in the United States, where risks are constantly rising: *When Rising Seas Transforms Risk Into Certainty, The Climate Issue* (2017), states:

> Along parts of the East Coast, the entire system of insuring coastal property is beginning to break down. (NYT)

Again, despite these articles mentioned above, the discussion and portrayal of flood insurance as a topic remains limited, even in the NYT material. Of the over 800 articles in the corpus, only a handful take up this crucial issue and talk about it in more detailed terms. When it is mentioned, however, the NYT is certainly the outlet that gives the most context on the debate around the program, its effectiveness, and possible reforms (which have not taken place so far).

Discussion of Results and Conclusion

As one can see in both the quantitative and qualitative analysis of disaster discourse in this case study, mainstream media in the United States only rarely reports on crucial issues around disaster risk reduction policy. This becomes especially visible in the case of flood insurance. As described, an effective flood insurance system would be crucial to address and mitigate risk due to natural phenomena. The National Flood Insurance Program in the United States has, however, not worked effectively almost since its inception. Several reforms have improved the system over the years, but the basic issues of the program remain the same, including the fact that it is not a mandatory insurance, the premiums are subsidized and too low, and it cannot update information on flooding and other related (climate) factors in a timely manner.

Yet, what we can see reflected in the media discourses on natural disasters in the United States is that American audiences are rarely exposed to the (long history of) underlying issues with the flood insurance program. Discussions around this issue are infrequent and media discourses rather reflect on the immediate effects of disasters, as well as affected households who are struggling to get by on disaster aid which the federal government should provide. This often lacks a reflection of the fact that there is a vicious cycle of calling on the federal government and rebuilding repetitive loss properties in affected regions over and over again without proper adaption and risk mitigation on the ground.

Most of the time, the media reports that most of the affected households do not hold flood insurance. Why this is the case, and the fact that this is actually part of the problem of disaster and risk mitigation in the United States, is not discussed. Instead, the partisan fights over disaster aid and reauthorization of the NFIP in the context of the debt ceiling are mentioned several times both by the conservative and mainstream outlets, reflecting the high degree of political polarization in the media sphere as well as in virtually every area of policy-making. While budgetary questions are of course an important information to convey, there is, again, a lack of context about how planned reforms of the program never seem to go anywhere and why one disaster aid package after the other is sent to vulnerable regions without addressing the root causes of this vulnerability in the first place.

Overall, these findings are consistent with other research on media discourses in disaster situations which found that technical issues such as flood insurance are rarely covered by media outlets (Strother 2018) and that media mainly focuses on the immediate effects of disasters in one specific region without a reflection of causal explanations of disasters (Houston et al. 2012). Given the fact that most media audiences have traditionally turned to TV to consume the news rather than written outlets (Pew-Research-Center 2016), this effect is even more pronounced.

Thus, for most Americans, it can be assumed, that there is little exposure to crucial elements of disaster risk reduction policy, which ultimately means that as voters, Americans remain uninformed about this issue and are thus possibly unwilling to support the often expensive adaption and mitigation measures in vulnerable regions. Yet, as Thomas Drabek argues, when "leaders cannot mobilize reasonable levels of consensus [among themselves and their voters] about the risk associated with natural and technological hazards, the policy-making process grinds to a halt" (Drabek 2010, 207).

Mainstream media's failure to provide information on disaster risk reduction policy might contribute to this lack of risk perception which is ultimately shaped by several factors such as the national culture and ideology that emphasizes self-reliance and small government or personal experience and knowledge about underlying risk factors. Yet, especially this personal experience can be misleading: "Denial and evidence to support a sense of invulnerability are operative […]." (Drabek 2010, 207) when risk perceptions are formed. Additionally, "Familiarity [with risks] breeds acceptance" (Drabek 2010, 208). So, the more people are exposed to risk on a regular basis, the more they become accepting of it, without questioning the vulnerability of their situation on a higher level. Thus, to effectively address vulnerability to disaster, one has to reach a certain level of knowledge about risk factors and mitigation measures in a community: "Mitigation actions of whatever type are less likely to be accepted by those whose risk perceptions have been blunted by years of indifference" (Drabek 2010, 209).

But given the results of this case study, mainstream media in the United States seems unlikely to be a positive force in this endeavor. The structure and practice of the US media system seem ill suited in this regard. As Thomas Patterson puts it:

> The problem is that the press is not a political institution and has no stake in organizing public opinion. The media are in the news business, and their inadequacy as a linking mechanism becomes obvious once the nature of news is understood. The news simply is not an adequate guide to political choice. Its major themes are dictated by journalistic values, not political ones. (Patterson 2000, 264)

Thus, the commercial nature of the US media system is a further hindrance to truly well-informed citizenry in the area of disaster mitigation—which will become even more crucial in the decades to come when climate change will worsen the already high vulnerability to disasters across the United States.

Timothy Recuber argues that the common "consumption of catastrophe" in the United States will likely remain the same, unfortunately, marked by an often detached engagement by audiences: "(...) the consumerism of security will resemble a kind of 'political anesthesia' (Szasz 2007, 195) that diminishes one's sense of risk and reduces the sense of urgency to make political demands about the alleviation of potential hazards" (Recuber 2016, 175). If the notion of this "political anesthesia" is accurate, it is unlikely that the United States will see essential changes in disaster risk reduction policy in the near future.

References

Bergstrand, Kelly, Brian Mayer, Babette Brumback, and Yi Zhang. 2015. Assessing the Relationship Between Social Vulnerability and Community Resilience to Hazards. *Social Indicators Research* 122 (2): 391–409.

Brezina, Vaclav. 2018. *Statistics in Corpus Linguistics. A Practical Guide*. Cambridge, UK: Cambridge University Press.

Brezina, V., M. Timperley, and T. McEnery. 2018. #LancsBox v. 3-4.x [software]. Accessed Nov 4, 2022. http://corpora.lancs.ac.uk/lancsbox.

Castells, Manuel. 2010. The Rise of the Network Society. Second Edition ed. 3 vols. Vol. I. *The Information Age: Economy, Society, and Culture*. Chichester: Wiley-Blackwell.

Committee on Increasing National Resilience to Hazards and Disasters. 2012. *Disaster Resilience: A National Imperative*. Edited by National Academy of Science. Washington, DC: The National Academies Press.

Cox, Robin S., Long C. Bonita, Megan I. Jones, and Risa J. Handler. 2008. Sequestering of Suffering: Critical Discourse Analysis of Natural Disaster Media Coverage. *Journal of Health Psychology* 13 (4): 469–480.

Curran, James, Shanto Iyengar, Anker Brink Lund, and Inka Salovaara-Moring. 2009. Media System, Public Knowledge and Democracy: A Comparative Study. *European Journal of Communication* 24 (1): 5–26.

Cushion, Stephen. 2019. The Political Impact of Media. In *Media and Society*, ed. James Curran and David Hesmondhalgh, 303–321. Bloomsbury Academic.

Cutter, Susan L. 2016a. The Landscape of Disaster Resilience Indicators in the USA. *Natural Hazards* 80:741–758.

Cutter, Susan L. 2016b. Social Vulnerability and Community Resilience Measurement and Tools. *Natural Disaster Resilience: An Urgent Need and Opportunity for Chile*.

Cutter, Susan L., and Christina Finch. 2008. Temporal and Spatial Changes in Social Vulnerability to Natural Hazards. *Proceedings of the National Academcy of Sciences of the United States of America* 105:2301–2306.

Cutter, Susan L., Lindsey Barnes, Melissa Berry, Christopher Burton, Elijah Evans, Eric Tate, and Jennifer Webb. 2008. A Place-based Model for Understanding Community Resilience to Natural Disasters. *Global Environmental Change* 18:598–606.

Davis, Justin M., and T. Nathaniel French. 2008. Blaming Victims and Survivors: An Analysis of Post-Katrina Print News Coverage. *Southern Communication Journal* 73 (3): 243–257.

Drabek, Thomas E. 2010. *The Human Side of Disaster*. Boca Raton: CRC Press, Taylor & Francis Group.

FEMA. 2010. *Flooding – Our Nation's Most Frequent and Costly Natural Disaster*. Accessed Nov 7, 2023. https://www.fbiic.gov/public/2010/mar/FloodingHistoryandCausesFS.PDF.

FEMA. 2022. *NFIP Debt*. Federal Emergency Management Agency. Accessed May 9, 2023. https://www.fema.gov/case-study/nfip-debt.

Houston, Brian J., Betty Pfefferbaum, and Cathy Ellen Rosenholtz. 2012. Disaster News: Framing and Frame Changing in Coverage of Major U.S. Natural Disasters, 2000–2010. *Journalism & Mass Communication Quarterly* 89 (4): 606–623.

Knowles, Scott Gabriel, and Howard C. Kunreuther. 2014. Troubled Waters: The National Flood Insurance Program in Historical Perspective. *The Journal of Policy History* 26 (3): 327–353.

Levendusky, Matt. 2014. Are Fox and MSNBC polarizing America? *The Washington Post*.

McRobbie, Angela, and Sarah L. Thornton. 1995. 'Moral Panic' for Multi-Mediated Social Worlds. *The British Journal of Sociology* 46 (4): 559–574.

Moeller, Susan D. 1999. *Compassion Fatigue: How the Media Sell Disease, Famine, War and Death*. New York: Routledge.

Norris, Fran H., Susan P. Stevens, Betty Pfefferbaum, Karen F. Wyche, and Rose L. Pfefferbaum. 2008. Community Resilience as a Metaphor, Theory, Set of Capacities, and Strategy for Disaster Readiness. *American Journal of Community Psychology* 41:127–150.

Patterson, Thomas E. 2000. The United States: News in a Free-Market Society. In *Democracy and the Media: A Comparative Perspective*, ed. Richard Gunther and Anthony Mughan, 241–265. Cambridge, UK: Cambridge University Press.

Pew Research Center. 2016. *The Modern News Consumer*. Pew Research Center. Accessed Jan 12, 2023. https://www.pewresearch.org/journalism/2016/07/07/pathways-to-news/.

Pew Research Center. 2020. *U.S. Media Polarization and the 2020 Election: A Nation Divided*. Pew Research Center. Accessed July 21, 2020. https://www.journalism.org/2020/01/24/u-s-media-polarization-and-the-2020-election-a-nation-divided/.

Pew Research Center. 2022. "News Platform Fact Sheet." Pew Research Center. Accessed Nov 8, 2023. https://www.pewresearch.org/journalism/fact-sheet/news-platform-fact-sheet/#:~:text=Americans%20turn%20to%20radio%20and,they%20prefer%20TV%20(33%25).

Rauscher, Natalie. 2024. In the Face of Disaster: Narratives of Community Vulnerability and Resilience in Media Coverage of Natural Catastrophes in the USA. In *Vulnerability – Real, Imagined, and Displayed Fragility in Language and Society*, ed. Silvia Bonacchi, 167–187. V&R Unipress.

Rauscher, Natalie, and Welf Werner. 2022. Why Has Catastrophe Mitigation Failed in the U.S.? *Zeitschrift für Politikberatung/Policy Advice and Political Consulting* 4 (2016).

Recuber, Timothy. 2016. *Consuming Catastrophe: Mass Culture in America's Decade of Disaster*. Philadelphia: Temple University Press.

Rhodes, Anne, and Max Besbris. 2022. *Soaking the Middle Class: Suburban Inequality and Recovery from Disaster*. New York: Russell Sage Foundation.

Shiller, Robert J. 2017. Narrative Economics. *The American Economic Review* 107 (4): 967–1004.

Strother, Logan. 2018. The National Flood Insurance Program: A Case Study in Policy Failure, Reform, and Retrenchment. *Policy Studies Journal* 46 (2): 452–480.

Szasz, Andrew. 2007. *Shopping Our Way to Safety: How We Changed from Protection the Environment to Protecting Ourselves*. University of Minnesota Press.

Tierney, Kathleen. 2019. *Disasters: A Sociological Approach*. Cambridge: Polity Press.

UNDRR. 2017. *Understanding Disaster Risk*. Accessed Oct 25, 2022. https://www.preventionweb.net/understanding-disaster-risk/key-concepts/resilience.

Wakeham, Joshua. 2015. Uncertainty: History of the Concept. In *International Encyclopedia of the Social & Behavioral Sciences*.

Zakour, Michael J., and David F. Gillespie. 2013. *Community Disaster Vulnerability*. New York: Springer Science and Business Media.

Impact and Response

Multinational Insurers and Catastrophic Loss: Responses to the San Francisco Disaster of 1906

Robin Pearson and Mikael Lönnborg

Abstract An important share of the huge economic burden caused by natural disasters is carried by the insurance industry. This chapter examines the impact on international insurance of one of the greatest disasters in American history: the San Francisco earthquake and fire of 1906. It has three aims: first, to explore the response of foreign fire insurers to the catastrophe, and their decisions to stay or quit the Californian market; second, to examine whether modern theory helps explain those decisions; third, to explore how the experience of insurers operating in San Francisco illuminates one of the central themes of this book—the particularity of American responses to natural disasters. We identify four factors—economic, market, company-specific, and institutional—whose interplay determined the effect of this disaster on the multinational insurance companies operating in California in 1906 and their post-disaster strategic decisions. In particular, powerful informal institutional pressures generated by the singular American approach to natural disasters narrowed the scope for reducing costs by those insurers who wished to remain in the market and drove other foreign insurance companies to quit the market altogether.

We thank the Skandia Insurance and Swiss Reinsurance companies for permission to cite material from their private archives. The views in this chapter are those of the authors and may not represent the views of either company. This research has been supported by an award made to Robin Pearson by the UK Economic and Research Council (R 000 234,859). Mikael Lönnborg's research was supported by Jan Wallander and Tom Hedelius Stiftelse (P12–0110). We are also indebted to the editors of this book and the participants of conferences in Heidelberg, Zurich and London for their valuable comments on earlier versions of this paper.

R. Pearson (✉)
University of Hull, Kingston upon Hull, UK
e-mail: R.Pearson@hull.ac.uk

M. Lönnborg
Södertörn University, Huddinge, Sweden
e-mail: mikael.lonnborg@sh.se

N. Rauscher, W. Werner (eds.), *Natural Disasters in the United States*, Risk, Governance and Society 22, https://doi.org/10.1007/978-3-031-96436-7_10

Introduction

> "San Francisco is merely waiting for the payment of her insurance to go right ahead"
> Eugene L. Lezinsky.[1]

The scale and frequency of natural disasters have increased enormously since the 1980s, a trend commonly associated with climate change.[2] Such catastrophes often have a significant economic impact on the regions they affect. An important share of this burden is carried by the insurance industry. Consequently, there has been a growing interest in the problem of managing catastrophe risk and pricing it for insurance purposes.[3]

The topicality of disasters has also drawn scholars to the history of catastrophes and perceptions of risk.[4] This chapter examines the impact on international insurance of one of the greatest natural disasters in American history: the San Francisco earthquake and fire of 1906. The total insurance payouts of $200 m covered 57 percent of the estimated losses in San Francisco, dwarfing the sums spent by other public and private bodies. Yet, the extensive literature on the disaster devotes relatively little space to the response of insurance. Those few works that have examined this have largely focused on the experience of single companies.[5] This chapter, for the first time, compares the responses of foreign insurers from several countries, places these companies in the context of the Californian insurance market, and investigates the factors influencing their decisions, drawing on modern economic, insurance, and institutional theory. Our focus is on foreign insurers, who accounted for 42 percent of the money paid out in claims. Most US fire insurers in 1906 had few or no operations outside North America and could not disperse San Francisco risks globally to the extent that the British and Europeans could. Thus, the factors affecting their post-disaster decisions were somewhat different, although their story remains to be written.[6]

---

[1] *New York Times*, 26 July 1906. Lezinsky was a member of the Merchants' Association of San Francisco, a major booster of the city.

[2] Between 1989 and 1995 $75bn (1996 US$) was lost through hurricanes, floods, tornadoes and earthquakes, compared to $51bn between 1950 and 1988. The 30 years prior to 1989 were thus a poor predictor of the loss to come in the following seven. Anderson, "Catastrophe insurance," 76; Insurance Services Office (ISO), *Managing catastrophe risk*, p. 4. We thank Ira Blatt, formerly of the ISO, for this reference.

[3] Cossette, Duchesne and Marceau, "Modeling catastrophes"; Froot, *Financing of catastrophe risk.*

[4] Papers or sessions were devoted to catastrophe history at the following conferences: Economic History Association 1999; Risk and disaster: an historical approach, Zurich 2006; Risk and the insurance business in history, Seville 2019; Natural Catastrophes in the United States, Heidelberg 2022.

[5] Trebilcock, *Phoenix Assurance*; Swiss Re, *A shake in insurance*; Rohland, "Earthquake versus fire."

[6] The most comprehensive study of early US fire insurance remains Baranoff. "Shaped by risk," but just one paragraph (p. 210) is devoted to the San Francisco disaster.

This chapter has three aims: first, to explore the response of foreign fire insurers to the catastrophe, and their decisions to stay or quit the Californian market; second, to examine whether modern theory helps explain those decisions; third, to explore how the experience of insurers operating in San Francisco illuminates one of the central themes of this book—the particularity of American reactions to natural disasters. Changing American perspectives on risk are not the primary focus of this chapter—that would require a much broader study than we can realize here.[7] Our evidence, however, does allow us to offer some reflections on popular responses to the disaster of 1906 and on American attitudes toward insurance protection in the wake of the catastrophe.

Section 1 draws on the theoretical literature from insurance, economics, disaster studies, and international business, to identify four sets of factors—economic, market, company-specific, and institutional—that can influence property insurers' responses to natural disasters. Section 2 describes the Californian fire insurance market before and after the earthquake. Section 3 summarizes the disaster and its aftermath. In Sects. 4 and 5, we analyze the responses of the foreign insurance and reinsurance companies operating in San Francisco in 1906. Section 6 presents our findings, and the final section describes in a brief postscript some of the longer-term consequences of the disaster.

Factors Impacting Insurers' Responses to Natural Disasters

Economic Factors

Most natural disasters have direct economic effects—loss of production, unemployment, falling profits, rental, and wage incomes—but there are also indirect effects channeled through insurance. A sudden surge in claims drains an insurance company's reserves and reduces its capacity to underwrite.[8] Premium rates rise, increasing the cost of living for local residents and threatening the relative economic competitiveness of the affected region.[9] The potency of indirect effects is partly related to the magnitude of the disaster. Insurance losses in major catastrophes are amplified through increases in the price of building materials and labor; delays to repairs that inflate property damage; claims inflation by policyholders as loss adjusters struggle to inspect properties; and the unanticipated expansion of insurance liability as the terms of policies are extended, often under public pressure, to cover

[7] For a pioneering study of risk culture and risk management in nineteenth and twentieth America, see Mohun, *Risk*.

[8] Underwriting capacity is the amount of insurance that can be written by a company for a given level of reserves set aside to cover existing insurance liabilities.

[9] Ratajczak, "What catastrophes do."

additional sources of loss.[10] Under these circumstances, the underwriter's fundamental assumption that losses are generated independently of each other breaks down. The higher the intensity of the catastrophe, the higher the interdependency of insured risks and the greater the reduction in underwriting capacity. Reduced capacity may encourage insurers to exit the disaster region. Preliminary estimations of damage, and the volume and quality of information underpinning those estimates, help determine expectations of the speed of recovery and may influence insurers' decisions to stay or quit.[11] Other factors may include the extent and form (grants versus loans) of external aid after the disaster, the degree of altruistic cooperative behavior among victims and donors, and the pre-disaster level of development, as the latter can determine the technologies of reconstruction.[12]

Company-Specific Factors

The composition and distribution of loss are a function of the intensity of the catastrophe and of the building types in the area affected.[13] The extent and composition of pre-disaster insurance coverage shape how losses impact individual insurers.[14] The probability of a company surviving is improved if it holds a balanced portfolio of independent non-catastrophe risks alongside those interdependent risks subject to catastrophe hazard and if it can lay off parts of the latter through reinsurance or spread them using the capital market.[15]

Reconstruction may take place at different rates for different types of capital. Using data from the 1964 Alaska earthquake and other disasters, Dacy and Kunreuther argued that public capital has a steeper recovery curve than business and residential capital.[16] Because of the different marginal productivity of each type of capital, and assuming there is a post-disaster scarcity of funds and labor, rationally allocated relief will go, first, to rebuild the public transport, service, and administrative infrastructure, second, to rebuild business property, and, third, to the majority of residential property. This will affect insurers in different ways. The insurance of public real estate is likely to be smaller in volume and more narrowly concentrated in a few large firms, than the insurance of other types of property. Investment in rebuilding public capital will therefore have a limited impact on the post-disaster

---

[10] Grossi and Muir-Wood, *San Francisco earthquake*.

[11] Dacy and Kunreuther, *Economics of natural disasters*, p. 77.

[12] De Alessi, "Postdisaster cooperation."

[13] Dacy and Kunreuther, *Economics of natural disasters*, p. 72

[14] Cossette, Duchesne and Marceau, "Modeling catastrophes."

[15] Powers and Ren, "Catastrophe risk." Reinsurance is the insurers' insurance, by which a direct insurance company (the cedant) "cedes" or "lays off" part of a risk that it has insured to another company (the reinsurer). This was an option in 1906, but diffusing risks via the capital market was not. It was another 80 years before catastrophe bonds were invented.

[16] Dacy and Kunreuther, *Economics of natural disasters*, pp. 73–74.

insurance market. Investment in rebuilding private business capital will have a greater effect, though smaller insurers will benefit least, as they are the most vulnerable to a few big losses on commercial and industrial property. Large domestic insurers and foreign multinationals with a wide geographical spread of liabilities will be in the best position to take advantage of the increased demand to insure commercial real estate. As a result, during the initial recovery stages, one would expect insurance operations to be confined to the few large companies most likely to have public buildings in their portfolio of risks. Entry or re-entry by most insurers will be delayed until the rebuilding of ordinary residential property is underway and a broad base of insurance demand is established.

Market Factors

The point in the underwriting cycle of insurance supply and demand at which a catastrophe strikes may also determine its impact. At the bottom of this underwriting cycle, in "soft" markets when profits, premiums, and demand are low, a catastrophe has only a limited impact on the share values of insurance companies. The capital market discounts catastrophic losses and values companies with large assets highly, because these are the ones most likely to survive. In an upswing, when profits are rising and demand for insurance is outstripping supply, capital markets treat catastrophic losses as good news. The price inelasticity of demand for insurance after a disaster, and the temporary reduction in underwriting capacity as insurers' funds are depleted, together result in increasing premiums and profits. By contrast, in "hard" markets at the top of the cycle, when premiums and profits peak and demand is saturated, a catastrophic loss impacts negatively on the value of companies.[17]

Institutional Factors

Institutional theory provides a further set of factors that may influence insurers' responses to natural disasters.[18] This states that the particular institutions, or "rules of the game," in any given country can determine the success or failure of international firms in that country. Institutions are defined as, on the one hand, formal rules

[17] Sougiannis, "Cyclical valuation." Cf., however, Christensen's recent finding that market reactions to earnings announcements are stronger in periods of heightened uncertainty about firm value arising from big catastrophes, regardless of the timing of the latter. Christensen, "Effects of uncertainty."

[18] The literature is vast. The seminal study is North, *Institutions*. For an overview in international business, see Dau et al., "Informal institutions." For the theory applied to the sociology of organisations, see Scott, *Institutions and organizations*.

and regulations, enforced, for example, by the state, and, on the other hand, informal unwritten rules that include cultural and ethical values and expected norms of behavior. Formal and informal institutions can reduce uncertainty and transaction costs for firms operating away from home markets because institutions constrain the range of acceptable actions and reduce opportunistic behavior. Institutions, however, only help reduce costs if firms adjust their behavior to fall in line with them. Firms that fail to adjust face increased uncertainty and costs because of institutional constraints, and this may diminish a firm's willingness to make a long-term commitment to a market.

A multinational's strategic choices are shaped by the interaction between the institutions of its home country, which influence the corporate culture of the firm's head office, and the institutions and markets of the host countries where its foreign operations are conducted, which impact the behavior of local agents and branch offices. This interplay of forces can create a misalignment of interests between different units within a multinational organization.[19] This problem may become particularly acute after a major shock to the business such as a natural disaster.

The institutional theory also suggests that foreign firms will adopt whatever practices they believe the formal and informal rules in the host country deem appropriate or legitimate, sometimes regardless of whether these practices increase efficiency or reduce costs. This has been called "the logic of appropriateness."[20] Multinationals, however, are not merely passive actors in the face of institutional pressures. Their strategic responses can range from compliance to resistance or abandoning a market altogether. Compliance may increase a firm's probability of survival and gaining legitimacy; resistance may help to maintain a firm's autonomy over its decision-making; quitting a market allows a firm to circumvent institutional pressures to conform. Firms will be more likely to challenge the rules of the institutional environment when they can demonstrate the rationality or moral probity of their own conduct, when anticipated losses in efficiency from conforming to those rules are high, when the threat to the firm's autonomy in decision-making is high, and when the perceived cost of market exit is low.[21]

In sum, modern theory suggests that the effect of a catastrophic loss on the insurance industry, and, by extension, on the economy of a disaster region, is determined by the interplay of economic, market, company-specific, and institutional factors, as shown in Fig. 1. We return to these points below when we discuss Californian insurance before and after 1906.

[19] Jackson and Deeg, "Comparing capitalisms"; Kristensen and Zeitlin, *Local players*.

[20] March and Olsen, "Elaborating the 'new institutionalism.'"

[21] Oliver, "Strategic responses."

Fig. 1 Factors determining a multinational insurer's response to natural disasters

The Californian Insurance Market Before 1906

Between the American Civil War and the First World War, California developed rapidly as a market for fire insurance. Premiums earned in the state rose from $2.4 m in 1872 to $16.6 m by 1914 (Fig. 2). Although the rest of the United States was catching up, per capita incomes and per capita premiums remained higher in California. With the population growing at over twice the national rate, it is easy to see why the state proved attractive to insurance companies (Table 1).[22] California

[22] Californian population growth was 175 per cent, 1880–1910, compared to 83 per cent for the United States. Calculated from Mitchell, *Historical statistics*.

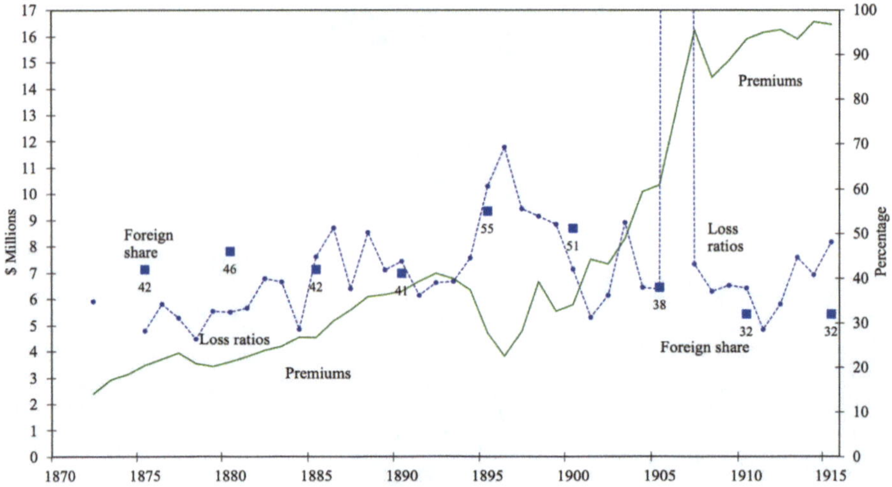

Fig. 2 Fire insurance in California 1872–1915. (Premiums graphed on left axis, loss ratios graphed on right axis. Loss ratios are claims paid as percentage of premiums earned. Values shown indicate share of market held by Foreign companies). Source: Calculated from Annual Reports of the Californian Insurance Commissioner. Note: The aggregate loss ratio reported for 1906 is off the chart at 1094%

Table 1 Insurance penetration: California and the United States 1880–1920

	Per caput premiums $		Per caput	Per caput
	USA	California	premium ratio	income ratio
			California/USA	Pacific region/USA
1880	0.9	4.2	4.7	2.04
1901	2.2	5.1	2.3	1.63
1910	3.1	6.7	2.2	
1920	7.9			1.35

Source: Income ratios calculated from Gallman, "Economic Growth," Table 1.15

had a savings rate much higher than the US average, and it accounted for an increasing proportion of US property income—rising to six percent by 1910.[23]

In 1906, over 90 percent of property owners in San Francisco held an insurance policy.[24] The high levels of insurance penetration were due to a wealthy population and the high premium rates charged. Before the disaster, fire insurance was already more expensive in California than in the United States as a whole (Table 2). Shortly after the earthquake, average premium rates rose to 1.7 times those in the United States. Thereafter through to 1913, rates in California fell by 28 percent, while in the United States, they fell by only nine percent. Insurers who quit the state after the disaster missed out on the short-term upswing in rates between 1906 and 1908. They also avoided the subsequent decline in rates and premium growth.

[23] Odell, "Capital markets"; Klein, "Personal income," Table 9.

[24] Grossi and Muir-Wood, *San Francisco earthquake*, p. 9.

Table 2 Average fire insurance premium rates in the United States and California, 1896–1914

	Stock companies only		
	USA	California	Ratio
1896	1.10	1.13	1.03
1897			
1898	1.01	1.47	1.46
1899	1.00	1.46	1.46
1900			
1901	1.07	1.45	1.35
1902	1.16	1.51	1.30
1903	1.19	1.58	1.33
1904	1.19	1.48	1.24
1905	1.17	1.57	1.34
1906			
1907			
1908	1.17	1.96	1.68
1909	1.17	1.74	1.49
1910	1.12	1.68	1.50
1911	1.11	1.59	1.43
1912	1.10	1.55	1.41
1913	1.06	1.42	1.34
1914	0.97	1.44	1.48

($ premiums per $100 of risks written)
Source: Average rates from *Distribution by States of Fire Insurance* (New York, 1896–1915), corrected for arithmetic errors in the source data for California 1901, 1903, 1905, and 1914

Fire underwriting in California was generally profitable before 1906. As Fig. 2 shows, loss ratios (claims paid as a percentage of premiums earned) only twice rose above 50 percent between 1872 and 1894, and in the years before the disaster they averaged only 40 percent. After the disaster, loss ratios soon returned to pre-earthquake levels, averaging 39 percent between 1907 and 1915. Figure 2 thus shows that on the eve of the disaster fire insurers were operating in a rising underwriting cycle with strong growth and satisfactory profits. The cyclical theory of underwriting outlined above suggests, first, that this sent a positive signal about the prospects of those companies that survived the wave of disaster claims, and, second, that falling loss ratios and limited supply induced re-entry into the market from 1907.

The number of foreign insurance companies with agencies on the Pacific Coast (excluding reinsurers) peaked at 37 between 1885 and 1900 (Table 3). Numbers fell during the 1900s before recovering a little before the First World War. Their share of the Californian market fluctuated between 40 and 49 percent between 1875 and 1890, much higher than their share of the national market (Fig. 2).[25] It rose to 55 percent following the crash of 1893 and the exit of many American insurers. From 1896, as profits improved and US companies returned, foreign market share

[25] Between 1872 and 1880, 26 per cent of net fire premiums in the United States were earned by foreign companies. Calculated from data in *Insurance and Commercial Magazine* 11 (1881), p. 118.

Table 3 Number of insurance companies underwriting fire insurance in California 1875–1915

	Total	Californian	County mutuals	Other US states	Foreign
1875	74	7		50	17
1880	98	8		60	30
1885	110	11		63	36
1890	122	11		74	37
1895	87	3		50	34
1900	106	2	9	58	37
1905	116	3	10	72	31
1910	118	3	17	75	23
1915	152	4	20	99	29

Notes: Californian companies operated statewide, and the county mutuals were local. The 1915 total for other US states includes eight mutual companies, separately enumerated for the first time
Source: Calculated from Annual Reports of the Californian Insurance Commissioner

declined. By the late 1900s, foreign insurers accounted for about one-third of premiums earned in California. In the years before the disaster, therefore, foreign insurers in California were experiencing declining market share and tougher competition. Nevertheless, nowhere else in America were property owners still so dependent upon foreign multinationals for their insurance protection than in the Golden State.

The Earthquake and Fire

The earthquake struck on the morning of 18 April 1906, devastating central San Francisco. Total losses amounted to about $350 m, or in present values about $76.3bn.[26] Over 28,000 buildings, about half of them residential, were destroyed, and fatalities may have exceeded 3000.[27] Only one-third of the damage was the direct result of the earthquake, the rest was caused by fire. The quake destroyed water pipes and hampered fire-fighting. Fires burned for three days and leveled a broad swathe of the city running from the waterfront in a south-westerly direction on both sides of Market Street, engulfing the Mission district and City Hall, and reaching as far south as 20th Street. The intense heat twisted steel frames and destroyed 80 percent of "fireproof" safes, making it difficult to ascertain property

[26] $76.3bn is the relative cost, as measured by GDP per capita, in 2023 US $ of $350 m losses in 1906. Calculated from measuringworth.com

[27] Some writers suggest fatalities between 500 and 700 and a total damage of $500 m. Recent estimates place deaths at over 3000. Cf. www.eyewitnesstohistory.com/sfeq.htm, "San Francisco earthquake"; Douty, *Economics of localized disasters*, p. 81; Phillips, "Coping with financial catastrophe"; Fradkin, *Great earthquake*, p. 234; Whitney, *Insurance settlements*, p. 10; Grossi and Muir-Wood, *San Francisco earthquake*, p. 4; Davies, "Points of origin," 279.

titles and insurance liabilities.[28] Wooden buildings generally survived the tremors better than stone and brick structures, only to provide fuel for the fires that followed.[29] The burned districts had housed up to 250,000 of the city's population of 425,000.[30] Around 100,000 left San Francisco to find shelter and work elsewhere, though many returned after a few months. Most of the remaining homeless were looked after by friends and relatives or were housed in tent camps run by the military in the city's squares and parks.[31] The estimated value of real and personal property in the city in the year before the earthquake was $749 m. The 108 fire insurance companies licensed to operate there insured about $239 m of this, of which at least $200 m was eventually paid out in claims.[32]

The rebuilding of San Francisco broadly fits Dacy and Kunreuther's reconstruction model described above. Aside from the erection of tents and temporary barracks for the homeless, recovery work began first on public utilities, transport, postal, and municipal services.[33] The focus then extended to commercial buildings in the district north of Market Street, while the poorer residential areas south of Market Street were neglected. Within two months, large department stores had begun to move onto lots on Fillmore Street, the Mission District and Van Ness Avenue, and the city's major financial institutions had reopened for business.[34]

It is unsurprising that public and commercial property recovered first. The sequence of rising rents and land values that attracted developers to the burned district could not commence until evacuees returned and demand for permanent housing increased. This took several months. At the end of July 1906, just 66 permanent buildings were under construction, while 4500 temporary structures had been erected. Many decisions to rebuild were taken during the second half of 1906 after

[28] When the San Francisco office of the Swedish company, Svea, burned down, all its policies were destroyed. Svea's overseas manager was quickly despatched to the city, followed by the company's surveyor carrying two large bags of policy copies. Bring, "Historik."

[29] State Earthquake Investigation Commission, *California earthquake*, vol.1, pp. 354–358, and atlas, map 20.

[30] Douty, *Economics of localized disasters*, p. 82. Initial estimates of the homeless were lower. 150,000 was reported by the *New York Times* 25 July 1906.

[31] Douty, *Economics of localized disasters*, pp. 124–131. Estimates of the number who fled the city vary. City boosters tried to refute stories of a mass exodus. One observer stated that 125,000 had left the city in first 30 days after the quake, but most had returned by the end of July. *New York Times* 27 July 1906. Fradkin states over 70,000 left the city. Fradkin, *Great earthquake*, pp. 207–209.

[32] The valuation and liability estimates are those reported in the *New York Times*, 20 April 1906. The final payout figure is that given by Trebilcock, *Phoenix Assurance*, p. 268. Fradkin puts the payout higher at $220–225 m. Fradkin, *Great earthquake*, p. 234.

[33] The following is based on contemporary reports in the *New York Times*, and the accounts by Fradkin, *Great earthquake*, pp. 208–247, and Douty, *Economics of localized disasters*.

[34] *New York Times* 26 July 1906. An eyewitness reported that temporary stores were first put up on "every available lot" on Fillmore Street, and then subsequently on Van Ness Avenue. Dicker, "San Francisco earthquake," p. 62.

which the pace of reconstruction picked up.[35] When developers finally moved into the burned district, they tended to convert land from residential to commercial or industrial use and to increase the density of residential housing by rebuilding fewer single-family dwellings compared to other types of homes.[36] Rebuilding was largely unregulated. New structures were mostly made from the same materials used in those destroyed.[37] The city issued a revised building code in May 1906, but this was widely ignored.[38] There was some street widening, some rebuilding of commercial property in concrete and steel, and an auxiliary water supply was created, but there was no long-term planning.

Most rebuilding was funded through private capital and insurance payments. $73 m had been spent by April 1907 on the rebuilding and repair of property and infrastructure. Just $1.6 m of this came from the city authorities.[39] The process of reconstruction, largely complete by 1910, boosted real estate and shipping services and the lumber industry in the Pacific Northwest.[40] There was, however, a downside to the inflow of insurance capital. Foreign insurers mostly paid their San Francisco claims by drawing on funds held at home, to maintain the liquidity of their American operations. This led to an outflow of gold to the United States and rising interest rates in Britain and Europe, which became factors in the financial panic of 1907.[41] Global economic integration thus ensured that the effects of the disaster reverberated through financial as well as insurance markets around the world.

[35] Douty, *Economics of localized disasters*, pp. 217, 222. At Christmas 1906 a reporter still found "a city of ruins," *New York Times* 6 January 1907. One year after the earthquake there were still 12,000 living in temporary tents and shacks. The last official refugee camp closed in June 1908. Fradkin, *Great earthquake*, pp. 224–225. Kroll-Smith, *Recovering inequality*.

[36] Razed blocks experienced a 62 per cent increase in density relative to unburned blocks by 1914. Siodla, "Razing San Francisco," 56. By 1931 non-residential land share had risen by 12 per cent in burned areas relative to unburned areas. Siodla, "Clean slate," 10.

[37] Wood frame buildings accounted for 88 per cent of those destroyed by fire in 1906, and 89 per cent of those constructed between 1906 and 1914. Calculated from Siodla, "Razing San Francisco," Table 2.

[38] Douty, *Economics of localized disasters*, pp. 207–211. Zoning and building codes enforced by inspection did not emerge until the 1920s in San Francisco. Siodla, "Razing San Francisco," 50–51.

[39] *New York Times* 19 April, 21 April 1907.

[40] Douty, Economics of localized disasters, p. 223.

[41] Noyes, "Panic of 1907"; Odell and Weidenmeier, "Real shock."

Foreign Insurers' Responses to the Disaster: The Europeans

Foreign insurers accounted for about 42 percent of the total insurance loss in San Francisco.[42] Most insurers paid claims worth multiples of their annual net premium income, with the smallest firms hit hardest.[43] The disaster continued to affect subsequent years' results as delayed or disputed claims were finally settled. The Alliance of London, for instance, suffered a 486 percent loss on its US operations in 1907.[44]

Companies reacted in various ways. Three days after the earthquake American and foreign insurers met in Oakland to establish a joint loss adjustment committee. A proposal to reduce all claims payments by a "horizontal cut" of 25 percent became the object of dispute. At a further meeting in New York at the end of May, 32 companies voted against the proposal, including 10 British. They were quickly lauded by the American press as the "dollar for dollar" companies. Sixty-eight voted for the cut, including four German, two Canadian, one Austrian, one Swedish, and one British. These were disparaged as the "six-bit" companies. A further 21 companies abstained or were absent.[45]

By this time, several German insurers had resolved to deny liability and abandon California. Before the earthquake, they had built up portfolios of business often by diluting policy conditions and accepting poorer risks. Their agents in San Francisco were instructed not to adjust losses precipitously and to impose the biggest deductions possible on sums paid to claimants.[46] Most German and Swiss companies persisted in asserting their indemnity from damage by fires directly caused by the earthquake or by explosives used by firefighters to demolish buildings.[47] Eighteen insurers and reinsurers, meeting at Frankfurt in May 1906, called for all companies to "refuse to pay any losses at all, as all originate from the earthquake and the intention of the parties contracting the insurance did not include payment of losses from that cause."[48] This outraged American public opinion. When claimants sued, German companies refused to acknowledge the verdicts of Californian juries and sought more favorable judgments before German courts. Some had their US assets seized by the California insurance commissioner after they refused to accept liability.[49] Some were pursued to Germany by lawyers representing Californian policyholders.

---

[42] Calculated from *Post Magazine* 68 (1907): 433.

[43] The correlation of US premium income to US loss ratios for British companies was −0.6. Calculated from *Australian Insurance and Banking Record* 30 (1906): 729.

[44] *Post Magazine* 70 (1909): 226.

[45] The full list is given in Whitney, *Insurance settlements*, pp. 19–24.

[46] Swiss Re Corporate Archive (hereafter SRCA), Swiss Re, Protocolle der Ausschuβsitzungen, 26 July 1906.

[47] The use of explosives by fire-fighters only accelerated the flames, see Fradkin, *Great earthquake*, pp. 74–79.

[48] London Metropolitan Archive (hereafter LMA), Ms. 18,248/6, Sun Insurance, US letter books, 16 May 1906.

[49] SRCA, Swiss Re, Protocolle der Ausschuβsitzungen, 26 July, 4 September, 5 October 1906.

When American lawyers succeeded in getting a favorable ruling in a Hamburg court against two local firms, Transatlantic and North German, the companies were quickly wound up. Another company, Rhine and Moselle, refused to settle any claims.[50]

This resistance was supported by the reinsurers, who, being involved in a business-to-business activity, did not have to confront American policyholders directly. At Frankfurt, a working party of directors from German and Swiss reinsurance companies resolved to draft a "perfect" earthquake indemnity to be inserted into every policy.[51] Using their connections to some 600 insurance companies worldwide, they analyzed hundreds of exclusionary clauses and drew up a model earthquake clause that they believed was legally watertight. Associations of fire insurers in several European countries then agreed upon a standard clause. In August 1906, following another massive earthquake at Valparaiso, Chile, the Imperial German Insurance Bureau queried whether the earthquake clauses of German companies were sufficiently unambiguous. In response, the leading German reinsurer, Munich Re, renewed the call to adopt the standard earthquake clause.

As discussed below, the efforts to introduce the earthquake clause ran into stiff opposition in the United States. Nor were all European insurers aligned with the Germans. The three Swedish companies operating in San Francisco demonstrated how strategic responses were contingent on individual pre-disaster patterns of business. Two of them, Skandia and Skåne, reinsured for English companies, while Svea operated its own agency in San Francisco for direct underwriting. All suffered heavy losses in the disaster, but each drew different lessons from it.[52]

Skåne's reinsurance income from the United States was small, and the company was only lightly committed to the American market. It was its greater loss from the Valparaiso earthquake that finally propelled Skåne to exit both South and North America.[53] The decision to quit was also influenced by the departure of its ceding partner, Alliance of London, following the latter's huge losses.

Skandia operated a general agency in New York that entailed substantial sunken costs. Withdrawal would have jeopardized this investment. Moreover, the company was convinced that its British reinsurance partner, Royal, would underwrite more cautiously after the disaster, thereby reducing Skandia's exposure in America, and

[50] *New York Times* 22 January 1907. On the cases against German insurers in US and German courts, not all of which ruled in favour of the plaintiffs, see James, "Six bits or bust," 13–14, 19–21.

[51] The following is based on Röder, "Urban catastrophes"; idem, *Rechtsbildung*. See also Rohland, "Earthquake versus fire."

[52] Skandia paid out 2.6 m Kr, about two-thirds of its US premium income. Svea paid out 5 m Kr, 116 per cent of its US premiums. Skåne paid out 1.2 m Kr., or 120 per cent of its non-Scandinavian premium income. Data derived from Skandia corporate archives, June 1906, p. 4 ff.; Ljungqvist, "Historik," p. 53 f.

[53] Lönnborg, "Skandiakoncernens," 240 f. Larsson and Lönnborg, *Omvandlingar*, p. 84 f.

that Californian rates would soon rise. Skandia, therefore, resolved to stay, which would help it continue to spread its risk portfolio widely.[54]

Svea was also dependent on the United States to achieve risk spreading.[55] US premiums accounted for about half of Svea's net insurance income. Average rates in San Francisco were eight times higher than those in Sweden, and Svea's directors expected rates in Sweden to decline. Competition in California was expected to ease as insurers left the market so that a portfolio of first-class risks might be accumulated while premiums were rising. Finally, the costs of exit were likely to be significant. After the disaster, it would be difficult to find a buyer for the business. Any decision to run off existing policies might result in rising losses through a deterioration in moral hazard.

Svea's board also considered the disadvantages of staying. There was the continued risk of major conflagrations. If Svea adopted the "earthquake clause" in the teeth of American and British opposition, that might impair its competitiveness. Competition in California would eventually recover and result in more generous policy conditions and increasing losses. Other difficulties of underwriting in the United States included: a volatile regulatory environment, hostile juries that made foreign companies vulnerable to costly lawsuits, competition for the services of local agents that drove up costs, the difficulty of monitoring agents over long distances, and what Svea believed was a innate moral hazard among American policyholders that derived from their "mentality"—a euphemism for recklessness.

Notwithstanding these concerns, Svea retained its US operations but took some remedial actions. Funds were transferred from Sweden to meet the San Francisco claims, to avoid weakening the company's financial position in the United States.[56] The maximum amount written on single policies was reduced, and more reinsurance was purchased. An earthquake clause was introduced for the western United States, although, as other firms dispensed with the clause, Svea began to restrict the areas to which it applied, and in 1909 it was abandoned altogether. Finally, more detailed agents' instructions were issued and a new scheme was devised to train recruits at the Gothenburg head office, before sending them to work overseas.

[54] The CEO of Skandia, Karl Herlitz, remarked on this decision: "It seems to me ... that what has happened now does not constitute a valid reason for departing from the policy followed in recent years, which is based on the indisputably theoretically correct proposition, that under otherwise equal conditions all insurance becomes safer, the greater the number of risks that it includes. That the San Francisco disaster will exert a beneficial influence on competition can be taken for granted and it would certainly be unwise if Skandia did not seek to be part of reaping its benefits." Herlitz, *Meddelande*, p. 6. [Authors' translation].

[55] The following is based on the separate reports of the overseas manager, C. A. Hjelm, and the company's president, Ernst Bring, given in Skandia corporate archives, Brand- och livförsäkrings Svea, Minutes of the board of directors, 6 April 1907, 5 July 1907.

[56] Lönnborg, Internationalisering.

Foreign Insurers' Responses to the Disaster: The British

British companies accounted for 83 percent of all losses affecting foreign insurers in San Francisco.[57] The traditional narrative of their response to the disaster is a heroic account of company representatives sifting through hot rubble in search of lost policy records in the days following the fire and of bold decisions ordering prompt and full settlement of any claims on the spot.[58] This story was given additional stimulus by informal institutional pressures. These were generated by, for instance, the "Policyholders Protective League," formed as a pressure group in June 1906 to lobby for full payment of claims; the Real Estate Board of San Francisco, which passed resolutions demanding that the disaster be referred to as a "great fire" rather than as an earthquake; and the "Committee of 50," a self-appointed body of business and civic leaders set up to govern the city immediately after the earthquake, which also pressed insurers for prompt and generous payment of claims.[59] In addition, the local press repeatedly fingered those companies regarded as "shaving" their payments or "welching" on their obligations to Californian policyholders.[60] The *San Francisco Express* published a "roll of honor" featuring 39 insurance companies, among them 11 British, which had "met their obligations at once, honorably and without vexatious delays."[61] Other papers compiled a weekly "black list" of companies that did not pay claims in full. The city's chamber of commerce published a report on how insurers had treated policyholders, listing individual companies by name.[62]

There is something in the traditional account. Fifteen out of 18 British insurers reimbursed claimants at 75 percent or higher, while four out of eight European offices paid nothing at all.[63] Some believed that generosity would generate more business.[64] The responses of British insurers, however, were more complex than this account suggests. Like the European firms, several British companies hesitated before admitting liability. Atlas Assurance initially advised its San Francisco manager not to pay any claims that derived from earthquake damage, or from quake-damaged buildings subsequently destroyed by fire or dynamited.[65] By the end of May, however, Atlas had joined the "dollar for dollar" companies. Commercial

[57] Calculated from Australian Insurance and Banking Record 30 (1906): 729.

[58] Liveing, *Century of insurance*, p. 100; Francis, *London and Lancashire*, pp. 61–62; North British and Mercantile, *Centenary*, p. 56; Dickson, *Sun Insurance*, p. 231; Supple, *Royal Exchange*, p. 249.

[59] Fradkin, *Great earthquake*, p. 226; Hudson, "California National Guard,"

[60] San Francisco Examiner 21 Oct. 1906; Post Magazine 68 (1907): 139.

[61] North British and Mercantile, *Centenary*, p. 56; Trebilcock, *Phoenix Assurance*, p. 277.

[62] Whitney, Insurance settlements; San Francisco Examiner 21 Oct. 1906; Svenska Dagbladet 6 Dec. 1906.

[63] Trebilcock, *Phoenix Assurance*, table 3.9, p. 276.

[64] Supple, *Royal Exchange*, p. 249; LMA, Ms. 18,248/6, 8, Sun Insurance, US letter books, 16 May 1906.

[65] Fradkin, *Great earthquake*, pp. 236–238.

Union was served with a lawsuit for rejecting a claim on the basis of an earthquake clause in its policy, but after the Ninth Circuit Court found in favor of the plaintiff, the company decided to make "generous compromises" with all its Californian claimants.[66]

Another London company, Alliance, at first resolved to reject claims for losses resulting directly from the earthquake and to "make the best possible compromise" where such proof was not available. Its agent in San Francisco was informed that "the directors will be much disappointed if the reduction by compromise is not greatly in excess of 30 percent." The agent, however, persuaded Alliance to settle small claims quickly and generously, as "the advantage to the company of full settlement will outweigh any pecuniary saving by enforced compromise as most of the claimants are poor, many in refugee camps, and liberal treatment of them is likely to favorably influence results in other cases." Aware that Californian juries were in no mood to hear technical arguments about liability, Alliance became "anxious to avoid inevitable adverse decisions before dealing with important cases." The strategy was to settle smaller claims after which "other cases will be rigidly dealt with." Several months later, this approach had tied Alliance down in litigation with large claimants, with little prospect of a successful outcome. By this stage, it had already resolved to wind up its entire US operation.[67]

The attitude of British companies to an improved earthquake clause also changed. Shortly after the disaster, the London Fire Offices Committee (FOC) began to draft a new standard clause. To the consternation of the Germans, however, the British suddenly switched course. Under pressure from its members wishing to recommence underwriting in San Francisco as quickly as possible, the FOC abandoned the earthquake clause and instead resolved to charge a supplementary premium for earthquake risk. The Germans rejected this, arguing that no rate could be high enough to offset the potential hazard.[68] Like Svea of Gothenburg, those British companies that did adopt an earthquake clause soon began to ignore it in the face of growing competition on the Pacific Coast.[69]

There were several reasons for this about-turn. There were significant local pressures to insure earthquake-prone risks. As the chapter by Birken in this volume shows, Californians were keen to portray the disaster not as an earthquake but as a conflagration, the risk of which could be controlled through preventative measures and improved building regulation. British insurers were also influenced by new scientific studies of seismic activity that suggested the probability of a similar catastrophe occurring in the near future was low. By contrast, the Germans remained unconvinced that the earthquake hazard was anything other than totally

[66] *New York Times* 16 June 1906, 25 January 1907; James, "Six bits or bust": 17–18; Liveing, *Century of insurance*, p. 100.

[67] LMA, Ms. 12162A/22, Alliance Assurance, Directors' minutes, 23 July, 1 Aug., 31 Oct. 1906; James, "Six bits or bust": 22.

[68] *Assecuranz Jahrbuch* 29 (1908): 101–102, 115–118.

[69] Trebilcock, *Phoenix Assurance*, p. 280.

unpredictable. To try to insure it would be merely a "quixotic gamble" (*abenteurliches Hazardspiel*).[70]

There were also strategic reasons enabling British companies to abandon the standard earthquake clause. As noted above, much of the pressure on companies to stand by their indemnity clauses had come from reinsurers, who were not subject to the same informal institutional pressures in California as their cedents. In the months after the disaster, Swiss Reinsurance Company, for example, pressed the five companies, including two British, for whom it reinsured business in California to deny liability and to obtain the highest deduction possible on any claims paid.[71] The Swiss backtracked, however, when it became apparent that its British partners were going to pay out regardless of any technical indemnity and that unanimity over an earthquake clause could not be reached. Swiss Re needed to maintain the goodwill of cedents and had no desire to sour relations with its English partners in the United States. Instead of trying to enforce the earthquake clause, it resolved to reduce its liabilities on the Pacific Coast but to continue operating there, not least because American reinsurances were so intimately linked to its business around the globe.[72] In 1910, Swiss Re established a US subsidiary, cementing its decision to remain in America.[73]

While European reinsurers worried about the impact of the disaster on their cedent partners, British companies who settled claims generously became concerned about the capacity and willingness of their reinsurers to pay up promptly and in full. The largest British companies reinsured relatively more of their risks than their smaller counterparts and were consequently more dependent upon reinsurers to keep their net losses down.[74] Sun Insurance, for instance, worried about disputes with its reinsurers and resolved that the latter were entitled to their full share of salvage and any discounts for prompt payment. A close watch was kept on the financial condition of the German reinsurance companies, and not just Sun's partners. By October 1906, there was palpable relief that most of its reinsurers were beginning to pay up.[75]

The British experience of the recovering market varied greatly. Some reaped the benefits of increased business at higher rates after the Pacific Coast Board of

---

[70] *Assecuranz Jahrbuch* 28 (1907): 112–113.

[71] An account of Swiss Re's response is given by Rohland, "Earthquake versus fire."

[72] By 1913, 17 per cent of Swiss Re's fire reinsurance business derived from American companies and 23 per cent from British companies. A high (but unknown) percentage of the latter reinsurances were on US risks. SRCA, Guggenbühl, Hausstatistiken (unpublished typescript, Zürich, 1939), p. 62.

[73] SRCA, Swiss Re., Protokolle der Verwaltungsratssitzungen, 2 May, 21 June 1907, 30 Jan. 1908, 14 July 1910.

[74] For British companies the correlation between the proportion of gross insurances affected by losses at San Francisco that was reinsured, and the proportion paid out in claims was −0.82. Calculated from *Australian Insurance and Banking Record* 30 (1906): 729.

[75] LMA, Ms. 18,248/7, Sun Insurance, US letter books, 6 June, 13 July, 3 Oct., 24 Oct. 1906. On the eve of the earthquake, 37 per cent of the risks insured by Sun in San Francisco were reinsured.

Underwriters (PCBU) introduced a 25 percent hike in rates, although the new tariff did not hold for more than a year.[76] Falling loss ratios and rebuilding in San Francisco encouraged the entry of non-tariff insurers such as Lloyd's syndicates and unlicensed companies.[77] In November 1906, Sun observed that the initial effect of the disaster had been "to put all business in the strongest companies," but that this effect was now "ebbing" as policyholders gravitated to other companies.[78] Even companies that anticipated further growth in California did not embark on unbridled expansion. Phoenix Assurance imposed new limits on their American liabilities and only remained in the United States after a fierce debate in the boardroom. Its gross premiums in the United States fell by 17 percent between 1906 and 1908 before recovering.[79] North British and Mercantile, 40 percent of whose losses for the year 1906 were accounted for by the disaster, closed down its San Francisco office in November 1907. Thereafter, its Pacific Coast agencies were managed through its New York office.[80] Some British companies quit the region entirely. Norwich Union ceased to underwrite in California 12 months after the earthquake. Royal Exchange Assurance left California in May 1908, selling its business there to the Home Insurance Company of New York. For the next five years, its US premiums remained below their pre-earthquake level.[81] Alliance closed down all US operations at the end of 1907, running off its liabilities within five years, and was in turn followed out of the United States by its Swedish reinsurer, Skåne.[82]

Conclusions

Why did foreign insurers respond in such a variety of ways to the San Francisco disaster and can such responses be explained by the theoretical framework presented in Fig. 1? Dacy and Kunreuther's recovery production model helps describe the stages of reconstruction in San Francisco, but it provides a poor explanation for insurance company decisions. Figure 3 charts the value of private building contracts issued monthly in San Francisco from May 1906 to April 1910 by type of construction. It shows that repairs and alterations to damaged property (measured by total contracts minus frame and brick contracts) accounted for a much higher proportion of all construction in the second half of 1906 than subsequently and that the value

[76] *Post Magazine* 68 (1907): 203.

[77] *Post Magazine* 68 (1907): 268.

[78] LMA, Ms. 18,248/8, Sun Insurance, US letter books, 13 Nov. 1906

[79] Trebilcock, *Phoenix Assurance*, pp. 279–282.

[80] *Post Magazine* 68 (1907): 621. Loss figures calculated from Aviva archive, North British and Mercantile, General Court Minutes, fol.35, Comparative statement of profit and loss for 1905 and 1906.

[81] *Post Magazine* 68 (1907): 308; *ibid.* 69 (1908): 439; Supple, *Royal Exchange*, p. 250.

[82] LMA, Ms. 12162A/26, Alliance Assurance, Directors' minutes, 11 Nov.1912; Larsson and Lönnborg, *Omvandlingar*.

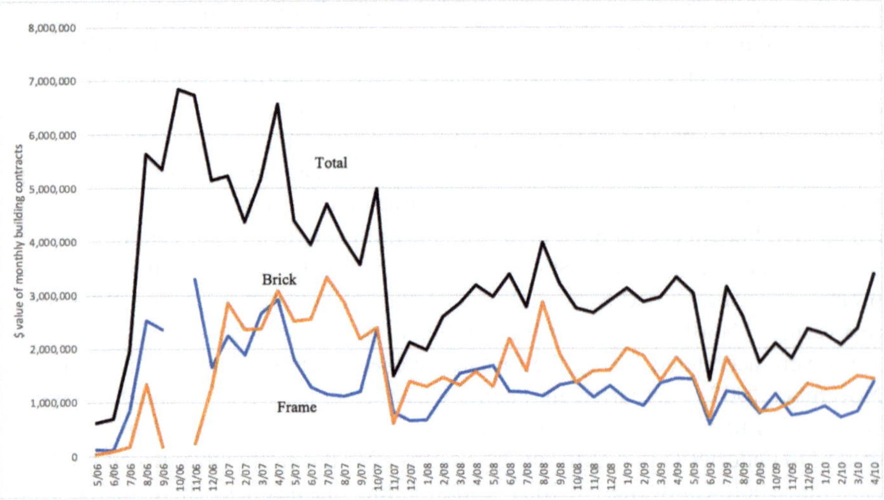

Fig. 3 San Francisco private building contracts, monthly, by type of construction, $ values, May 1906–April 1910. Source: Douty, *Economics of localized disasters*, appendix table A, pp. 393–394

of contracts for brick buildings did not overtake that for wood-frame property before the end of 1906.[83] In other words, the end of the repair phase and the peak of reconstruction in the safest materials occurred after most foreign insurers had decided to quit or stay and after most reinsurers had begun to pay their cedents' claims.[84] This evidence of timing, and the paucity of any discussion of reconstruction in insurance boardrooms, suggest that the rebuilding process itself, the nature of the materials used, the lack of effectiveness of building codes, and the increasing density of rebuilt housing, had, at most, a marginal impact on insurers' strategic decisions.

Risk-related market and economic factors, rather than the rebuilding process, were more important. A catastrophe of such high intensity, generating unprecedented losses and a reduction in financial strength, coupled with highly interdependent insurance in the disaster region, heightened concern about future risks and spelled bankruptcy or exit for some foreign companies. The underwriting cycle also played a role in determining insurers' decisions. The disaster struck at a point when premium revenue was growing strongly, when loss ratios were declining and foreign market share was shrinking. This helps explain why, in the aftermath, many insurers expected premium rates and demand to rise together, and that competition would subsequently intensify with the re-entry of large American firms. For those

[83] Alterations averaged 13 per cent of all building contracts by value from the earthquake through to the end of 1906, and peaked in August and December at nearly one-quarter. In 1907 and 1908 however, alterations averaged only four per cent of the total. Calculated from Douty, *Economics of localized disasters*, appendix table A, pp. 393–394.

[84] Alliance resolved in October 1906 to withdraw, although its US winding up operations were not completed until the end of 1907. A similar lag occurred in other companies quiting the market. Other companies such as Sun and Svea made their decisions to stay before the end of 1906.

companies that resolved to stay in California, any concern about the immeasurability of earthquake risk was outweighed by the belief that this was historically a profitable high-volume market worth being involved in.

Institutional factors were also important in shaping insurers' responses. The formal rules under which insurance companies operated in California before 1906, such as the requirement to make a deposit of funds, obtain a license, and submit an annual report, were not onerous compared to those elsewhere and were "in general quite vague" and easily evaded.[85] The state insurance department was chronically underfunded and incapable of closely monitoring the hundreds of companies active in the state.[86] In 1897, California's attorney general declared that the state insurance commissioner had no discretionary authority to expel any insurance company from the state, and was required to license any that complied with the law.[87] Only after the disaster did the state finally pass a law that introduced a more rigorous procedure for companies applying to wind up operations and gave the commissioner the power to revoke the license of any insurer that sought to move Californian lawsuits to other jurisdictions.[88] This new regulatory framework, however, arrived too late to affect the post-disaster strategic decisions of foreign insurers, which, as noted above, had mostly been made by the end of 1906.

Informal institutional pressures were much more significant. Birken's chapter in this volume shows how city boosters, media commentators, and politicians renarrated the disaster to expunge the earthquake and retell it as a manageable fire event, rather than an unpredictable natural disaster, and as an opportunity for the renewal of the city. California's insurance commissioner himself claimed that the earthquake had caused minimal damage, no more than three percent of the total loss.[89] This attitude also echoes the accounts given by Garrido and Rauscher elsewhere in this volume of the particular American approach to natural disasters: the emphasis on displays of local sacrifice and heroism and the ultimate triumph of local communities over adversity; the reluctance to draw appropriate lessons from disasters and to develop resilience strategies in order to reduce future exposure.

Historians who have explored American attitudes to risk have argued that these were shaped by two opposing cultures: a "vernacular risk culture" of chance that believed in a contingent universe where luck mattered and that was at ease with randomness and irrationality; a positivist culture of control that believed that risk could be managed and even eradicated through science, technology, education, and regulation. The latter culture rose to prominence during the Gilded Age, but never

---

[85] This was the verdict of the insurance commissioner, E. Myron Wolf, in 1907. State of California, *38th Annual Report of the Insurance Commissioner,* p. 48. A similar approach was seen in Californian regulation of banking, where entry was virtually unrestricted between 1864 and 1908. Lister, "Bank behaviour."

[86] For complaints of underfunding, see State of California, 27th Annual Report of the Insurance Commissioner, p. 1; 38th Annual Report of the Insurance Commissioner, p. 51.

[87] *Insurance Critic* 25 (1897): 425.

[88] State of California, 39th Annual Report of the Insurance Commissioner, pp. 29–32.

[89] Steinberg, *Acts of God,* chapter 2, fn 37.

entirely replaced the former. Notwithstanding the great efforts made by professional and commercial groups, including engineers and insurers, from the 1880s to improve safety at work, at home, in public places, and on the railways, the old trope about the "reckless and careless habits" of the natives, continued to influence some foreign underwriters in their views of American risk, as we have seen above in the case of the Svea of Gothenburg.[90] The hectic and unplanned way in which San Francisco was rebuilt was hardly designed to alter such views.

Those foreign companies with a large stake in the United States, however, could not easily allow any prejudices about the supposed American predilection for risk-taking to determine their strategic decisions in post-earthquake California. They quickly became acutely aware of the bad publicity and hostile juries that awaited them if they challenged the heroic narrative of the disaster, and tried to defy popular expectations of correct and moral conduct by contesting the claims of either the poorer policyholders camped out in the city parks or those of wealthy business interests. This considerably narrowed the scope for reducing costs for those insurers wishing to remain in the market. Such institutional pressures also meant that in the weeks and months following the earthquake insurance companies were dynamic in their responses to the rapidly changing market conditions in their industry. This undoubtedly increased the unpredictability for Californian customers seeking insurance, and yet the statistical evidence presented in Tables 1 and 2 and Fig. 2 indicate that, despite the higher premium rates being charged, premium volume and insurance penetration (insurance per capita) continued to grow in the Golden State. This does not suggest that Californians began to reevaluate their trust in insurance providers or that they embarked on greater risk-taking and self-insurance. On the contrary, their appetite to seek formal insurance protection and willingness to pay high premiums remained undiminished by the disaster.

The evidence suggests that the post-disaster responses of insurers were also contingent upon company-specific pre-disaster patterns of business, strategic alliances with other insurers and reinsurers, capital strength, managerial preferences, and planning decisions. The importance of a company's access to reinsurance in offsetting the interdependency of risks insured in the disaster region was reflected in the concern of the British about the solidity of their European reinsurers and in the way reinsurers' losses reverberated around the world. San Francisco losses forced several German reinsurers to suspend operations in Russia, which in turn drove at least one Russian insurance company into bankruptcy.[91] Insurers in Denmark, Austria,

[90] The phrase was used in a British article about fire risks in the United States, *Post Magazine* 32 (1871): 341, one of many contemporary examples of such sentiments. On the safety movement in America and attitudes to risk-taking, see Mohun, *Risk*; Tebeau, *Eating smoke*; Wermiel, *Fireproof building*; Lears, *Something for nothing*.

[91] St Petersburg Insurance Company went bankrupt due to a loss of 790,000 roubles on San Francisco retrocessions accepted from Cologne Reinsurance Company. *Assecuranz Jahrbuch* 29 (1908): 332, 336. A retrocession is a reinsurance risk insured, or "retroceded," by a reinsurer with another insurer.

and Switzerland also suffered soaring losses due to reinsurance or retrocession liabilities from California.[92]

The degree to which an insurance company depended on its reinsurance partners, and vice-versa, underlined the need for each side constantly to monitor the solidity of their reinsurance arrangements. Power relations in reinsurance markets varied greatly by country and company. The largest British insurance exporters had considerable bargaining strength, which helps explain why they felt able to resist German demands for a standard earthquake clause and to adopt a liberal rather than legalistic approach to claims settlements in California that satisfied local expectations. With the British so dominant in global fire insurance before 1914, reinsurers from small home markets such as Switzerland, Sweden, or Russia depended disproportionately on British businesses to help spread their risks.[93] Swiss Re drew back from insisting upon the imposition of the earthquake clause because of the harm this was doing to its relations with British companies. Smaller European companies such as Skandia, however, also worried about the quality of the risks ceded to them by their British partners. By contrast, with a large domestic market to fall back upon, German reinsurers, such as Munich Re, had less need for British custom.[94]

Company-specific lock-in effects also influenced decisions to stay or quit. The high ratio of expenses to income in the United States meant that a large volume of insurance was required in order to achieve a good return. Small portfolios were susceptible to sudden losses from a few major fires that could consume profit streams. Low premium volume, coupled with the concentration of risks in a limited geographical area, reduced the opportunities to take advantage of the law of large numbers. Thus, there were major incentives for foreign insurers in the United States to sell as much insurance as possible. The higher the share of a market in a company's total revenue, the more it was locked into that market, since a large reduction in income would diminish economies of scale. This, together with the institutional memory of surviving past disasters, were important reasons for some companies such as Svea of Gothenburg and Phoenix of London deciding to remain in the United States after the San Francisco disaster.[95]

Exit entailed costs, which reinforced lock-in effects, especially for the larger foreign insurers. Existing policies might be run off and agencies closed, but losses tended to rise quickly during run-offs since the monitoring of claims by agents became less rigorous. Moreover, the number and size of claims tended to increase when policyholders learned that a company was about to quit a market. Another exit strategy was to reinsure the portfolio and close down agencies once the reinsurance contract expired, but in the aftermath of a disaster, it could be difficult to obtain full

[92] *Ibid.* 28 (1907): 23; *ibid.* 29 (1908): 231, 316–317.

[93] On the eve of the First World War, nearly one quarter of Swiss Re's income from fire reinsurance derived from British insurers. See footnote 72 above.

[94] Borscheid, "Vertrauensgewinn."

[95] Thirty-six per cent of net premiums earned by Phoenix in 1905 came from the United States. This figure was over 50 per cent in other companies that stayed, such as Commercial Union. Trebilcock, *Phoenix Assurance*, p. 230, Table 2.20.

value for a company's risks. One alternative was to reduce exposure in the most hazardous segments of a market by introducing more restrictive policy conditions. The fate of the earthquake clause, however, demonstrated how counter-productive this could be in a post-disaster market where institutional pressures were high and competition was increasing.

Postscript

The San Francisco disaster did not reverse the decline in the foreign share of the US property insurance market.[96] Before 1906, foreign insurers operating on the Pacific Coast were already making greater use of reinsurance. After the disaster, new European reinsurers stepped in quickly to meet the demand of insurers still operating in America.[97] Notwithstanding the many disputes over the "welching" companies, the disaster was generally a triumph for the international insurance industry. Unlike the Chicago fire of 1871, when hundreds of insurance companies had gone bust, relatively few companies failed after San Francisco even though the level of loss was far greater. The difference was the enormous development of reinsurance capacity since 1871, which enabled insurers to distribute their Californian risks around the world.

Nor did the disaster lead to fundamental changes in the social structure and spatial distribution of San Francisco. Thanks to the speed and scale of insurance payments, relatively little real estate changed hands after the disaster and broad patterns of residence, work, and ethnicity remained intact.[98] The disaster also did not produce major alterations in Californian property insurance. Most changes in premium rates and policy conditions were short-lived and the local market returned to one characterized by generous policy terms and little scope for insurers to protect themselves with the aid of restrictive clauses. In sum, long-run economic forces, and that particular dualistic American attitude to risk noted above, reasserted themselves.

References

Anderson, D. R. 2000. Catastrophe Insurance and Compensation: Remembering Basic Principles. *Society of Chartered Property and Casualty Underwriters. CPCU Journal* 53:76–89.

Baranoff, D., 'Shaped by Risk: The American fire Insurance Industry 1790–1920', unpublished PhD thesis, John Hopkins University, 2003.

Borscheid, P. 2001. Vertrauensgewinn und Vertrauensverlust: Das Auslandsgeschäft der deutschen Versicherungswirtschaft 1870–1945. *Vierteljahrschrift für Sozial- und Wirtschaftsgeschichte* 88:311–345.

[96] Cf. Figure 1 in Pearson, "British and European insurance enterprise."

[97] *Post Magazine* 67 (1906): 767, 923; *ibid.* 69 (1908): 399.

[98] Davies, "Points of origin"; Douty, *Economics of localized disasters*, pp. 205, 213–216; Siodla, "Clean slate," 12.

Bring, E., 'Historik', in *Brand- och Lifförsäkrings AB Svea. Minnesskrift 1866–1916* (Göteborg, 1917).
Christensen, T. E. 2002. The Effects of Uncertainty on the Informativeness of Earnings: Evidence from the Insurance Industry in the Wake of Catastrophic Events. *Journal of Business, Finance and Accounting* 29:223–255.
Cossette, H., T. Duchesne, and É. Marceau. 2003. Modeling Catastrophes and Their Impact on Insurance Portfolios. *North American Actuarial Journal* 7:1–22.
Dacy, D. C. and Kunreuther, H., *The Economics of Natural Disasters: Implications for Federal Policy* (New York, 1969).
Dau, L. A., A. S. Chacar, M. A. Lyles, and J. Li. 2022. Informal Institutions and International Business: Towards an Integrative Research Agenda. *Journal of International Business Studies* 53:985–1010.
Davies, A. R., 'Points of Origin: The Social Impact of the 1906 San Francisco Earthquake and Fire', in G. Bankoff, U. Lübken and J. Sand eds., *Flammable Cities: Urban Conflagration and the Making of the Modern World* (Madison, 2012), pp. 273–294.
De Alessi, L. 1975. Toward an Analysis of Postdisaster Cooperation. *American Economic Review* 65:127–138.
Dicker, L. M. 1980. The San Francisco Earthquake and Fire: Photographs and Manuscripts from the California Historical Society Library. *California History* 59:34–65.
Dickson, P. G. M., *The Sun Insurance Office, 1710–1960* (London, 1960).
Douty, C. M., *The Economics of Localized Disasters: The 1906 San Francisco Catastrophe* (New York, 1977).
Fradkin, P. L., *The Great Earthquake and Firestorms of 1906: How San Francisco Nearly Destroyed Itself* (Berkeley and Los Angeles, 2005).
Francis, E. V., *London and Lancashire History* (London, 1962).
Froot, K. A., ed., *The financing of Catastrophe Risk* (Chicago, 1999).
Gallman, R. E., 'Economic Growth and Structural Change in the Long Nineteenth Century', in S. L. Engerman and R. E. Gallman, eds., *The Cambridge History of the United States: volume II – the Long Nineteenth Century* (Cambridge, 2000), pp. 1–56.
Grossi, P. and Muir-Wood, R., *The 1906 San Francisco Earthquake and Fire: Perspectives on a Modern Super Cat* (Newark, 2006).
Herlitz, K., '*Meddelande från Försäkringsaktiebolaget Skandia*', June 1906.
Hudson, J. J. 1976. The California National Guard in the San Francisco Earthquake and Fire of 1906. *California Historical Quarterly* 55:137–149.
Jackson, G., and R. Deeg. 2008. Comparing Capitalisms: Understanding Institutional Diversity and Its Implications for International Business. *Journal of International Business Studies* 39:540–561.
James, R. A. Six Bits or Bust: Insurance Litigation Over the 1906 San Francisco Earthquake and Fire. *Western Legal History* 24 (2011): 1–39.
Insurance Services Office, *Managing Catastrophe Risk*, (ISO Insurance Issues Series, May 1996).
Klein, A., 'Personal Income of US States: Estimates for the Period 1880–1910', University of Warwick, Warwick Economic Research Papers, no. 916 (September 2009).
Kristensen, P. H. and Zeitlin, J., *Local Players in Global Games: The Strategic Constitution of a Multinational Corporation* (Oxford, 2005).
Kroll-Smith, S., *Recovering Inequality: Hurricane Katrina, the San Francisco Earthquake of 1906, and the Aftermath of Disaster* (Austin, 2018).
Larsson, M. and Lönnborg, M., *Omvandlingar. Försäkringsbolaget Skandia 1990–2016 [Transitions. The Insurance Company Skandia 1990–2016*] (Stockholm, 2019).
Lears, Jackson, *Something for Nothing: Luck in America* (London, 2003).
Lister, R. C. 1983. Bank Behaviour, Regulation and Economic Development: California 1864–1910. *Journal of Economic History* 43:284–286.
Liveing, E., *A Century of Insurance: the Commercial Union Group of Insurance Companies, 1861–1961* (London, 1961).
Ljungqvist, G., 'Historik [History]', in *Försäkringsaktiebolaget Skandia 1855–1930 [The insurance company Skandia 1855–1930]*, part 1, (Stockholm 1930).

Lönnborg, M., *Internationalisering av svenska försäkringsbolag. Drivkrafter, organisering och utveckling 1855–1913* (Uppsala, 1999).

Lönnborg, M. 2002. Skandiakoncernens Internationella verksamhet 1887–1995 [The international business of Skandia 1887–1995]. *Scandinavian Insurance Quarterly* 2:237–256.

March, J. G. and Olsen, J. P., 'Elaborating the 'New Institutionalism', in Sarah A. Binder, R. A. W. Rhodes and Bert A. Rockman eds., *The Oxford Handbook of Political Institutions* (Oxford, 2008), pp. 3–20.

Mitchell, B. R., *International Historical Statistics: The Americas 1750–2000* (5th edition, Basingstoke, 2003).

Mohun, Arwen P., *Risk: Negotiating Safety in American Society* (Baltimore, 2013).

North, D. C., *Institutions, Institutional Change and Economic Performance* (Cambridge, 1990).

North British and Mercantile Insurance Company, *Centenary, 1809–1909* (Edinburgh, 1909).

Noyes, A. D. 1909. A Year After the Panic of 1907. *Quarterly J. of Economics* 23:185–212.

Odell, K. A. 1989. Integration of regional and interregional capital markets: evidence from the Pacific Coast, 1883-1913. *Journal of Economic History* 49:297–310.

Odell, K. A. and Weidenmeier, M. D., 'Real shock, monetary aftershock: the San Francisco earthquake and the panic of 1907', *NBER Working Paper*, no. 9176, (Washington, 2002).

Oliver, C. 1991. Strategic responses to institutional processes. *Academy of Management Review* 16:145–179.

Pearson, R. 1997. British and European insurance enterprise in American markets, 1850-1914. *Business and Economic History* 26:438–451.

Phillips, R. J. 2003. Coping with financial catastrophe: the San Francisco clearinghouse during the earthquake of 1906. *Research in Economic History* 21:79–104.

Powers, M. R., and J. Ren. 2003. Catastrophe risk and insurer solvency: a diffusion-jump analysis. *Assurances et Gestion des Risques* 71:239–264.

Ratajczak, D. 1994. What catastrophes do to the economy. *Journal of the American Society of CLU & ChFC* 48:22–24.

Rohland, E., 'Earthquake versus fire: The struggle over insurance in the aftermath of the 1906 San Francisco disaster', in Janku, A., Schenk, G. J. and Mauelshagen, F., eds., *Disasters in Context: Science, Religion and Politics* (New York, 2011), pp. 174–194.

Röder, T. J., *Rechtsbildung im wirtschaftlichen 'Weltverkehr': die internationale Standardisierung von Vertragsbedingungen, 1871–1914* (Frankfurt am Main, 2005).

Röder, T. J., 'Urban catastrophes: patterns of perception and reaction', unpublished paper presented to the EABH conference on Risk and Disaster, Swiss Re Centre for Global Dialogue, Zurich, 2006.

Siodla, J. Razing San Francisco: the 1906 disaster as a natural experiment in urban redevelopment. *Journal of Urban Economics* 89 (2015): 48–61.

Siodla, J. Clean slate: land-use changes in San Francisco after the 1906 disaster. *Explorations in Economic History* 65 (2017): 1–16.

Sougiannis, T. 1997. The cyclical valuation of the catastrophic losses in the insurance industry. *Journal of Financial Statement Analysis* 3:32–43.

Steinberg, T., *Acts of God: The Unnatural History of Natural Disaster in America* (New York, 2006).

Supple, B. 1970. *The Royal Exchange Assurance: a history of British insurance, 1720–1970*. Cambridge.

State of California, *27th Annual Report of the Insurance Commissioner of the State of California* (Sacramento, 1895).

State of California, *38th Annual Report of the Insurance Commissioner for the State of California* (Sacramento, 1907).

State of California, *39th Annual Report of the Insurance Commissioner for the State of California* (Sacramento, 1908).

State Earthquake Investigation Commission, *California earthquake of April 18 1906: report* (Washington, 1908), 2 volumes.

Swiss Re, *A Shake in Insurance History - the 1906 San Francisco Earthquake* (Zurich, 2006).

Tebeau, Mark, *Eating Smoke: Fire in Urban America, 1800–1950* (Baltimore, 2003).
Trebilcock, C., *Phoenix Assurance and the development of British insurance*, vol. 2, (Cambridge, 1998).
Wermiel, Sara E., *The Fireproof Building: Technology and Public Safety in the Nineteenth-Century American City* (Baltimore, 2000).
Whitney, A. W., *On insurance settlements incident to the 1906 San Francisco fire* (1906, reprinted Pasadena, 1972).
"The San Francisco earthquake, 1906, eyewitness – history through the eyes of those who lived it." www.eyewitnesstohistory.com/sfeq.htm

Oil Ports and Hurricanes Along the Texas Gulf Coast

Alan Lessoff

Abstract This chapter considers vulnerability to hurricanes as a factor that over the twentieth century shaped the Texas Gulf Coast as a region of oil and gas ports. Projects to open Texas's shallow bays and estuaries to ocean shipping preceded the 1901 Spindletop strike, which set in motion Texas's transformation into a global hub for the energy sector. Nevertheless, organizers of the early Texas oil industry quickly identified three major deep-water channels built in the 1910s–1920s—the 52-mile Houston Ship Channel, the Sabine-Neches Waterway at Beaumont-Port Arthur, and the 20-mile Corpus Christi Ship Channel—as exemplary locations for the processing and shipping stages of their enterprise. These waterways, along with shorter channels added later at Freeport, Matagorda Bay, and Brownsville, offered ample space for freight yards, pipelines, storage tanks, refineries, loading docks, petrochemical plants, equipment fabricators, and other support industries. Storm protection provided an argument for the arduous, expensive extension of navigation inland. Yet industries and port authorities repeatedly rationalized exposed locations close to the coast when business strategy seemed to demand this. As recent storms from Ike in 2008 to Harvey in 2017 and Beryl in 2024 portend, the concentration of Texas's industrial petroleumscape along dredged channels has created conditions for environmental cataclysm. The continent's largest, most strategic petroleum and petrochemical complex becomes ever more vulnerable to storm surges and flooding, whose frequency and intensity increase because of carbon-driven climate

The author thanks Professor Carola Hein of the TU-Delft for arranging the funding that underwrote this work and for welcoming him into her research group through the coronavirus pandemic and beyond. The author also thanks his Illinois State University colleague Keith Pluymers for pointing him toward one of this chapter's driving ideas. Portions of this essay build upon Alan Lessoff, "The Texas Coast: Ship Channel Network of the Petroleum Age," *Urban Planning* 8, n. 3 (2023): 330–345, published by Cogitatio Press under a Creative Commons BY license. The author thanks Tiago Cardoso and Stephen Ramos for help with that earlier work.

A. Lessoff (✉)
Illinois State University, Normal, Il, USA
e-mail: ahlesso@ilstu.edu

N. Rauscher, W. Werner (eds.), *Natural Disasters in the United States*, Risk, Governance and Society 22, https://doi.org/10.1007/978-3-031-96436-7_11

change. Oil and gas made the modern Gulf Coast but threatened what it made. The globalized character of the oil industry, of course, means that this bleak dynamic looms over port cities on every continent. Much US discussion of the exposure of people and property to coastal disasters focuses on residential or resort development in Florida and elsewhere. Along the Texas Gulf Coast, the key factor drawing investment and population into harm's way has been the fossil fuel energy regime and its commercial and industrial infrastructure.

"Horrific Future Scenario"[1]

On August 25, 2017, Hurricane Harvey made landfall as a category 4 storm on San Jose Island, one of seven barrier islands that line the Texas coast. The picturesque coastal towns of Rockport and Fulton bore the brunt as the storm moved ashore. Corpus Christi, 30 miles to the south, was spared massive damage. Casualty figures from Harvey for Texas vary from 68 to 103, depending on how they are measured. In either case, Harvey would cause the most hurricane-related deaths in the state since September 14, 1919, when a 12-foot surge washed over central Corpus Christi. The official toll from that earlier disaster was 284, though the number of people killed may have been hundreds higher. As this chapter relates, the 1919 disaster catalyzed the construction of the Corpus Christi Ship Channel, which opened in 1926. Corpus Christi's was the third of Texas's great ship channels after Houston's, which opened in 1914 and the Sabine-Neches Canal, constructed mainly in the 1910s. By 1941, Corpus Christi would also receive a seawall 15 feet high and 2 miles long, counterpart to the 17-foot seawall extended along the gulf side of Galveston Island after its catastrophic 1900 storm[2] (see Fig. 1).

As Harvey made landfall, storm surges of 6 feet or more and gusts over 130 miles per hour battered Port Aransas at the north end of Mustang Island, entry point to the Corpus Christi Ship Channel. Winds and floods struck the coastal towns of Aransas Pass and Ingleside, as well as the Port of Corpus Christi's La Quinta Channel. Starting in the 2010s, these places had seen heavy investment in energy processing, storage, and export facilities related to the Eagle Ford shale oil field about halfway between Corpus Christi and San Antonio. Refineries and petrochemical plants along the ship channel's inland turning basins suffered minimal damage. Port Lavaca, about 85 miles to the north, another notoriously vulnerable point where industrial

[1] Jim Blackburn, *A Texan Plan for the Texas Coast* (College Station: Texas A&M University Press, 2017), 92.

[2] Eric S. Blake and David A. Zelinsky, *Tropical Cyclone Report: Hurricane Harvey (AL092017), 17 August—1 September 2017*, National Hurricane Center, May 9, 2018, 1–9; National Weather Service, "Major Hurricane Harvey—August 25–29, 2017," www.weather.gov/crp/hurricane_harvey (accessed Jan. 22, 2024); National Weather Service, "The Hurricane of 1919," www.weather.gov/crp/hurricane1919 (accessed Dec. 28, 2023).

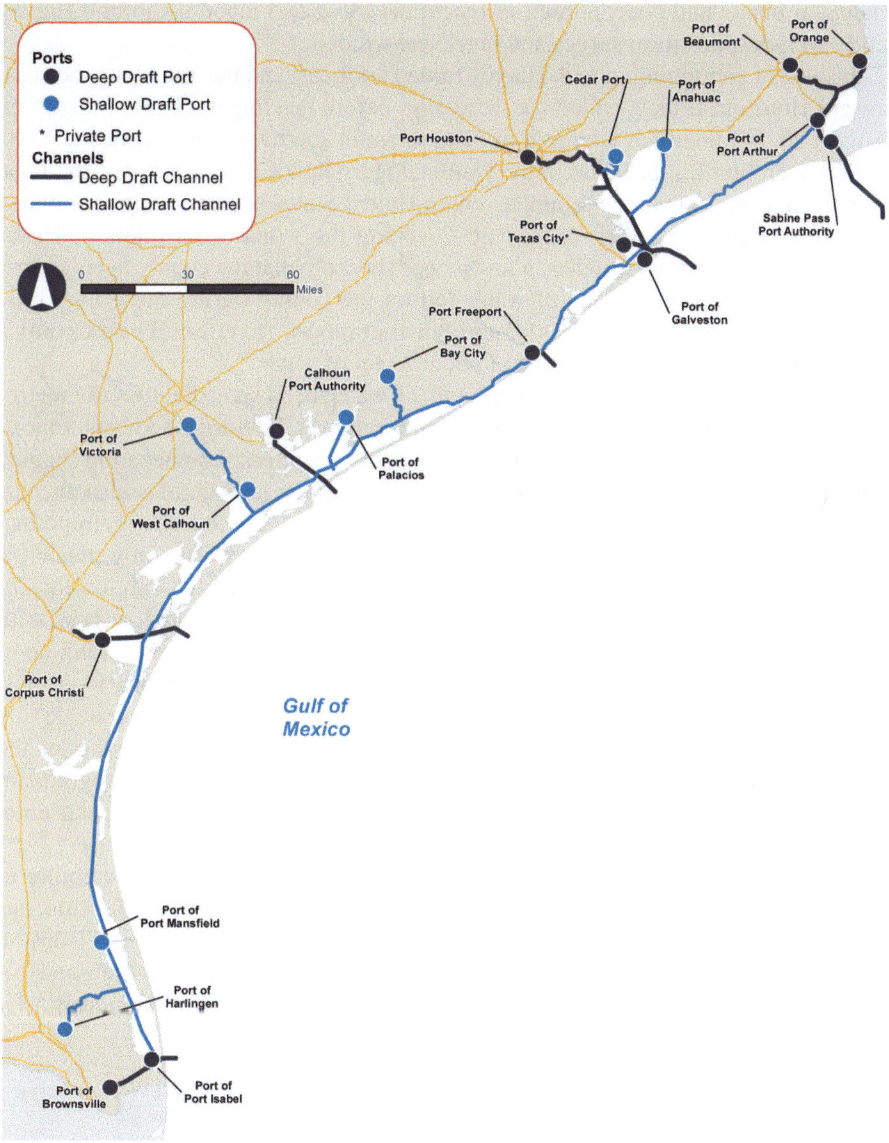

Fig. 1 Ship channels and ports along the Texas Gulf Coast. After the Houston Ship Channel opened in 1914, the Texas coast gained five more deep-water ship channels over the next half century. Other than Houston, the most significant for the oil-and-gas and petrochemical sectors are the Sabine-Neches Waterway in the upper right and the Matagorda Bay (Calhoun County) and Corpus Christi channels toward the middle south. The Gulf Intracoastal Waterway extends behind the chain of barrier islands, connecting these and smaller ports.]. Source: Texas Department of Transportation, https://ftp.dot.state.tx.us/pub/txdot-info/tpp/giww/map_tx_ports.pdf

chemical plants had concentrated in recent decades, also made it through Harvey fairly well despite storm surges and hurricane winds.[3]

Instead of continuing inland, Harvey turned northeast and tracked along the coast toward Houston. It then stalled for three days before heading back into the gulf on August 28, eventually to land as a tropical storm in southwestern Louisiana. While stalled over Texas, it became, in the National Hurricane Center's words, "the most significant tropical cyclone rainfall event in United States history." Rains continued for five straight days, with August 26–27 being the worst. Rain totals in Harris County averaged over 45 inches, a few inches short of what the county averages per year. About 1 trillion gallons of water fell on the county during these five days, "enough," remarked a local flood control district report, "to cover Harris County's 1777 square miles with an average of about 3 feet of water."[4]

Sited on a low-lying coastal plain, with 23 watersheds spread across the metropolitan area that feed into creeks, bayous, and wetlands, Houston had been prone to flooding since its founding. State laws that supported aggressive annexation enabled the city to expand to nearly 640 square miles. Suburban jurisdictions also encouraged headlong development, so that by the time Harvey arrived, the metropolitan statistical area covered 9444 square miles over nine counties. The only major US city to have no zoning regulations, Houston permitted residential subdivisions to spread through the landscape, with development loosely coordinated, at best, with flood plans and drainage systems. Commercial and office complexes sprang up in satellite business districts or a miscellany of strip-style structures along roadways. This diffuse, multicenter, auto-oriented pattern suits a city with such a strategic role in fueling internal combustion engines. Other chapters in this book consider the perverse consequences of flood control systems and flood insurance programs in magnifying damage from urban flooding while hampering recovery. For those not yet alert to the problem, Harvey revealed Houston to exemplify it.[5]

Deforestation and the filling of wetlands diminished the landscape's ability to absorb water passing through even in typical rainfall years. "As Houston has expanded in almost all directions," explained Jim Blackburn, the renowned Houston environmental lawyer and policy expert, "we covered hundreds of thousands of acres with concrete and highly efficient drainage where there was previously very

[3] Blake and Zelinsky, *Tropical Cyclone Report: Hurricane Harvey*, 10; National Weather Service, "Major Hurricane Harvey—August 25–29, 2017."

[4] Blake and Zelinsky, *Tropical Cyclone Report: Hurricane Harvey*, 6; also Merritt Kennedy, "Harvey The 'Most Significant Tropical Cyclone Rainfall Event In U.S. History,'" Jan. 25, 2018, www.npr.org (accessed Jan. 23, 2024). Harris County Flood Control District, *Hurricane Harvey: Impact and Response in Harris County*, May 2018, 9; Jim Blackburn, *Houston Flooding 3.5 Years after Harvey* (Houston: Baker Institute for Public Policy, Rice University, 2021), 3.

[5] Anna Rhodes and Max Besbris, *Soaking the Middle Class: Suburban Inequality and Recovery from Disaster* (New York: Russell Sage Foundation, 2022); Jake Bittle, "They Were Proud Houston Homeowners. Then It All Fell Apart," *New York Times*, Feb. 4, 2023; Shawn Boburg and Beth Reinhard, "Houston's 'Wild West' Growth," *Washington Post*, Aug. 29, 2017. Jim Blackburn and Philip B. Bedient, *Houston a Year after Harvey: Where We Are and Where We Need to Be* (Houston: Baker Institute for Public Policy, Rice University, 2018), 4–12.

little. We converted natural sponges" into impermeable surfaces "generating much greater runoff than before." Blackburn reiterated concerns expressed regularly over the preceding decade, for example in a 2015 Delft Technical University study. The Dutch researchers had warned that metropolitan Houston's "vast swaths of impervious surfaces and widespread loss of wetlands" meant that during heavy storms, "normally languid" bayous—especially those lined with concrete and "channelized" into drainage canals—would "quickly morph into raging rivers," sending flood water downstream at high speed.[6]

Indeed, during Harvey, the threat of dam failures compelled the Army Corps to release water from reservoirs. Seemingly safe areas that had grown around managed bayous, canals, and ponds found themselves flooded. Interstates 10 and 45—landmarks that defined the petroleum metropolis as well as transportation arteries for it—were both inundated. Over the next years, as Houston struggled to overcome the social as well as physical damage that Harvey left, lesser episodes of the same dynamic repeatedly undid the recovery of vulnerable neighborhoods. "Our primary drainage mechanism throughout this city is our streets," remarked a municipal public works executive in the aftermath of Hurricane Beryl in July 2024, "for better or worse." When Beryl hit, thousands of Houstonians were already stuck in shelters, driven from their homes in mid-May when severe thunderstorms brought flooding and power outages. The recurrent floods, lamented an environmental justice activist, had left behind "a kind of numbness."[7]

Total damage estimates from Harvey hovered around $125 billion, which made Houston's 2017 catastrophe vie with Katrina's wrecking of New Orleans in 2005 as the costliest tropical cyclone disaster in US history. Because Harvey's flooding came from rainfall rather than storm surge, from inland, that is, rather than the sea, Houston-based experts at times downplayed how much the 2017 disaster could reveal about the distinctive dangers posed by the complex of petroleum storage and transport operations, refineries, and petrochemical plants along the Houston Ship Channel. Any metropolitan area built in a similar manner in such an environment, the reasoning went, would have experienced similar flooding, regardless of the industrial base.[8] The connection between Harvey and oil becomes more manifest when one's view broadens to include what came close to happening at Corpus Christi or Port Lavaca. Harvey weakened by the time it reached the Texas–Louisiana border. Even so, flooding left the enormous Motiva refinery, owned by Saudi

[6] Blackburn, *Houston Flooding 3.5 Years after Harvey*, 3; Antonia Sebastian, "Flood Mitigation in Multi-Hazard Coastal Environments," in *Delft Delta Design: Houston Galveston Bay Region, Texas U.S.A.*, ed. Bauke Kothuis, et al. (Delft: Delft University Publishers, 2015), 69.

[7] Quotations: Brady Dennis and Scott Dance, "Houston Has a Flooding Problem; Beryl Was More Proof It's Hard to Fix," *Washington Post*, July 10, 2024. J. David Goodman, "'This Storm Has Broken People': After Beryl, Some Consider Leaving," *New York Times*, July 16, 2024. Orrin H. Pilkey and Keith C. Pilkey, *Sea Level Rise: A Slow Tsunami on America's Shores* (Durham, NC: Duke University Press, 2019), 124–125.

[8] Blake and Zelinsky, *Tropical Cyclone Report: Hurricane Harvey*, 9; Lessoff, "Texas Coast," 340–341.

Aramco and located on the Intracoastal Waterway as it runs through Port Arthur, damaged and coated in oil. In any case, Houston's petrochemical complex did become a half-hidden factor in the Harvey disaster, for example when floods dumped toxic soil and poisoned groundwater into houses in so-called fenceline neighborhoods.[9]

Whatever implications they might draw from Harvey, Houston-based researchers and officials had for years attempted to draw attention to concerns at the heart of this chapter. The convergence of the Spindletop discovery of 1901 and the construction of the Houston Ship Channel between 1902 and 1914 meant that Texas's system of dredged inland ports evolved in tandem with the state's central role in the North American and worldwide petroleum and petrochemicals industries. The symbiosis of ship channels and petroleum enabled Houston to become the self-proclaimed energy capital, home of "the oldest, largest, and most diverse complex of oil- and natural gas-related activities in the world," as historians Joseph Pratt and Martin Melosi assert. Vulnerability to storms was only one factor motivating inland channel construction. When business strategy seemed to demand it, port authorities and energy firms rationalized the siting of oil, gas, and petrochemical operations at inherently exposed locations such as Corpus Christi's La Quinta Channel, Port Lavaca and Matagorda Bay, or Texas City and Baytown near Houston. By the twenty-first century, the extractive energy regime that made the Texas coast into a thriving industrial port region now posed an inexorable threat.[10] Much US discussion of the exposure of people and property to coastal disasters focuses on residential or resort development, such as along the Florida coast. This applies to portions of the Texas Gulf Coast, such as Galveston Island or North and South Padre Island. Yet by far, the main factor drawing investment and population into harm's way has been the fossil fuel energy regime and its commercial and industrial infrastructure.

Well before Harvey, researchers were modeling scenarios for a catastrophe that seemed increasingly inevitable: a direct hit by a climate-change-fueled hurricane on one of the region's agglomerations of petroleum facilities and petrochemical plants. These studies took as their starting point not Harvey, but Hurricane Ike, which had trudged over the Bahamas, Haiti, and Cuba before landing at Galveston on September 13, 2008. Although by wind speed only a category 2 hurricane, Ike pushed an enormous storm surge. The 10–15-foot surge that hit the gulf side of Galveston Island could not surmount the seawall. Yet floods in excess of 12 feet poured into Galveston from the bayside to the west. A surge of up to 20 feet inundated the Bolivar Peninsula, northeast across the pass from Galveston. Storm surges

[9] Pilkey and Pilkey, *Sea Level Rise*, 105; Alex Stapleton, dir., *God Save Texas*, episode 2: *The Price of Oil*, HBO Max documentary, 2024.

[10] Martin V. Melosi and Joseph A. Pratt, "The Energy Capital of the World? Oil-Led Development in Twentieth-Century Houston," in *Energy Capitals: Local Impact; Global Influences*, ed. Joseph A. Pratt, Martin V. Melosi, and Kathleen A. Brosnan (Pittsburgh: University of Pittsburgh Press, 2014), 30; Carola Hein, "Port City Regions as Nodes in the Global Petroleumscape," *Sartonia* 33 (2020): 193–224. Ashley Dawson, *Extreme Cities: The Peril and Promise of Urban Life in the Age of Climate Change* (New York: Verso, 2017), tells a similar story for the Louisiana coast.

over 12 feet were recorded as far north as the Sabine Pass and over 11 feet as far inland as Port Arthur, where at least one major chemical facility flooded. Observers such as Blackburn emphasized that for all the "misery" on Galveston and the Bolivar Peninsula, "the Houston region was spared the worst." Ike's eventual path across Galveston Bay meant that authorities could instruct Houston's two million people to shelter in place, avoiding a repeat of the botched evacuation that accompanied Hurricane Rita in 2005, when confusion and panic proved deadlier than the storm itself. The storm's shift northward meant minimal damage to the Houston area's oil facilities and chemical plants.[11]

Rice University's Severe Storm Prediction, Education and Evacuation from Disaster Center (SSPEED) calculated that if Ike had increased 15 percent in intensity to category 3 and made landfall at Galveston Island's southwestern end—a plausible alternative—it could well have pushed a 24-foot storm surge, enough to pour over the levees at Texas City, on the mainland across from Galveston, while also flooding Baytown, east across the bay from where the ship channel enters Buffalo Bayou. The storm in the model would then head up the ship channel into "the oil refining and chemical manufacturing center of the United States." Storm water would crush storage tanks or lift them from their foundations, while breaking apart pipelines and spreading "a deadly soup of toxic pollutants" across metropolitan Houston, as the eminent coastal scientist Orrin Pilkey remarks. Over 2220 oil or chemical storage tanks would experience at least some flooding, Blackburn explains, as would "at least six refineries and well over a hundred chemical plants." Up to 90 million gallons of oil and hazardous chemicals would spread across the urban area and into Galveston Bay, "the worst environmental disaster in US history," Blackburn adds, "far exceeding the Exxon Valdez oil spill [1989] and possibly the Deepwater Horizon event [2010]." These projections did not reckon in rising sea levels and stronger storms related to climate change, which could make surges at Houston of 34 feet "reasonably foreseeable" within the next half-century[12] (see Fig. 2).

What Blackburn understandably labeled a "horrific future scenario" prompted proposals for coastal protection systems. The best-known of these proposals, the so-called Ike Dike—modeled on Dutch projects such as Rotterdam's Maeslant Barrier—would consist of levees and gates running 55–80 miles along the Gulf of Mexico from the Bolivar Peninsula down Galveston Island and south across San Luis Pass (see Fig. 3). This project came with daunting costs along with considerable environmental disruption and might not even prevent the ship channel from flooding in a strong enough hurricane. Less costly alternatives all had drawbacks, including continued exposure of refineries and petrochemical plants outside probable dike routes. Nevertheless, port officials, civil engineers, and civic and business interests pressed ahead. Environmental writer Roy Scranton recounts attending a

[11] Blackburn, *Texan Plan*, 81–82. Robbie Berg, "Tropical Cyclone Report: Hurricane Ike, (AL092008), 1–14 September 1008," US National Hurricane Center, 23 January 2009, updated to 18 March 2014, 6–10. Harvey Rice, "Hurricane Ike Worst Storm in Decades," Houston *Chronicle*, Oct. 22, 2016.

[12] Pilkey and Pilkey, *Sea Level Rise*, 106. Blackburn, *Texan Plan*, 84, 92.

Fig. 2 Carol Highsmith, aerial photo of the Houston Ship Channel, 2014. From a series of aerial photos Highsmith took of the ship channel in May of that year. These underscore how densely oil facilities cluster along a waterway increasingly vulnerable to storm surge. Source: Courtesy the Lyda Hill Texas Collection of Photographs in Carol M. Highsmith's America Project, Prints and Photographs, US Library of Congress

five-hour hearing at Galveston of a Texas state legislative committee on coastal barrier systems. Public and private sector groups detailed "how the various projects might affect insurance rates, how long different projects might take to build, and how the [funding] pitch might be put to the U.S. Congress." Ruefully, Scranton noted, "What obstacles might have remained between this roomful of committed public servants and the building of one of the largest coastal projects in the world seemed for the moment insubstantial." For his part, Pilkey dismisses the Ike Dike as "a nineteenth-century solution for a twenty-first-century problem." A "new disaster" on the scale of Galveston's 1900 storm, still the deadliest natural disaster in US history, was "forthcoming," Pilkey insists. "Better to spend the money" on cleaning toxic waste along the Houston Ship Channel, ripping up concrete atop flood plains and marshes, and moving endangered communities, even if that meant abandoning swaths of metropolitan Houston altogether.[13]

Ruminations over global warming while living in and observing Houston inspired Timothy Morton, an environmental and cultural scholar at Rice, to coin the term *hyperobject*. "These huge, gigantic entities," Morton explained to Scranton as they

[13] Blackburn, *Texan Plan*, 92–95. Roy Scranton, "Anthropocene City: Houston as Hyperobject," in *More City than Water: A Houston Flood Atlas*, ed. Lacy M. Johnson and Cheryl Beckett (Austin: University of Texas Press, 2022), 42–43; Pilkey and Pilkey, *Sea Level Rise*, 106.

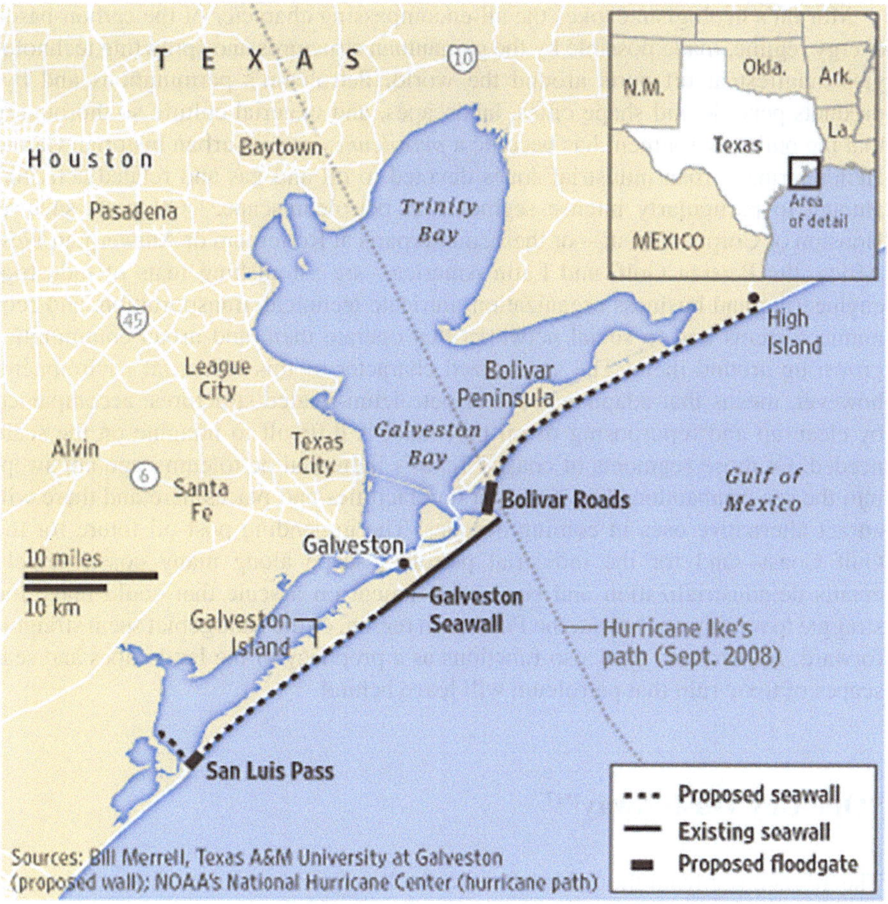

Fig. 3 The Ike Dike System as proposed in 2014. The Ike Dike system was devised by a research team working with Professor William J. Merrill, a marine scientist at Texas A&M University at Galveston. Modeled in part on Dutch coastal defense systems such as Rotterdam's Maeslant Barrier, the project drew upon Dutch experts as well as US ones. The proposal sparked much discussion and a variety of alternative schemes in the years since. Source: Professor William Merrill, https://www.tamug.edu/ikedike/index.html

rode past refineries on the public boat tour of the Houston Ship Channel, "are all symptoms of this even larger, much more disturbing thing that we can't point to directly."[14] Morton and Scranton were discussing, among other matters, the predicament that the Gulf Coast faces as the need to move away from carbon-based energy becomes more total and immediate.

[14] Scranton, "Anthropocene City," 44; Timothy Morton, *Hyperobjects: Philosophy and Ecology after the End of the World* (Minneapolis: University of Minnesota Press, 2013).

Morton's neologism evokes the all-encompassing character of the carbon-based energy regime, made possible by the gargantuan structures and sprawling technologies manifest at oil ports around the world. Petroleum's permutations and by-products pervade and shape cities, landscapes, and material culture so thoroughly that the built environment has become a *petroleumscape*, in urban historian Carola Hein's terms. Urban-industrial zones devoted to oil and gas and related activities amount to particularly intense segments of petroleumscape.[15] Oil ports such as Houston or Corpus Christi—or their counterparts at Rotterdam or Antwerp, in West Africa, the Persian Gulf, and Latin America—are astonishing feats of industrial engineering and business organization, intricate technical, transportation, and economic systems tied to social networks that operate them and urban communities grown up around them. The specialized character of this industrial environment, however, means that adaptive reuse of petroleum-related structures, accompanied by clean-up and repurposing of oil port land, is difficult to imagine on the scale needed. Of those segments of coastal Texas's industrial petroleumscape not swept into the sea or abandoned to it, only a few facilities and places here and there will attract alternative uses in coming decades. The impending post-oil future for the Gulf Coast—and for the industrial petroleumscape along many coasts—likely entails deindustrialization and site abandonment on a scale that could make the struggle to repurpose Detroit, the Pittsburgh region, or the Ruhrgebiet seem straightforward. *Hyperobject* thus also functions as a prophecy of the landscapes and seascapes of toxic ruin that petroleum will leave behind.[16]

"One Day Wash Away"[17]

The Texas coast from the Sabine Pass to the Rio Grande extends about 375 miles. From Galveston Island in the north to Padre Island—the world's longest barrier island—in the south, a string of barrier islands separates the mainland from the Gulf of Mexico. The lagoons, marshes, and bays inside the islands are uniformly shallow, with drafts of 8 or 10 feet at best and with sandbars forming at the passes. The Laguna Madre, which extends from Corpus Christi Bay down behind Padre Island to Tamaulipas, Mexico, averages around 3 feet deep. Another noteworthy feature of

[15] Carola Hein, "Oil Spaces: The Global Petroleumscape in the Rotterdam/the Hague Area," *Journal of Urban History* 44 (Sept. 2018): 887–929; Carola Hein and Alan Lessoff, "The Original North American Petroleumscape: Oil and Gas Empire, Petrochemical Nation" in *Oil Spaces: Exploring the Global Petroleumscape*, ed. Carola Hein (London: Routledge, 2022), 28–31.

[16] Hein, "Port City Regions," 220–221. On the limits of adaptive reuse of industrial sites, James Douet, ed., *Industrial Heritage Retooled: The TICCH Guide to Industrial Heritage Conservation* (Lancaster: Carnegie Publishing, 2012). On modernity as a generator of ruins: Nick Yablon, *Untimely Ruins: An Archaeology of American Urban Modernity, 1819–1919* (Chicago: University of Chicago Press, 2010); Camilo José Vergara, *American Ruins* (New York: Monacelli Press, 1999).

[17] Quoted in David G. McComb, *Galveston: A History* (Austin: University of Texas Press, 1986), 48.

the Texas coast is how low-lying it is. The reputed highest point, the Corpus Christi Bluff south of the Nueces River, rises to 40 feet. The salt dome at High Island on the Bolivar Peninsula is 38 feet. Only in a few other places does the shoreline rise more than 5 feet above sea level.[18] Taken together, these features—the shallow bays and estuaries, the low-lying shore, and the relative protection from Gulf of Mexico storms provided by the barrier islands—explain the waterscape's human history, including the Anglo-American preoccupation with re-engineering it.

Sustained attempts to create ports along the Texas coast preceded Anglo-American colonization, but not by much. The port of Galveston originated in the 1810s in the convoluted dealings of French privateers Louis Aury and Jean Lafitte. In the 1820s, Anglo-Americans made it their port of entry into Mexican-era Texas and then a naval stronghold during the Texas Revolution of 1835–1836. At the northern edge of Texas's northernmost barrier island, 2 miles from the mainland, Galveston was self-evidently not a great site. As elsewhere along the Gulf Coast, the bay was shallow and blocked by sand bars. The island rises less than 9 feet above sea level, which meant that it was manifestly vulnerable to inundation in hurricanes. At least 11 hurricanes struck Galveston in the nineteenth century. Jean Lafitte lost four ships in an 1818 storm. In 1837, shortly after the Texas Republic assumed control, a storm beached eight ships and rendered nearly every building unusable. "Many felt discouraged and left Galveston, thinking it would always be subject to such storms," recorded the Texas Republic's military commander on the island.[19]

Whatever the shortcomings of its physical site, Galveston had what geographers label a favorable *situation*. It was more accessible to ships than any place in the vicinity. Goods and people—including enslaved workers bound for inland plantations—could unload there for transshipment along the coast or upstream to Houston, about 50 miles to the northeast. Galveston became the main export point for Texas commodities, especially cotton, grain, and hides. With 4177 people in 1850, the first census under US rule, Galveston ranked as Texas's largest city, a status maintained off and on into the 1880s, when shifting transportation methods enabled Dallas, San Antonio, and then Houston to surpass it.[20]

Beyond vulnerability to storms, limited fresh water—essential for industry as well as domestic use—hampered Galveston's development. Railroads especially exposed the island port's constraints. As rail lines spread across Texas starting in the 1850s, Houston, with its broad, flat, well-watered site, became their Gulf Coast hub. By the early twentieth century, 17 railroads met at Houston's four depots. In 1859, a rail line opened between Houston and Virginia Point on the shore opposite Galveston. This connection required transshipment across the bay until a trestle

---

[18] John W. Tunnell, Jr., "Geography, Climate, and Hydrography," in *The Laguna Madre of Texas and Tamaulipas*, ed. John W. Tunnell Jr. and Frank W. Judd (College Station: Texas A&M University Press, 2002), 7–9; John J. Kosovich, "State of Texas—Highlighting Low-Lying Areas Derived from USGS Digital Elevation Data," US Geological Survey Scientific Investigations Map 3050, 2008, https://pubs.usgs.gov/sim/3050/ (accessed June 9, 2022).

[19] McComb, *Galveston*, 27, 34–41.

[20] McComb, *Galveston*, 66–68.

bridge opened in 1860. This bridge survived the Civil War only to be wiped out in an 1867 hurricane. Railroad companies rebuilt it and added two more bridges. All three of these crossings to Galveston were wrecked in the Great Storm of 1900.[21]

The certainty of storms weighed on investors. In 1876, one writer to the local press recounted negative views of Galveston heard in New York, Chicago, and other northern cities. The city would draw investment, the writer remarked, "if we could offer a reasonable argument that the island will not one day wash away."[22] Nevertheless, Galveston's ability to handle Texas and southwestern commodities continued to grow. By 1900, the island city ranked as the country's largest cotton port, as well as its third leading port for wheat exports. The city seemed so indispensable that the federal government made it a focus of port improvement activities. In 1880, the government established a Galveston District of the US Army Corps of Engineers, which oversaw a harbor-deepening program, including the construction of the world's longest jetties. Through these efforts, the harbor on the bay side of the island reached a depth of over 25 feet.[23]

The long-feared catastrophe came on September 8, 1900. The Great Storm pushed a surge of over 15 feet and flooded Galveston Island from both sides. Around 3600 houses were destroyed, including every building within six blocks of the beach. The destruction would have been more widespread, except for a 30-block barricade of debris that limited damage behind it. The estimated 6000 casualties on the island amounted to 15 percent of the city's population (see Fig. 4). Texas lore holds that the 1900 storm ended Galveston as a significant port. In fact, the Great Storm seemed to redouble public and private sector commitment. The Army Corps worked with local government to lift up 2156 buildings spread over 500 city blocks. In 1904, residents celebrated the completion of the seawall's first segment. By 1912, the island city was again the country's leading cotton port. The structures intended to protect that trade held up in their first test in 1915, when another powerful hurricane caused eight local deaths and only a fraction of 1900's damage.[24]

While Galveston would come to specialize in tourism and cruise ships, it remained a significant port for grain, fruits, and other commodities. The city's population reached a high in 1960 of 67,175, before falling back to under 60,000 by the twenty-first century. By 1960, though, Houston was approaching one million

[21] William C. Barnett, "A Tale of Two Texas Cities: Houston, the Industrial Metropolis, and Galveston, the Island Gateway," in *Energy Metropolis: An Environmental History of Houston and the Gulf Coast*, ed. Martin Melosi and Joseph A. Pratt (Pittsburgh: University of Pittsburgh Press, 2007), 192–193; McComb, *Galveston*, 49–58; David G. McComb, *Houston: The Bayou City* (Austin: University of Texas Press, 1969), 34–40.

[22] Quoted in McComb, *Galveston*, 48.

[23] McComb, *Galveston*, ch. 3–4; Earle B. Young, *Galveston and the Great West* (College Station: Texas A&M University Press, 1997); Lynn M. Alperin, *Custodians of the Coast: History of the United States Army Engineers at Galveston* (Galveston: United States Army Corps of Engineers, 1967), ch. 2.

[24] Patricia Bellis Bixel and Elizabeth Hayes Turner, *Galveston and the 1900 Storm* (Austin: University of Texas Press, 2000); Alan Lessoff, "Galveston Hurricane," *American Disasters*, ed. Ballard Campbell (New York: Checkmark Books, 2008), 185–187.

Fig. 4 "Galveston's Awful Calamity." Kurz & Allison lithograph of the Great Storm of September 1900. Source: Courtesy Prints and Photographs, US Library of Congress

people, and by 2020 it was challenging Chicago for the status of the country's third-largest city. Even before the 1900 storm, factors that would shift activity toward Houston were in place. The inland port's advantages over its island rival accelerated with the advent of oil and the opening of the ship channel in the decade and a half after the Great Storm.

Galveston was the most significant attempt during the nineteenth century to establish a saltwater port on the Texas Gulf Coast. Migration into northeastern Texas moved over land or by river from Louisiana and Arkansas. This lessened pressure to break through the Sabine Pass on the Louisiana–Texas border. Beaumont, on the Neches River about 30 miles from the Gulf, remained a tiny river port until the 1880s, when railroads and lumber mills caused the population to grow to about 9400 by 1900. Orange, along the Sabine River, also depended not on navigation but on railroads hauling the products of the Piney Woods. Port Arthur, founded as land speculation in 1894, remained in 1900 a struggling resort and port town of 900 people on Sabine Lake inside the pass.[25] By the early 1890s, however, a Beaumont

[25] David G. McComb, *The City in Texas: A History* (Austin: University of Texas Press, 2015), ch. 16.

lumber merchant named Patillo Higgins was trying to interest investors in a suspected pool of oil beneath Spindletop Hill 3 miles south of town.

Until the 1880s, Galveston's most plausible rivals were on Matagorda Bay about 130 miles to the south, earlier the site of France's ill-fated La Salle expedition. The bay offered a promising place to land settlers and supplies headed to central, southern, and western Texas. It provided an outlet for cattle and hides from ranches taking shape in the grasslands of central Texas or in South Texas's shrubland. Oxcarts and wagon trains plying the dangerous roads from northern Mexico headed there. After Comanches burned a previous attempt at a port in the vicinity in 1840, Anglo-Texans founded Lavaca on the western side of Matagorda Bay. This site's access to Victoria and beyond to some degree offset the harbor's shallowness and obstruction with sandbars. About 15 miles southeast, Indianola emerged on a more accessible, though dangerously low-lying site. In the 1850s, the influential Morgan Steamship line moved its mid-coast operations to Indianola, which thereby became Texas's second port, with a peak population of around 2000. Flooded by a hurricane in 1875, Indianola had partly recovered when another hurricane in 1886 swept what remained of the town into the bay. Nowadays, a historical marker identifies the site of this storied ghost town.[26]

"Sustained Safety"[27]

Incorporated in 1837, Houston served as the temporary capital until the Texas Republic's move to Austin in 1839. Founded by brothers J.K. and A.C. Allen, migrants from New York, Houston was located at the confluence of Buffalo Bayou and White Oak Bayou, 18 miles from where Buffalo Bayou flows into the San Jacinto River, which in turn connects to Galveston Bay. One can absorb this geography from atop the 567-foot San Jacinto Monument, which overlooks the mouth of Buffalo Bayou and thus the main section of the Houston Ship Channel. The monument marks the site where in April 1836, General Sam Houston's forces routed the Mexican army, gaining precarious independence for Texas.

The Allens and their partners claimed that Buffalo Bayou was tidal and navigable up to the site. That was true mainly in comparison to other shallow, clogged Texas river ports. Without dredging and the clearing of vegetation, only light steamboats could make their way through the bayou. Despite this, state and federal officials came to agree with local boosters about the potential of a shipping canal to Houston. Even as it pursued costly improvements to Galveston, the Army Corps reported

[26] Keith Guthrie, *Texas Forgotten Ports,* vol. 1*: Mid-Gulf Ports From Corpus Christi to Matagorda Bay* (Austin: Eakin Press, 1988); Brownson Malsch, *Indianola: The Mother of Western Texas*, rev. ed. (Austin: State House Press, 1988); Lonnie Ficklen Maywald, "Port Lavaca, TX," *Handbook of Texas Online*, www.tshaonline.org/handbook/entries/port-lavaca-tx (accessed Jan. 27, 2024).

[27] *The Port of Corpus Christi, Where Texas Meets the Sea, Opens to World Commerce*, pamphlet 1926, [6], copy in author's possession.

upon Houston as an "obvious" location for a ship channel because railroads and industry could readily concentrate there. In 1870, the federal government designated Houston a port of entry, after which Congress funded studies of a deep-water channel through Buffalo Bayou. While this was going on, the Morgan Line constructed a 9-foot channel to a railroad 6 miles from the city.[28]

From the Rhine-Main system to the Manchester Ship Canal to the Chicago Sanitary and Ship Canal, the half-century before World War I was an era of shipping canals and the channelization of rivers. Common sense and tradition guide people to think of ports as waterfront places to which merchants bring goods. Yet counterintuitively, industrialization and railroads were making it cost-effective to bring ships inland to freight yards and factory districts. Ample land along Houston's rail lines attracted a range of industries, often offshoots of the commodities in which pre-Spindletop Texas specialized: cotton compresses, cotton-seed oil mills, textile mills, sawmills, leather works, and meatpacking plants. The Southern Pacific system made Houston its eastern hub, operating shops and yards there that employed over 1200 men by the early 1890s. Railroad construction and repair provided Houston with an industrial base in metals and machinery, skills later valuable to the oil industry.[29] These considerations explain why in 1896, after years of studies and lobbying, Congress authorized a channel 52 miles from the Gulf, initially 25 feet deep. By 1902, Congress began funding the Houston project, which the city and its banks underwrote as well through bond sales.

Although the Houston Ship Channel had an origin separate from the Texas oil boom, the potential interconnection of these two economic and engineering dramas quickly became evident. In 1910, Houston, with 78,800 people, was still Texas's third-largest city. But its comprehensive rail network, its diverse industries, and its vibrant banking and commercial sectors made it an attractive base for oil operations. The motive to concentrate around Houston increased with the discovery of oil reserves in the city's direct hinterland. By the 1920s, major companies created by the Texas oil boom—Gulf, Texaco, and Humble—all had headquarters in Houston, which also became a locus for the American operations of foreign enterprises such as Shell. Despite the lore of Texas oil as a creation of organization-averse wildcatters, the industry's major firms almost all began with northeastern or international finance and management behind them. From experience with Cleveland, Philadelphia, New Jersey, or Rotterdam, organizers of the Texas oil industry recognized the ship channel as an excellent place to bring together production, storage, and transport. By 1930, the Port of Houston's nine refineries were capable of processing 194,000 barrels per day. By World War II, the Baytown refinery of the Standard Oil of New Jersey (ESSO) affiliate Humble was the largest in the United States, a status that the plant, later renamed after ESSO's successor Exxon/Mobil, retained into the twenty-first century. Along with refineries, oil companies

[28] Alperin, *Custodians of the Coast*, 96; McComb, *Houston*, 46–48.

[29] McComb, *Houston*, 91–110; Harold L. Platt, *City Building in the New South: The Growth of Public Services in Houston, Texas, 1830–1915* (Philadelphia: Temple University Press, 1983), 80–83.

developed enormous storage terminals at the Houston Ship Channel, fed by pipeline networks for crude and refined oil and, eventually, natural gas.[30]

The Port of Houston attracted oil industry equipment and rig manufacturers and industrial construction companies. The port's mix of petroleum operations and industrial services in turn attracted manufacturers of fertilizers, petrochemicals, and synthetic materials. By the 1980s, the coastal area from Beaumont to Houston accounted for half of the US petrochemical capacity. Cement works, metal producers, and inorganic chemical plants also concentrated along the ship channel. Such industries used high-energy manufacturing processes that benefited from steady, inexpensive supplies of natural gas.[31] Houston took its place alongside Galveston as a leading export point for agricultural commodities, especially cotton and grain. Still, the industrial infrastructure that collected along the Houston Ship Channel or at satellites like Baytown and Texas City shifted the balance toward oil and petroleum products, which accounted for 80 percent of the port's cargo by mid-century.[32]

The Corpus Christi Ship Channel arose from a similar mix of considerations, except that a hurricane directly prompted the decision to build a deep-water port in South Texas. Founded in 1839 by Henry Lawrence Kinney, a Pennsylvania-born trader and land dealer, Corpus Christi took its name from the magnificent crescent bay along which it stretches. The combined effect of the bay and the 40-foot bluff a few hundred yards from the shore inspired generations of promoters to follow Kinney in using the slogan "Naples of the Gulf" to advertise for settlers and tourists. Corpus Christi's semi-protected location 20 miles inland from Aransas Pass came with the typical Gulf Coast downside, a shallow bay accessed via a silt-prone channel. The original practice was to reload cargo and passengers onto lightering boats with drafts of maybe 4 feet for passage to town wharves. The federal government recognized Corpus Christi as a port in 1852, the same year the town gained incorporation. Yet in contrast to Houston, the Corps of Engineers long argued that Corpus Christi was probably not worth the effort, given the likely low level of traffic in remote South Texas. It would be easier to extend railroads to points near the coast, for example, Aransas Pass or Rockport. With federal support unlikely, Kinney acquired a dredge and by 1857 had cut a 5-foot channel across the bay. This was insufficient even for the Morgan Line's shallow-draft steamers, one reason for Morgan's ill-fated decision to concentrate mid-coast and South Texas operations at Indianola.[33]

One fallout from the collapse of Reconstruction in Texas in the mid-1870s was the imposition of Anglo-American control over South Texas borderlands, carried

[30] Lessoff, "Texas Coast," 336–337.

[31] Lessoff, "Texas Coast," 337–338.

[32] James J. Parsons, "Recent Industrial Development in the Gulf South," *Geographic Review* 40 (Jan. 1950): 72. Lessoff, "Texas Coast," 339.

[33] Alan Lessoff, *Where Texas Meets the Sea: Corpus Christi and Its History* (Austin: University of Texas Press, 2015), 29, 98, 272. Bill Walraven, "Port of Corpus Christi, 1500–1926," *Nueces County Historical Commission Bulletin* 4 (Nov. 1997): 6.

out by vigilante bands and the Texas Rangers. Since the Texas Revolution, this region had experienced cycles of raids and reprisals that veered between banditry and ethnonational border war. Settlement of the countryside under Anglo-American control gave renewed impetus to port improvements at Corpus Christi. The Anglo landholding families at the heart of South Texas ranching lore, the Kings, Klebergs, Kenedys, and Driscolls, from the start ran their operations as diversified land development enterprises, with cattle ranching only one dimension. Identifying Corpus Christi as the regional city and port their enterprises required, ranch families worked with railroad promoters to bring the four lines that converged on Corpus Christi by the 1910s. The concentration of railroads at a townsite capable of expansion was the formula that eventually won federal backing for the Houston Ship Channel.[34]

The troubles of South Texas's cattle business in the years after the Civil War, meanwhile, dramatized the case for deep-water access. Obstructions to shipping, combined with low cattle prices, induced packing plants along Corpus Christi Bay and on Padre Island to strip hide, tallow, and horns from slaughtered cattle, ship only these by-products, and throw the carcasses into the bay or leave them rotting on the beach. In 1871, Richard King, founder of King Ranch, organized a company to dredge an 8-foot channel connected to a railroad pier built into the bay. When a Morgan Line steamer reached Corpus Christi in May 1874, 3000 people turned out to celebrate. King's pass had already silted about the time of the 1886 storm that destroyed Indianola. Morgan decided to try Rockport, 30 miles to the north.[35]

One more attempt to open a channel with private capital occurred in the 1890s. A New Jersey businessman and former Union colonel named Elihu Ropes stopped in Corpus Christi on his way to California. The high bluff and sweeping bay gave him a vision of a combination port and resort, a "Chicago of the Southwest" and a "Southern Newport." With financing from New York investors, Ropes began dredging a channel straight through Mustang Island. He planned a railroad bridge near the site of the present Kennedy Causeway to what he grandly labeled Port Ropes. "The problem of deep water on the Texas coast, without government aid," remarked the *Houston Post*, "seems to be near solution." Ropes's activities set off a real estate bubble whose bust began even before the Panic of 1893. Defeated and pursued by creditors and lawsuits, Ropes returned to New York. The barrier island sand absorbed his broken-down dredge.[36]

---

[34] Lessoff, *Where Texas Meets the Sea*, 64–65, 91–94, 98–99.

[35] Walraven, "Port of Corpus Christi," 6; Mary Jo O'Rear, *Storm over the Bay: The People of Corpus Christi and Their Port* (College Station: Texas A&M University Press, 2009), 18; O'Rear, *Barrier to the Bays: The Islands and the Texas Coastal Bend and Their Pass* (College Station: Texas A&M University Press, 2017), 58–62; Murphy Givens and Jim Moloney, *Corpus Christi: A History* (Corpus Christi: Nueces Press, 2011), 122–123.

[36] *Deep Water and Corpus Christi*, pamphlet, c. 1892, Kilgore 6235; and "Deep Water at Last," flyer c. 1892, Kilgore 6237, Special Collections and Archives, Texas A&M University–Corpus Christi [hereafter TAMU-CC]; William E. Kent, "Ropes Pass, Texas," *Engineering Magazine* 3 (June 1892): 340–349. Lessoff, *Where Texas Meets the Sea*, 99–100.

Once construction began on the Houston and Sabine-Neches channels, South Texas politicians, backed by ranching as well as town interests, stepped up lobbying for an equivalent channel to Corpus Christi. The Army Corps continued to argue for less expensive improvements nearer the coast. The defect of the Army Corps' position, from the perspective of channel advocates, became evident in August 1916, when a storm wiped out much of its work at Aransas Pass.[37] As they watched the havoc at Indianola and Galveston, Corpus Christians assured themselves that their town was storm proof. This despite numerous first-hand accounts of damaging storms in the 1870s–1880s.[38]

Whatever "false sense of security" Corpus Christi derived from its location along a "shallow inshore" bay disappeared on September 14, 1919, when an "unprecedented" hurricane drove a huge storm surge over the barrier islands, across the bay, over the North Beach resort adjacent to the town, and into the low-lying business district. Given that the surge measured 12 feet when it reached Corpus Christi, contemporary analysts considered this hurricane "undoubtedly much more severe than the Galveston hurricane of 1900," which after all did its worst damage on an island[39] (see Fig. 5). Yet for all the wreckage at Corpus Christi, the storm "destroyed everything at Port Aransas and on Mustang, Harbor, and St. Joseph Islands," as ship channel advocates made sure to point out. And even as flooding spread corpses and piled debris in low-lying areas of the city, those who lived on the Corpus Christi Bluff went through the night hardly aware of the wretchedness hundreds of yards away. "In the interest of safety and property, a safe harbor should be established against the high land," wrote Roy Miller, the former mayor who edited the Corpus Christi *Caller*, the region's main newspaper.

A close associate of Richard J. Kleberg, Richard King's son-in-law and successor as head of the King Ranch, Miller treated the crisis as an opportunity to secure the long-sought ship channel and put together a broad, effective campaign. Miller went on to serve as Washington lobbyist for Texas navigation projects, especially the Gulf Intracoastal Waterway (GIWW), the shipping artery that by 1949 connected the Gulf Coast from the Rio Grande to Florida.[40]

When the Corpus Christi Ship Channel opened in September 1926, on the seventh anniversary of the deadly storm, backers stressed the "tremendous asset of high

[37] "Brief of the City of Corpus Christi, Texas, and the Corpus Christi Commercial Association, submitted to the Board of Engineers," Sept. 20, 1920, Kilgore 6205, TAMU-CC, 5-6; Port of Corpus Christi, *Port Book, 1976, 50th Anniversary*, 14. Richard A. Laune, "The Political Struggle for the Port of Corpus Christi," *Journal of South Texas* 18 (June 2005): 91–112.

[38] For first-hand accounts of nineteenth-century hurricanes at Corpus Christi: W. Armstrong Price, "Hurricanes affecting the Coast of Texas from Galveston to Rio Grande," Technical Memorandum no. 78, Beach Erosion Board, Corps of Engineers, 1956, Appendix A, copy in TAMU-CC. O'Rear, *Barrier to the Bays*, 134–142.

[39] Terrell Bartlett, "The Tidal Storm at Corpus Christi and Its Effect on Engineering Structures," *Engineering News-Record* 83 (Nov. 13–20, 1919): 848, appended to Price, "Hurricanes Affecting the Coast of Texas."

[40] "Brief of the City of Corpus Christi," 6; Miller quoted in Walraven, "Port of Corpus Christi," 7. O'Rear, *Storm over the Bay*, ch. 7–12. Alperin, *Custodians of the Coast*, 126–137, 151–156.

Fig. 5 Postcard of wreckage in front of Lichtenstein's Department Store, Corpus Christi after the devastating September 1919 hurricane. Source: Courtesy L. H. Gross Collection, Collection 48, Special Collections and Archives, Mary and Jeff Bell Library, Texas A&M University—Corpus Christi

land," which provided "sustained safety" and thereby solved "the greatest problem in port construction." With a 30-foot turning basin 21 miles from the Gulf "protected by an embankment fifteen feet above high tide and fifty feet thick at the top," the Port of Corpus Christi illustrated how "engineering ingenuity" could create "*a safe port*." As at Houston, boosters of the Corpus Christi Ship Channel emphasized the ability of an inland port to facilitate what later became known as multimodal transportation: the harbor site offered "ample" space "for the construction of all necessary piers, docks, and slips, and railway switches and terminals," along with "an unlimited quantity of fresh water."[41]

The blowout in 1916 of an exploratory well at White Point along Nueces Bay revealed substantial reserves of gas and oil within miles of the future inland harbor. Corpus Christi gained the capacity to handle oil in 1930. By 1935, oil had already surpassed cotton, which as at Houston was initially envisioned as the leading commodity. Again, as at Houston and elsewhere along the Gulf, South Texas business leaders soon recognized the value to the industry of natural gas, earlier despised as a drilling hazard and waste product. In the early 1930s, local commercial interests were so eager to win a $7 million inorganic chemical plant that they underwrote the deepening and extension of the channel, the construction of a second turning basin

[41] *The Port of Corpus Christi, Where Texas Meets the Sea, Opens to World Commerce*, pamphlet 1926, [6, italics original]; "Brief of the City of Corpus Christi," 7–8. Lessoff, *Where Texas Meets the Sea*, 67.

(of an eventual five), and reconstruction of the city's main dam and reservoir. Local investors also underwrote what became a formidable gas pipeline and supply company named Southern Minerals, to fulfill the "industrialist's dream" of "low-priced fuel at tidewater."[42] By the early 1960s, in addition to five major refineries, the Corpus Christi Ship Channel was lined by industrial chemical plants, metals operations, and a cement factory, along with food processors and grain and cotton operations. Through the rest of the century, oil and related activities accounted for 70–80 percent of activity at the Port of Corpus Christi.

Unlike Corpus Christi and Houston, oil was an overriding motive from the start for the Sabine-Neches Waterway to Port Arthur and Beaumont, studies for which began within a year of the Spindletop discovery. As constructed in the 1910s, the northeast Texas system ran up the Sabine River to Port Arthur and then up the Neches River to Beaumont, with another branch extended eastward to Orange, which became a petrochemical center on the Gulf Intracoastal Waterway. By mid-century, the Y-pattern Sabine-Neches Waterway extended over 50 miles, with a channel over 30 feet deep. Clusters of refineries meant that the region produced nearly as much oil as the Houston area and at times even exceeded Houston in cargo tonnage.[43]

Freeport, about 60 miles south of Houston, received a 3-mile, 32-foot ship channel starting in the 1920s. During World War II, this became the site of a Dow Chemical complex underwritten by the US government. Then, in the late 1940s, proximity to gas wells encouraged Alcoa to construct an enormous aluminum plant at Port Comfort on Matagorda Bay. Union Carbide and Dupont were soon located in the area for similar reasons. The presence of these high-energy industries persuaded federal and state officials to push a deep-water channel through Matagorda Bay between 1958 and 1966. The area, with its agglomeration of plants producing industrial chemicals that in turn generated highly toxic wastes, remained as storm-prone as it had been in the 1870s–80s when Indianola was twice washed away.[44]

"Let's Get Real About the Texas Coast"[45]

Rotterdam, Europe's largest port, extends over 12,600 hectares about 30 kilometers up the Maas and the Nieuwe Waterweg, past the Maeslant Barrier to Europoort, Hoek van Holland, and the North Sea. Beginning in the nineteenth century, but especially after World War II, Rotterdam developed into the central transit and

[42] "Industrial Banquet, Plaza Hotel, Corpus Christi," Oct. 26, 1934, program in Southern Alkali packet, Lessoff collection, TAMU-CC. Lessoff, *Where Texas Meets the Sea*, 103.

[43] Parsons, "Recent Industrial Development," 72–74; Alperin, *Custodians of the Coast*, ch. 3.

[44] Edwin J. Foscue, "Industrialization of the Texas Gulf Coast Region," *Southwestern Social Science Quarterly* 31 (June 1950): 12–13; Parsons, "Recent Industrial Development," 76; Alperin, *Custodians of the Coast*, 166–170.

[45] Blackburn, *Texan Plan*, 113.

processing node for Western Europe's energy sector, with tanker terminals, refineries, pipeline networks, and storage tanks supporting an industrial agglomeration analogous to that along the Houston Ship Channel.[46]

As one might imagine, given the threat climate change poses to this city and country, Port of Rotterdam documents in the 2020s highlighted clean energy and climate programs "over the entire chain of production, infrastructure, transport, and consumption." Yet the port also drew attention to how Russia's 2022 invasion of Ukraine undercut such efforts. With Rotterdam and Antwerp, Europe's second largest port, key entry points for substitutes for Russian natural gas, imports of liquified natural gas (LNG) rose 63.9 percent in 2022 and another 9.8 percent in the first half of 2023. Nearly two-thirds came from the United States. Rotterdam handled LNG imports at the Gate Terminal on the Maasvlakte, the reclaimed extension of the port into the North Sea. Opened in 2011, "Europe's LNG hub" offered "excellent accessibility" for LNG carriers. Even before February 2022, the equivalent of a third of Dutch natural gas consumption passed through the Gate Terminal's three specialized tank complexes. The Ukraine War prompted the construction of an additional tank complex of 180,000 cubic meters, scheduled to begin operation in 2026. While insisting that LNG already counted as "one of the most environmentally friendly" fossil fuels, Port of Rotterdam websites expressed hope that a sustainable "bio-LNG" could soon be manufactured that might "even be carbon positive in certain cases."[47]

That the United States could fill the gap left by the European Union's recoil from Russia represented a turn of events that seemed unlikely a few years earlier. Amid the 1970s oil crisis, the United States imposed export restrictions. By 2016, the accelerating exploitation of shale oil fields encouraged the lifting of these controls. By 2024, the United States was the world's largest gas supplier, a "remarkable shift," the New York *Times* remarked, "that has profited oil and gas companies and strengthened American influence abroad." While LNG export plants appeared near such Atlantic ports as Savannah and Baltimore, Texas and Louisiana, with their pre-existing energy operations and their proximity to fracking fields, became the locus of this new export economy. Texas alone came to handle over 90 percent of the country's crude oil exports and more than 70 percent of refined product and petrochemical exports.[48]

---

[46] Hein, "Oil Spaces," 905–922.

[47] https://reporting.portofrotterdam.com/halfjaarbericht-2023/verslag-van-de-directie/ (accessed Sept. 1, 2024); Port of Rotterdam, *Highlights, Annual Report*, 2022, 58; *Half-Year Report*, 2023, 4; "Gate Terminal: Access Gateway for Global Supply of LNG," https://www.gateterminal.com/en/gate-terminal/ (accessed August 27, 2024); "Gate Terminal Starts Construction of fourth LNG Tank at the Port of Rotterdam," news release, Aug. 24, 2023, https://www.portofrotterdam.com/en/news-and-press-releases/gate-terminal-starts-construction-of-4th-lng-tank-at-the-port-of-rotterdam (accessed August 27, 2024).

[48] Brad Plummer and Nadja Popovich, "How the U.S. Became the World's Biggest Gas Supplier," *New York Times*, Feb. 3, 2024; Galveston District, US Army Corps of Engineers, *Coastal Texas Protection and Restoration Feasibility Study*, draft report, Oct. 2020, 9.

The North Sea and Texas ports began strengthening ties well before the 2022 crisis. On February 25, 2021, a year minus a day before the Russian invasion, Rotterdam signed an agreement with the Port of Corpus Christi, self-proclaimed "Energy Port of the Americas." (Both Corpus Christi and the Sabine-Neches Waterway promoted themselves as the country's "energy gateway.") Rotterdam and Corpus Christi agreed to "co-develop … trade and commercial opportunities," while cooperating on "innovative technologies specifically related to navigational safety and environmental protection." Such deals represented another remarkable historic shift. Since the Texas oil bust of the 1980s, Corpus Christi's energy sector and its overall economy had stagnated. While still enormous and strategically crucial, the Port of Corpus Christi had functioned more and more as a satellite of Houston and other cities where the energy sector's corporate authority and technical expertise increasingly concentrated. Then, in the 2010s, multinational investment in the Eagle Ford shale oil field drew billions in new investment to the South Texas port. By 2013, Eagle Ford ranked as "the largest oil and natural gas development in the world," possibly "the largest single economic development" in Texas history.[49] This accounted for Corpus Christi's reinvigoration and emergence as the country's largest exporter of crude oil. Corpus Christi's LNG capacity, however, was what drew European attention. In 2018, the opening of a $13 billion gas liquification plant built by Houston-based Cheniere Energy made Corpus Christi the first port capable of exporting LNG to Europe.[50] Four years later, Vladimir Putin's army made American LNG supply a geopolitical priority for the European Union.

Even as Corpus Christi became a focus for the fracking-related revival of Texas oil and gas, the port also emerged as the preferred entry point for imported wind turbines to support Texas's vast wind farms. During the 2010s, Texas came to account for a quarter of US wind energy capacity. Such trends encouraged experts to speculate hopefully that Texas's oil ports could shift to renewable energy and become as formidable in wind, solar, or hydrogen as they had been in oil and gas.[51] Hydrogen, a familiar industrial product at Texas ports and another fuel that could be marketed in gas or liquified form, held a special attraction. Hydrogen "has great promise as a flexible energy carrier," argued Sean Strawbridge, Corpus Christi's port CEO, "and many American allies in Europe and Asia are seeking hydrogen exported from the US Gulf Coast to meet their energy transition—and energy security—objectives." For its part, Rotterdam identified "green hydrogen" as its preferred alternative to LNG. Rotterdam embarked on partnerships with Shell and other European companies to develop hydrogen as a standard fuel for the port and

[49] "Port of Corpus Christi and Port of Rotterdam Enter into Historic Agreement," media advisory, Feb. 25, 2021; Port of Corpus Christi, *Strategic Plan, 2014–2020*, Dec. 10, 2013, 7. Lessoff, *Where Texas Meets the Sea*, ch. 6.

[50] "Liquid Bulk," https://portofcc.com/capabilities/cargo/liquid-bulk/ (accessed Sept. 1, 2024).

[51] Kenneth B. Medlock III, "The Future of Houston as Energy Transitions," Baker Institute for Public Policy Report, May 13, 2021, Rice University, Houston.

its industries. Even so, the Dutch port projected that imports would still account 90 percent of the hydrogen handled or used there through at least 2050.[52]

"Let's get real about the Texas coast," insisted Jim Blackburn, the Houston environmental lawyer and policy expert, in response to speculation that a post-oil future would come to Texas's port network anytime soon. Too many in the region "rely on hydrocarbons for their livelihoods" to count on change happening fast enough. In a book published shortly before 2017's Hurricane Harvey, Blackburn recounted that "every major oil and gas company" already had "extensive knowledge about climate change and its carbon footprint," however much the industry might prevaricate or talk in bromides publicly. For the foreseeable future, the most likely "path forward" involved encouraging Houston to become the metropole for a "Texas Coastal Exchange" that dealt in carbon offsets, carbon sequestration, and other "ecological services." The financing and administration of such a market remained uncertain. Climate scientists and environmentalists deplored this approach as untested and likely inadequate.[53]

Corpus Christi's dealings with Rotterdam capture the attitude that Blackburn sought to explain. Certainly, wrote Strawbridge, the Corpus Christi CEO, in introducing the port's 2023–2026 plan, the "dire need for climate action" created a "tremendous opportunity for innovation and workforce development." Even so, "energy security, and the geopolitical stability that it affords, is an essential precursor to energy transition." This meant, in Strawbridge's words, that "continued export of energy in all forms, particularly the continued growth in the production and export of liquified natural gas (LNG) in the Coastal Bend, is critical to breaking the tenuous dependence of key American allies on energy produced by nefarious actors." When in January 2024, the Joe Biden administration announced a climate-related pause in approvals for new or expanded LNG export facilities, the reaction in Corpus Christi and other oil ports was to scoff at its ineffectuality. "We are confident we will continue to secure all regulatory approvals," a Cheniere spokesperson told the Corpus Christi *Caller-Times,* "within our expected timelines."[54]

Since the 1910s–1920s, when the symbiosis between ship channels and Texas oil became apparent, the public and private interests involved have been aware that the protection afforded by an inland harbor was relative. Every oil port had a long list of near misses, as well as an evacuation and recovery plan for when the hit finally came. In September 1933, as Corpus Christi was shifting emphasis to oil, a hurricane bearing down upon the city weakened to category 1 after doing major damage to Brownsville and the Rio Grande Valley. Five ships and around 50 tugs sheltered safely in the main turning basin. Still, the surge from this storm flowed over the breakwater constructed after 1919, once again flooding North Beach and the lower

[52] Port of Corpus Christi, 2026 *Strategic Plan*, 5; Port of Rotterdam Authority, *Highlights, Annual Report*, 2022, 38–43.

[53] Blackburn, *Texan Plan*, ch. 5–6.

[54] Port of Corpus Christi, 2026 *Strategic Plan*, 5; John R. Moses, "Cheniere Taking Biden Adminstration's LNG Export Terminal 'Pause' in Stride," Corpus Christi *Caller-Times*, Jan. 26, 2024.

city, though not until 1938 did voters authorize the long-contemplated seawall.[55] Texas City is infamous for the 1947 fertilizer ship explosion in which an estimated 576 people died. Yet this may be the most vulnerable of the Houston area's collections of oil facilities and petrochemical plants. Repeated hurricanes, including one in 1943 as the federal government was underwriting a war-related expansion of chemical manufacturing there, prompted the Army Corps to build and later raise the Texas City levees that Rice University models now predict will be overmatched by intensified storm surges[56] (see Fig. 6).

Much of the investment from Eagle Ford went to areas such as the La Quinta Channel or Ingleside that were well outside Corpus Christi's inland turning basins. One recalls that after the 1919 storm, port planners had abandoned these areas as too dangerous. The port now devised flood mitigation schemes for sites that could not meet the customary understanding of semi-protection. One scheme that gained wide notice involved the use of dredged material to create an absorbent marsh habitat across from the La Quinta Channel. The marsh, the port argued, would enhance the state's carbon sequestration capacity while, by the way, reducing emissions.[57]

Historians of cities, technology, and urban environments resort to the concept of *path dependence* to describe the situation in which Texas's oil ports find themselves. The concept expresses the tendency for arrangements, processes, institutions, or technologies, often good ideas at the time, to perpetuate themselves, imposing constraints on future decisions and actions. Indeed, since the 1970s, analysts have noted that the energy sector's price cycles and its other upheavals have, paradoxically, tended to reinforce the Texas coast's engagement with petroleum and petrochemicals. This occurred despite much discussion of regional diversification and imaginative attempts at it, for example, Houston's thriving aerospace sector and its renowned medical sector.[58] Path dependence may be too tame a concept to express the situation confronting the ship channel cities of the Texas Coast. The change needed is so broad in scale and scope that it amounts to an abandonment of a political economy, a culture, and a way of life. As this volume suggests, aspects of US vulnerability to coastal disaster arose from economic, political, or cultural conditions distinctive to the United States. The bleak fate foretold for the Texas coast occupies a central place in the global story of global warming.

[55] Mary Jo O'Rear, *Bulwark against the Bay: The People of Corpus Christi and Their Seawall* (College Station: Texas A&M University Press, 2017), 81–82, 106–108; Coleman McCampbell, *Texas Seaport* (New York: Exposition Press, 1952), 101.

[56] John Neal Phillips, *Sitting on a Keg of Dynamite: Father Bill, Texas City, and a Disaster Foretold* (Norman: University of Oklahoma Press, 2022), 68–70, 102–104; Alperin, *Custodians of the Coast*, 224–233.

[57] Mott McDonald and the Port of Corpus Christi, "La Quinta Aquatic Habitat Mitigation Project," presentation, Western Dredging Association, Chicago, June 2019 (westerndredging.org/phocadownload/2019_Chicago/La Quinta.pdf, accessed Aug. 29, 2024); Judith Powers, "La Quinta Aquatic Habitat Flourishing in Corpus Christi Bay," *Waterways Journal Weekly*, June 18, 2018 (www.waterwaysjournal.net, accessed Aug. 29, 2024).

[58] Lessoff, *Texas Coast*, 341–343.

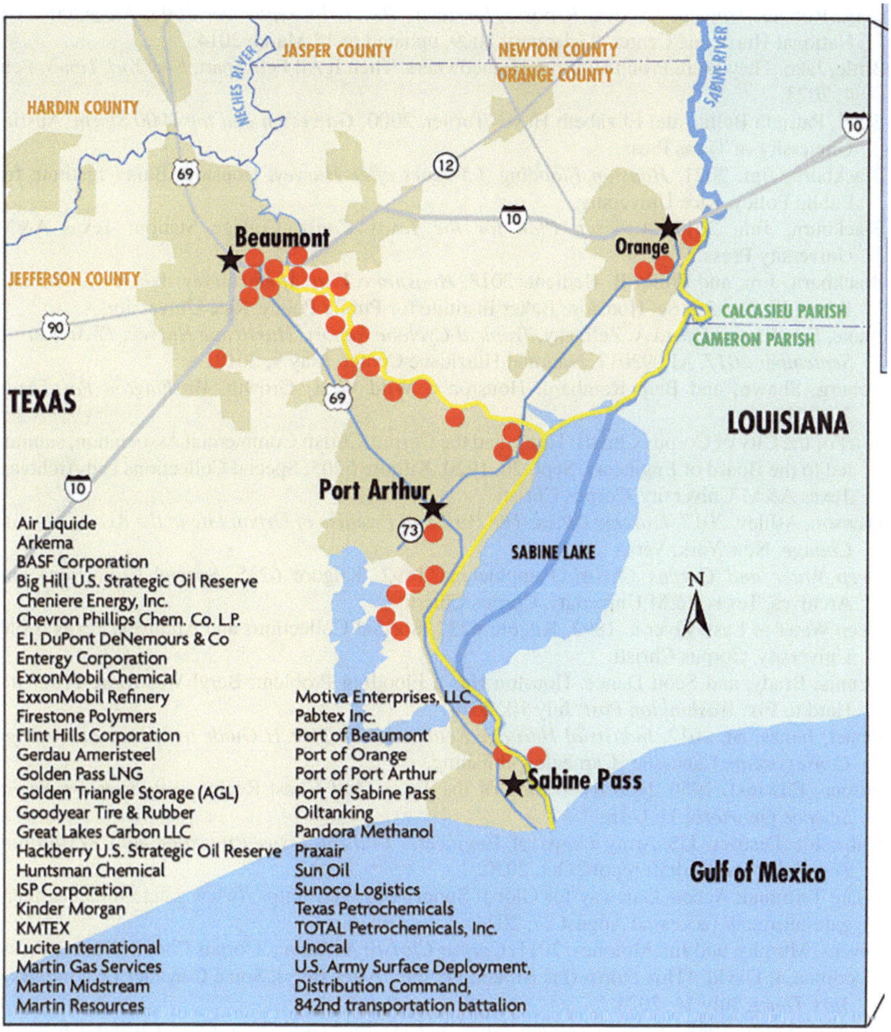

Fig. 6 Map of energy operations along the 57-mile Sabine-Neches Waterway on the southeast Texas coast. The Army Corps of Engineers and local officials devoted considerable attention to the combined threat of sea-level rise and intensifying gulf storms in this vulnerable area. Source: Courtesy Sabine-Neches Navigation District, https://navigationdistrict.org/fact_sheets/waterway-quick-facts/

References

Alperin, Lynn M. 1967. *Custodians of the Coast: History of the United States Army Engineers at Galveston*. Galveston: United States Army Corps of Engineers.

Bartlett, Terrell. The Tidal Storm at Corpus Christi and Its Effect on Engineering Structures. *Engineering News-Record* 83 (Nov. 13–20, 1919): 848–52.

Berg, Robbie. *Tropical Cyclone Report: Hurricane Ike, 1–14 September 1008*. AL092008. US National Hurricane Center, 23 January 2009, updated to 18 March 2014.

Bittle, Jake. They Were Proud Houston Homeowners. Then It All Fell Apart. *New York Times*, Feb. 4, 2023.

Bixel, Patricia Bellis, and Elizabeth Hayes Turner. 2000. *Galveston and the 1900 Storm*. Austin: University of Texas Press.

Blackburn, Jim. 2021. *Houston Flooding 3.5 Years after Harvey*. Houston: Baker Institute for Public Policy, Rice University.

Blackburn, Jim. 2017. *A Texan Plan for the Texas Coast*. College Station: Texas A&M University Press.

Blackburn, Jim, and Philip B. Bedient. 2018. *Houston a Year After Harvey: Where We Are and Where We Need to Be*. Houston: Baker Institute for Public Policy, Rice University.

Blake, Eric S., and David A. Zelinsky. *Tropical Cyclone Report: Hurricane Harvey, 17 August—1 September 2017*. AL092017. National Hurricane Center, May 9, 2018.

Boburg, Shawn, and Beth Reinhard. Houston's 'Wild West' Growth. *Washington Post*, Aug. 29, 2017.

Brief of the City of Corpus Christi, Texas, and the Corpus Christi Commercial Association, submitted to the Board of Engineers, Sept. 20, 1920. Kilgore 6205, Special Collections and Archives, Texas A&M University, Corpus Christi.

Dawson, Ashley. 2017. *Extreme Cities: The Peril and Promise of Urban Life in the Age of Climate Change*. New York: Verso.

Deep Water and Corpus Christi. Pamphlet, c. 1892. Kilgore 6235, Special Collections and Archives, Texas A&M University, Corpus Christi.

Deep Water at Last. Flyer c. 1892. Kilgore 6237, Special Collections and Archives, Texas A&M University, Corpus Christi.

Dennis, Brady, and Scott Dance. Houston Has a Flooding Problem; Beryl Was More Proof It's Hard to Fix. *Washington Post*, July 10, 2024.

Douet, James, ed. 2012. *Industrial Heritage Retooled: The TICCH Guide to Industrial Heritage Conservation*. Lancaster: Carnegie Publishing.

Foscue, Edwin J. 1950. Industrialization of the Texas Gulf Coast Region. *Southwestern Social Science Quarterly* 31:1–18.

Galveston District, US Army Corps of Engineers. *Coastal Texas Protection and Restoration Feasibility Study*, draft report, Oct. 2020.

"Gate Terminal: Access Gateway for Global Supply of LNG." https://www.gateterminal.com/en/gate-terminal/ (accessed August 27, 2024).

Givens, Murphy, and Jim Moloney. 2011. *Corpus Christi: A History*. Corpus Christi: Nueces Press.

Goodman, I. David. 'This Storm Has Broken People': After Beryl, Some Consider Leaving. *New York Times,* July 16, 2024.

Guthrie, Keith. 1988. *Texas Forgotten Ports, vol. 1: Mid-Gulf Ports From Corpus Christi to Matagorda Bay*. Austin: Eakin Press.

Harris County Flood Control District. *Hurricane Harvey: Impact and Response in Harris County*. Houston, May 2018.

Hein, Carola. 2018. Oil Spaces: The Global Petroleumscape in the Rotterdam/the Hague Area. *Journal of Urban History* 44:887–929.

Hein, Carola. 2020. Port City Regions as Nodes in the Global Petroleumscape. *Sartonia* 33:193–224.

Hein, Carola, ed. 2022. *Oil Spaces: Exploring the Global Petroleumscape*. London: Routledge.

Industrial Banquet, Plaza Hotel, Corpus Christi, Oct. 26, 1934, program in Southern Alkali packet, Lessoff collection, Special Collection and Archives, Texas A&M University—Corpus Christi.

Johnson, Lacy M., and Cheryl Beckett, eds. 2022. *More City than Water: A Houston Flood Atlas*. Austin: University of Texas Press.

Kennedy, Merritt. Harvey The 'Most Significant Tropical Cyclone Rainfall Event In U.S. History.' Jan. 25, 2018. www.npr.org (accessed Jan. 23, 2024).

Kent, William E. 1892. Ropes Pass, Texas. *Engineering Magazine* 3:340–349.
Kosovich, John J. *State of Texas—Highlighting Low-Lying Areas Derived from USGS Digital Elevation Data*. U.S. Geological Survey Scientific Investigations Map 3050, 2008. https://pubs.usgs.gov/sim/3050/ (accessed Sept. 1, 2024).
Kothuis, Bauke, et al., eds. 2015. *Delft Delta Design: Houston Galveston Bay Region, Texas U.S.A*. Delft: Delft University Publishers.
Laune, Richard A. 2005. The Political Struggle for the Port of Corpus Christi. *Journal of South Texas* 18:91–112.
Lessoff, Alan. 2008. Galveston Hurricane. In *American Disasters*, ed. Ballard Campbell, 185–187. New York: Checkmark Books.
Lessoff, Alan. 2023. The Texas Coast: Ship Channel Network of the Petroleum Age. *Urban Planning* 8:330–345.
Lessoff, Alan. 2015. *Where Texas Meets the Sea: Corpus Christi and Its History*. Austin: University of Texas Press.
Malsch, Brownson. 1988. *Indianola: The Mother of Western Texas*. Rev. ed. Austin: State House Press
Maywald, Lonnie Ficklen. Port Lavaca, TX. *Handbook of Texas Online*. www.tshaonline.org/handbook/entries/port-lavaca-tx (accessed Jan. 27, 2024).
McComb, David G. 2015. *The City in Texas: A History*. Austin: University of Texas Press.
McComb, David G. 1986. *Galveston: A History*. Austin: University of Texas Press.
McComb, David G. 1969. *Houston: The Bayou City*. Austin: University of Texas Press.
Melosi, Martin V., and Joseph A. Pratt, eds. 2007. *Energy Metropolis: An Environmental History of Houston and the Gulf Coast*. Pittsburgh: University of Pittsburgh Press.
McCampbell, Coleman. 1952. *Texas Seaport*. New York: Exposition Press.
Medlock, Kenneth B. III. *The Future of Houston as Energy Transitions*. Baker Institute for Public Policy Report, May 13, 2021, Rice University, Houston.
Morton, Timothy. 2013. *Hyperobjects: Philosophy and Ecology after the End of the World*. Minneapolis: University of Minnesota Press.
Moses, John R. Cheniere Taking Biden Adminstration's LNG Export Terminal 'Pause' in Stride. *Corpus Christi Caller-Times*, Jan. 26, 2024.
Mott McDonald and the Port of Corpus Christi. La Quinta Aquatic Habitat Mitigation Project. Presentation, Western Dredging Association, Chicago, June 2019. westerndredging.org/phocadownload/2019_Chicago/La Quinta.pdf, accessed August 29, 2024.
National Weather Service. *The Hurricane of 1919*. www.weather.gov/crp/hurricane1919 (accessed Dec. 28, 2023).
National Weather Service. *Major Hurricane Harvey—August 25–29, 2017*. www.weather.gov/crp/hurricane_harvey (accessed Jan. 22, 2024).
O'Rear, Mary Jo. 2022. *Barrier to the Bays: The Islands and the Texas Coastal Bend and Their Pass*. College Station: Texas A&M University Press.
O'Rear, Mary Jo. 2017. *Bulwark against the Bay: The People of Corpus Christi and Their Seawall*. College Station: Texas A&M University Press.
O'Rear, Mary Jo. 2009. *Storm Over the Bay: The People of Corpus Christi and Their Port*. College Station: Texas A&M University Press.
Parsons, James J. 1950. Recent Industrial Development in the Gulf South. *Geographical Review* 40:67–83.
Phillips, John Neal. 2022. *Sitting on a Keg of Dynamite: Father Bill, Texas City, and a Disaster Foretold*. Norman: University of Oklahoma Press.
Pilkey, Orrin H., and Keith C. Pilkey. 2019. *Sea Level Rise: A Slow Tsunami on America's Shores*. Durham, NC: Duke University Press.
Platt, Harold L. 1983. *City Building in the New South: The Growth of Public Services in Houston, Texas, 1830–1915*. Philadelphia: Temple University Press.
Plummer, Brad, and Nadja Popovich. How the U.S. Became the World's Biggest Gas Supplier. *New York Times*, Feb. 3, 2024.

Port of Corpus Christi. *Liquid Bulk*. https://portofcc.com/capabilities/cargo/liquid-bulk/ (accessed Sept. 1, 2024).
Port of Corpus Christi. *Port Book, 1976, 50th Anniversary*.
Port of Corpus Christi. Port of Corpus Christi and Port of Rotterdam Enter into Historic Agreement. Media advisory, Feb. 25, 2021.
Port of Corpus Christi. *Strategic Plan, 2014–2020*, Dec. 10, 2013.
Port of Corpus Christi. *Strategic Plan,* 2026. PDF download, https://portofcc.com/about/port/about-us/.
The Port of Corpus Christi, Where Texas Meets the Sea, Opens to World Commerce. Pamphlet 1926, copy in author's possession.
Port of Rotterdam. Gate Terminal Starts Construction of 4th LNG Tank at the Port of Rotterdam. News Release. Aug. 24, 2023a. https://www.portofrotterdam.com/en/news-and-press-releases/gate-terminal-starts-construction-of-4th-lng-tank-at-the-port-of-rotterdam (accessed August 27, 2024).
Port of Rotterdam. *Halfjaarbericht 2023: Verslag van de Directie*. https://reporting.portofrotterdam.com/halfjaarbericht-2023/verslag-van-de-directie/ (accessed Sept. 1, 2024).
Port of Rotterdam. *Half-Year Report*, 2023b.
Port of Rotterdam. *Highlights, Annual Report,* 2022.
Powers, Judith. La Quinta Aquatic Habitat Flourishing in Corpus Christi Bay. *Waterways Journal Weekly*, June 18, 2018. www.waterwaysjournal.net, accessed Aug. 29, 2024.
Pratt, Joseph A., Martin V. Melosi, and Kathleen A. Brosnan, eds. 2014. *Energy Capitals: Local Impact; Global Influence*. Pittsburgh: University of Pittsburgh Press.
Price, W. Armstrong. *Hurricanes affecting the Coast of Texas from Galveston to Rio Grande*. Technical Memorandum no. 78, Beach Erosion Board, Corps of Engineers, 1956. Copy in Special Collection and Archives, Texas A&M University—Corpus Christi.
Rhodes, Anna, and Max Besbris. 2022. *Soaking the Middle Class: Suburban Inequality and Recovery from Disaster*. New York: Russell Sage Foundation.
Rice, Harvey. Hurricane Ike Worst Storm in Decades. *Houston Chronicle*, Oct. 22, 2016.
Stapleton, Alex, dir. *God Save Texas*, episode 2: *The Price of Oil*. HBO Max documentary, 2024.
Tunnell, John W., Jr., and Frank W. Judd, eds. 2002. *The Laguna Madre of Texas and Tamaulipas*. College Station: Texas A&M University Press.
Vergara, Camilo José. 1999. *American Ruins*. New York: Monacelli Press.
Walraven, Bill. 1997. Port of Corpus Christi, 1500–1926. *Nueces County Historical Commission Bulletin* 4:5–9.
Yablon, Nick. 2010. *Untimely Ruins: An Archaeology of American Urban Modernity, 1819–1919*. Chicago: University of Chicago Press.
Young, Earle B. 1997. *Galveston and the Great West*. College Station: Texas A&M University Press.

“Becoming Green”: Resilient and Green Building as Risk Mitigation in Atlanta, Georgia

Katharina Wood

“A nation that destroys its soils destroys itself.”

—Franklin D. Roosevelt qtd. in Ratcliffe 2017

Abstract Buildings constructed with resilience in mind can serve as spaces of refuge during natural disasters and mitigate the effects of these events through effective stormwater management or energy independence. The “Kendeda” living building in Atlanta, Georgia, was built to moderate these risks and serve as an example of regenerative and green building practices in the American South. It is the first building that was awarded the living building challenge certification in the South West in order to showcase that regenerative and climate-sensitive building is possible even in hot and humid climates. This article investigates in which ways the Kendeda building contributes to climate-sensitive and resilient building practices and how building standards can reduce disaster risks and promote equity. Whereas the green building standards LEED (Leadership in Energy and Environmental Design) and LBC (Living Building Challenge) are critically evaluated, the case study highlights that on the one hand green building standards provide tangible solutions with regard to disaster prevention and resilience. On the other hand, the study establishes that single “eco-cathedrals” [Eco-cathedrals are a concept introduced by urban planner Hans Schulte to refer to flagship projects that serve an inspirational function and as cultural nodes (Günter 2010)]. will not suffice as a mitigation strategy, but more widespread application of green building standards is necessary to achieve climate goals and create resilient communities. This chapter conceptualizes the term “eco-cathedral” in a more ambidextrous way: On the one hand, eco-cathedrals serve as inspiring flagship projects. On the other hand, they often stand alone, serving more

K. Wood (✉)
Technical University Dortmund, Dortmund, Germany
e-mail: katharina.wood@tu-dortmund.de

N. Rauscher, W. Werner (eds.), *Natural Disasters in the United States*, Risk, Governance and Society 22, https://doi.org/10.1007/978-3-031-96436-7_12

of a representational function, and their achievements are oftentimes not transferred to more low-cost projects.

Introduction

Every year, Georgia is affected by several forms of natural disasters such as an average of 30 tornadoes, wildfires, floods, extreme heat and droughts, or tropical storms, and hurricanes. Due to global warming, these disasters occur more often and have stronger impacts (Stott 2016; McBean 2004). In a CPI-adjusted[1] event-count analysis based on billion-dollar disasters in the United States between 1980 and 2023, 2023 leads the count with 23 major disasters in comparison to an average of eight major events throughout the studied period (NOAA 2023). The city of Atlanta, GA, is especially at risk of flooding and severe storms (Wright et al. 2012). The case study at hand investigates one building confronting, renegotiating, and mitigating these risks: The Georgia Tech "Kendeda" living building in Atlanta. The Kendeda building is a highly innovative building showcasing state-of-the-art trends in green, resilient, and regenerative building.

As a first step, this study introduces the relevance of building codes and the green building standards LEED (Leadership in Energy and Environmental Design) and LBC (living building challenge) and contextualizes resistance to building codes and standards with regard to the culture of American individualism. The respective green building standards have been shaped by different green concepts such as resilience, circularity, or sustainability to varying degrees. In order to understand and reflect on resiliency efforts in regard to the prevention of natural disasters, the ways in which resilience materializes in the Kendeda building are assessed. Accordingly, the concept of resilience is defined from the perspective of different disciplines in engineering, the social sciences, and humanities and then applied to the aspirations of the Kendeda building in the case study. In addition, the case study looks beyond materiality at narrative approaches toward the building and how it is perceived. In a final step, the significance of the Kendeda buildings' achievements is scrutinized through the previously established contexts of what building standards, and their respective buildings can or cannot provide as mitigation strategies for Atlanta's communities. This study employs techniques from the field of "rapid ethnography." Baines and Cunningham explain that

> [r]apid ethnography (RE) is a form of multi-method ethnography involving data collection from numerous sources over a relatively short period of time including interviews, participant observations, document review and sometimes surveys and focus groups. The goal of this method is to produce a multi-layered, textured analysis of a slice of life in an organization, environment or some part of society, intertwining the structure and consciousness with the larger socio-political context shaping social relations. (Baines and Cunningham 2013, p. 74)

[1] CPI = consumer price index.

For this reason, expert interviews as well as anecdotal student interviews are included in the analysis of the Kendeda building case study. The data for the case study was collected during a three-month participant observation internship with the architectural and engineering firm "Epsten Group" in Atlanta in 2022. The company represents a relevant and well-known stakeholder in the realm of green building certification in the Atlanta metropolitan area and beyond. This case study is engaged in the transdisciplinary field of urban studies. By researching green building standards and buildings as cultural artifacts from a literary and cultural studies perspective, new questions of sustainability and climate adaptation are focalized. Among these questions are the following: For whom are flagship sustainable buildings built? Through which conceptual origins is their building culture established? And in which ways do these buildings serve specific purposes like refuge or natural disaster mitigation? The case study of the Kendeda building gives answers to these questions from a qualitative standpoint that can serve to open up larger discourses through inductive reasoning.

Building Codes and Standards in the United States

As far back as the nineteenth century, structures like London's Crystal Palace and Milan's Galleria Vittorio Emanuele II used passive systems,[2] such as roof ventilators and underground air-cooling chambers to moderate indoor temperature (Cassidy 2003, p. 4). In the early twentieth century, skyscrapers like New York's Flatiron Building and the New York Times Building employed deep-set windows to shade the sun. Also, the first photovoltaics were developed in the 1800s. Nevertheless, the contemporary green building movement did not gain traction until the 1970s.[3] Around the time the "glass box" style high rise had become the icon of the American city, a forward-thinking group of architects, environmentalists, and ecologists were inspired by the growing environmental movement and the high fuel costs that were prevalent during the 1970s (Cassidy 2003, p. 4). The genesis of these two scenarios

[2] Passive cooling systems require little to no external energy or movement to operate.

[3] Certainly, traditions of building inspired by nature are much older than the green building movement that established itself in the 1970s in Europe and North America. Nature-inspired building techniques for example used on Nouméa by Kanak people developed in a process of trial and error over centuries in an evolutionary process (Nachtigall 2003 170). These architectures, are influential in contemporary green building discourses especially in the field of building biology and bionics (German: Baubiologie, Baubionik). The green building movement that started in the 1970s was inspired by the European and North American environmental movement as a reaction to the environmental burden brought forth by the on-going industrialization processes and environmental pollution. At this time, certain building materials like asbestos were widely used and perceived as ingenious materials by construction professionals. A broad awareness for the health risks of building materials was developed over many years as a result of the mix between a re-gained environmental consciousness and building expertise.

ultimately resulted in the modern green building movement. The first Earth Day, celebrated in 1970, granted even more credibility to this new building concept. Additionally, with the Clean Air (1970) and Clean Water Acts (1972), the United States began to regulate air and water pollutants.

The first explicit green building program was launched in Austin, Texas under the title Austin Green Building Program in 1991. Designing and implementing green building designs is arguably one of the most crucial steps to greatly reducing greenhouse gas emissions. According to the US Green Building Council, buildings are responsible for 39% of the United States yearly carbon emissions (USGBC 2024a, "Building Decarbonization"). With the increasing recognition of human's influence on global climate change and the necessity to reduce reliance on oil and other non-renewable energy sources, combined with the fact that buildings account for a large portion of carbon dioxide emissions, it may seem logical that the green building movement and green cities have become more prominent over the past five decades.

The green building industry, furthermore, measures sustainability through the use of rating tools. Certification systems provide several benefits, such as risk management, optimization, and quality control. Generally, certifications have important benefits with regard to the planning, use, and reuse phases. They ensure that a wide arrangement of criteria is included from day one. The certification system "Leadership in Energy and Environmental Design" (LEED) was developed as a pilot rating system by the Green Building Council (USGBC) in 1998. In 2023, version 5 is being drafted, targeting especially carbon emissions, equity, health, and resilience (USGBC, "What's new in LEED v5"). LEED standard is a widely recognized internationally acclaimed green building standard. Version 4 incentivizes resilience criteria in the aspects of "sustainablee One of the benefits of LEED certification is that it has come to be internationally renowned and has developed an international reputation in the construction industry. This is especially interesting for firms that sell their products globally or want to establish such a global reputation. The living building challenge (LBC), on the other hand, is one of the most ambitious green building standards in the United States that is gaining more and more recognition—although it is used far less than LEED due to its stronger requirements that include the monitoring of actual energy use over the course of an entire year. The LBC was created in 2006 by the non-profit International Living Future Institute.

The Living Building Certification groups its aspirations and focus areas into different petals of a flower. The petals are: "place," "water," "energy," "materials," "health & happiness," "equity," and "beauty." (Fig. 1)

According to the International Living Future Institute, the metaphor of the flower is used because "the ideal built environment should function as cleanly and efficiently as a flower" (International Living Future Institute 2023). According to the co-founder of the standard, Jason McLennan, "flowers are an accurate representation of a truly regenerative building which receives all of its energy from the sun, nutrients from the soil, and water from the sky. Similar to a flower, they simultaneously shelter other organisms and support the surrounding ecosystem. They also serve as beauty and inspiration and adapt to their surroundings" (Thomas 2016). In contrast to LBC, LEED is more easily adaptable to current market standards.

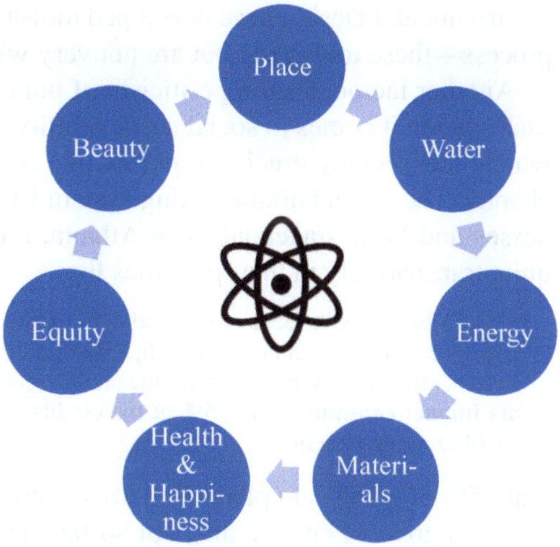

Fig. 1 Living Building Certification Petals, Source: Own Diagram

Criticisms of Building Codes and Standards

Kaid Benfield, co-founder of the LEED for Neighborhood Development system,[4] critically recognizes

> but, man, there are a lot of warts in this system. For starters, LEED has been criticized for being insufficiently demanding of its applicants. I believe that, to the extent this criticism is well-placed, it stems from a belief held by many involved with the Green Building Council, some of them representatives of the building industry, that the standards should be set only a little bit above what industry is likely to do anyway (Benfield 2016).

Critics of LEED follow this line of argumentation, arguing that LEED is not aspirational enough. A USA Today research of 7100 LEED-certified commercial buildings demonstrates "that designers target the easiest and cheapest green points by trying to create pleasant and healthful office spaces; using common building materials; or taking steps with an unknown effect, such as providing preferred parking for fuel-efficient cars, bike racks and showers, and posting educational displays about the building" (USA Today 2012). In contrast, the US Green Building Council tries to counter negative images through "myth-busting"—especially the myth that the system is not "innovative" (US Green Building Council). The USGBC argues that it has an innovation category. Nonetheless, this category is minor and therefore only a small incentive overall to be more innovative in LEED-certified projects.

One of the systems' blind spots remains social sustainability and equity. In his article "LEED after Ten Years," architect Michael Zaretsky evaluates that by "ignoring the social issues of sustainable design, LEED purports a message that a 'green' building is not required to address social and economic issues" (Michael Zaretsky 2010, p. 193). In response to this criticism, networks like SEED: Social Economic

[4] LEED for Neighborhood Development is a version of LEED. LEED has several versions for different use cases.

Environmental Design have developed tools that promote social equity in the design process—these tools exist but are not very widely adopted.

Another factor of strong criticism of building codes and standards in the United States when it comes to social sustainability is its blind spot when it comes to race, which is especially crucial in places like Atlanta, Georgia—the case study in this chapter. The green building rating system LEED, which is mainly adopted by businesses and large stakeholders in Atlanta, reflects the lack of social sustainability quite transparently in their priorities list:

> Of all LEED credits, 35% of the credits in LEED are related to climate change, 20% of the credits directly impact human health, 15% of the credits impact water resources, 10% of the credits affect biodiversity, 10% of the credits relate to the green economy, **5% of the credits impact community** and 5% of the credits impact natural resources. (USGBC 2023a, "LEED rating system")

Only 5% of credits are planned to affect "community." There are several pilot credits in progress to change this, but so far there is no incentive to incorporate these with more weight in the rating system. The way the green building rating system LEED is built illustrates the concepts of "greenness." It is conceptualized through which is a mix of **ecological sustainability** (mostly energy efficiency) and **ecological resilience**. The question of "sustainability for whom?" plays a very minor role, even though the health of the building's inhabitants plays a relatively large role as it is included in 20% of the rating's criteria. Green buildings are conceptualized to be healthy buildings for those who use them through the use of low-emitting materials, daylight views, and human-friendly air, acoustic, and thermal qualities.

Nevertheless, the question remains: "Who inhabits these buildings?" The vast amount of LEED certification in Atlanta happens in the affluent corporate sector while the rest of South-West Atlanta, where the LEED Platinum Mercedes Benz Stadium is situated, suffers from energy insecurity and burdens, as well as poorly insulated buildings. There is a history of disinvestment in the Western and Southern areas of the Atlanta metropolitan area. Racial segregation is the most relevant historical influence on this disparity. West Atlanta has fewer green spaces, and its main waterways are polluted. Researchers and partly Atlanta-based activists Milligan, McCreary, and Jelks further conclude that "Atlanta's urban hydrology presents a landscape of racial disparity" (Milligan et al., 2021, p. 1600). Proctor Creek, which runs through West Atlanta, is at high risk of stormwater flooding and sewage overflows polluting the waterway in the majority Black and economically depressed neighborhood. High levels of bacteria in the overflowing water regularly cause problems for the local residents (US EPA 2014, "Urban Waters and the Proctor Creek Watershed/Atlanta (Georgia)").

Gentrification has further become a more pressing issue and challenge in recent years in the Atlanta metropolitan area. Especially "green gentrification" surrounding the "Atlanta Beltline"[5] has fallen short of providing affordable housing projects

[5] The Beltline is a multi-use trail in Atlanta on former railroad tracks with adjacent green spaces, culture venues, and event spaces. It connects multiple neighborhoods and is used for leisure activities. The idea for the project was developed in a master's thesis by Georgia Tech student Ryan Gravel, and its development is ongoing.

that had originally been promised by the project's initiators. The Beltline has made the city more attractive to what Richard Florida coins as the "creative classes" and also increased the property values of adjacent spaces, adding to the pressures on legacy residents. The question of accessibility and equal distribution of green and quality housing therefore remains urgent in Atlanta. From a normative standpoint, green and resilient buildings should be available and accessible not only to affluent groups. Abe Kruger, founder of SkCollaborative, and other advocates in Atlanta promote the inclusion of sustainability criteria, especially for affordable housing and mixed-income housing. Fortunately, the LIHTCs (Low Income Housing Tax Credits) in Georgia legislate that new affordable housing projects must adhere to one of several green building standards such as LEED. This promises to alleviate some of the energy burden for the most vulnerable populations. Nevertheless, Atlanta has a history of reducing affordable housing while not remodeling or replacing it adequately. This for example happened in the case of the demolition of Techwood Homes prior to the 1996 Olympics, one of the first public housing projects in the United States. The Kendeda building stands in contrast to these questions of equity as more of a flagship project that is on the one side accessible to specifically educated campus audiences but remains a flagship project with little outreach toward these Westside communities (see Chapter "All of the Above: Energy, Environmental Politics, and US Climate Change Policy since the 1970s Oil Crises").

Generally speaking, both the LEED and LBC rating systems could be more equipped to deal with social and negotiated resilience or adaptation urbanism (see Chapter "Nature by Numbers: The National Flood Insurance Program"). LEED has a strong focus on health-related indoor environmental qualities that enhance user satisfaction such as thermal, visual, and acoustic comfort (USGBC 2023a, "LEED rating system"). LBC, on the other hand, goes a lot further with more community-oriented resilience criteria such as "Human Scale and Humane Places," "Universal Access to Nature and Place," "Equitable Investment," and "JUST Organizations." For LBC, all buildings must ensure access for people with disabilities and permit access to fresh air, light, and water. The criterion of JUST organizations evaluates the equity of business practices in the project under the criteria of diversity, equity, safety, worker benefit, and stewardship (Georgia Tech, "Equity Potal").

Apart from warranted criticism about some aspects of the building standards such as social equity considerations, insurance data has shown that stronger building codes have a positive effect on natural disaster mitigation. For example, buildings can limit the effects of windstorm losses by 40–60% (Czaijkowski 2019, p. 311). During disaster events, "the most vivid images presented [...] involve buildings" (Czaijkowski 2019, p.303), as Economist Jeffrey Czaijkowski argues. Losing a home can become an existential crisis for the former inhabitants, not simply monetarily but also with regard to identity, memory, and attachment to possessions and place. In his analysis of building code implementation in the United States, Czaijkowski finds that stricter codes do not only have safety benefits but also economic benefits as buildings built under them sell for 10% higher prices (Czaijkowski 2019, p. 306). Yet, adaptation and code prescriptions vary strongly by state and region (Czaijkowski 2019, p.307). He also finds that "36% of states do not adhere to uniform statewide residential building code standards" (Czaijkowski 2019, p.

307). If buildings that meet more elaborate codes and standards withstand natural disasters with higher resiliency rates and are even sold for higher premiums, why aren't these standards adopted more widely? I argue that opposition to more regulation through stricter building codes or certification systems culturally stems from US-American affection for cultural concepts such as self-reliance and individualism. As social scientist Toby Miller describes, "American individualism is a bad source of ideas for addressing global warming and environmental degradation, [...] because it misreads the real role of government in our lives" (Miller and Maxwell 2017).

Strikingly, global warming is perceived differently by humans with more individualistic cultural values in contrast to more egalitarian views (Smith and Leiserowitz 2012, p. 1029). Environment and forestry scholars Smith and Leiserowitz describe that individualism and egalitarianism are strong predictors with regard to pro-environmental behaviors (Smith and Leiserowitz 2012, p. 1030). On this note, environmental philosopher Baird Callicott emphasizes that "the hyper-individualistic and rationalistic paradigms—originating in the late eighteenth century and dominating moral philosophy—cannot capture the moral concerns evoked by the prospect of global climate change" (Callicott 2011, p. 101). Few solutions to a warming climate can be found at the individual level. Solutions to climate change demand laws, policies, and regulations at higher regional, national, or international levels (Callicott 2011, p. 115–116). One possible measure for climate change mitigation would be the implementation of green building certification systems at these levels. These certification systems contribute to resilience on the economic, social, but most strongly on ecological levels. How do these standards contribute to disaster mitigation? First of all, certifications encourage a structured and well-organized planning phase that considers a manifold catalog of criteria related to ecological resilience. Second, sustainable building certification systems tackle the problem from two directions: prevention and adaptation.

Defining Resilience: Setting the Scene

In order to reflect the achievements of the Kendeda Living Building, which will be detailed below, the concept of resilience will be reflected through the lens of different disciplines. This opens a space for reflecting the impact of resilience in urban planning as a formulaic and form-giving concept not only in sustainability studies but also for the construction of the Kendeda building itself.

The term resilience is used in a multitude of disciplines such as systems thinking, psychology, ecology, geography, or sociology. Conceptualizations of "resilient cities" have evolved as "normative, desired goals in both academic and policy arenas" (Meerow and Newell 2019, p. 311), and Béné et al. describe resilience as a relevant "policy narrative" that drives urban planning paradigms (Béné et al., 2018. p. 116). The Rockefeller Foundation's 100 Resilient Cities (100RC) program has also popularized the idea of urban resilience manifesting specific "urban resilience imaginaries and practices" (Webber et al., 2021, p. 344). Urban planner Rega and engineer Bonifazi claim that "not only resilience has gained prominence in science and policy, but it has also emerged first as a complement to, then increasingly as a substitute for, sustainability" (Rega and Bonifazi 2020, p. 2). Rega and Bonifazi call this the "resilience turn" in spatial planning (Rega and Bonifazi 2020, p. 2). They further argue that "the promise held by sustainability that better living conditions for all were within reach (no matter how slow the progress) has been losing grip in the past decades of economic and social turbulences, and growing awareness of global environmental uncertainties" (Rega and Bonifazi 2020, p. 2). Whereas sustainability has come to be regarded more and more as an "empty signifier," resilience holds the potential to describe green aspirations in a more clear-cut manner.

In order to further conceptualize the term, Rega and Bonifazi distill three major conceptual approaches to resilience. First of all, "engineering resilience" stemming from the discipline of engineering, and theorized by Holling in the 1970s, focuses on the ability of systems to bounce back to their state of origin when confronted with major shocks (Rega and Bonifazi 2020, p. 4). Second, "ecological resilience" refers to "the magnitude of the disturbance that can be absorbed before the system changes its structure" (Holling 1996, p. 33). Third, "socio-ecological resilience," transformational or adaptive resilience can be defined as a "dynamic non-equilibrium, it signals that systems undergo constant changes and have no stable state. Here, resilience is the ability of the system not only to bounce back but also to adapt and transform" (Rega and Bonifazi 2020, p. 2).[6] Urban geographers Meerow and Newell critically state that systems thinking has prevailed in definitions of resilience focusing on "quantitative modeling rather than the interactions between individual components and dynamics within the boundaries of a system" (Meerow and Newell 2019, p. 310). Exactly, these interactions and dynamics are where transdisciplinary research is beneficial. In an often-cited report for the Environmental Advisory Council to the Swedish Government, scientists Folke et al. circumscribe resilience for socio-ecological systems as "(i) the magnitude of shock that the system can absorb and remain within a given state; (ii) the degree to which the system is capable of self-organization; and (iii) the degree to which the system can build capacity for learning and adaptation" (Folke et al., 2002, p. 438). Their approach combines different branches of resilience thinking, focusing not only on bouncing back but also on processes of self-organization and learning, thereby incorporating more social components.

With regard to "urban resilience," Meerow and Newell argue for the implementation and inclusion of the "five Ws of urban resilience": Resilience for whom, what, when, where, and why (Meerow and Newell 2019, p. 310). This includes perspectives of the humanities and social sciences on the concept of resilience asking about beneficiaries and affected groups. According to the authors, "this approach engenders a politics of resilience that includes grappling with trade-offs and scalar

[6] These different schools of resilience thinking and conceptualizations have been reviewed and theorized by many scholars. In their article "negotiated resilience," Harris et al. provide an extensive overview of critiques regarding the term "resilience" (Harris et al., 2017, p. 7–9).

complexities and delineating how political context and power dynamics shape resilience policies, with inevitable winners and losers" (Meerow and Newell 2019, p. 311). Those on the losing end of resilience measures are often forgotten, while the winners celebrate their successes. This critique is also evident in Evans and Reid's critique of the use of resilience in policy discourses. Evans and Reid argue in *Resilient Life: The Art of Living Dangerously* (2014) that adopting the discourse of resilience is a neoliberal practice of putting responsibility on the individual to adapt in the face of dangers rather than on policymakers to make decisions that protect and secure healthy living conditions for humans and other species.[7] According to their analysis, adopting a resilience mindset places the "burden of the crises directly on the shoulders of the globally impoverished" (Evans and Reid 2015, p. 47). Evans and Reid warn that resilience can be practiced as "neoliberal interventionism" disregarding political and economic responsibilities (Evans and Reid 2015, p. 47).

In order to synthesize critiques and opportunities of thinking in terms of resilience, Geerlin et al. propose the term "adaptation urbanism."

> [C]urrent forms of resilience planning often operate in operational silos and can result in a type of market-oriented resilience [...] that improves safety for some but perpetuates many negative status quo trends. Adaptation urbanism in practice is designed to create a more holistic process of engaging residents in a long-term method of rebuilding cities to make them safer for a changing climate, more energy efficient through better urbanism, and more livable and equitable for all residents. This is the adaptation urbanism resilience process that is designed to focus policymakers and the public on a process for creating safer, desirable, more equitable, and sustainable places. (Geerlin et al., 2021, p. 62)

Adaptation urbanism is a powerful tool to tackle issues of resilience in a more equitable fashion. Introducing the concept of "negotiated resilience," Harris et al. argue that "[r]esilience thinking often overlooks issues of differentiated power attributed to political ideology, finance, class, race, or knowledge" (Harris et al., 2017, p. 2). Therefore, Harris et al. conceptualize the idea of "negotiated resilience" as a process with a focus on inclusivity and participation that always entails a contested and political character (Harris et al., 2017, pp.10–12). Drawing on research from social scientists Renn and Schweizer (2010), Harris et al. proclaim that "engagement will necessarily be uneven, political, and contested, and will point to complex trade-offs rather than win-win-win scenarios as often uncritically implied by the term" (Harris et al., 2017, p. 2). In practice, this can be exemplified by the city of Ghent's (Belgium) move toward resilience through removing pavement, facilitating greening, and installing bike lanes (Stad Gent 2023). Belgian politician Filip Watteuw explains in several interviews that the plans for creating a car-free city center were met with resistance and that trade-offs were inevitable (Faltermann 2021).

Environmental law scholar Shari Shapiro, considering research by Hassler and Kohler, further defines resilience for the built environment through the following two concepts (Shapiro 2016, p. 491):

---

[7] In this context, see also Rauscher's contribution to this volume and her reflections on the turn to resilience and away from discourses of vulnerability.

Resilience is a term that can 'collapse into meaninglessness [...] from having too many meanings' (Hassler and Kohler 2014, p. 119). Here, resilience incorporates two concepts: built-in resilience and disaster risk reduction. Lee Bosher defines 'built-in resilience' as 'a quality of a built environment's capability (in physical, institutional, economic and social terms) to keep adapting to existing and emergent threats' (Hassler and Kohler 2014, p. 123). By contrast, 'disaster risk reduction' is preventing threats from happening in the first place (Hassler and Kohler 2014).

Green building certifications, as I will argue in the following, contribute to built-in resilience as well as disaster risk reduction. Green building certifications further provide concrete criteria to measure resilience and circumvent and confront the problem of elusiveness—but they oftentimes fail at negotiated resilience and adaptation urbanism. Concepts of engineering resilience are pursued more thoroughly than the involvement of multiple stakeholders in achieving resilient communities. In the following case study of the Kendeda building, these different forms of resilience will be analyzed in practice with regard to their effects on creating resilient communities and/or mitigating climate-related risks.

Case Study: "Kendeda" Living Building in Atlanta, Georgia

Campus landscapes can serve as living laboratories for reducing carbon footprints, conserving water and aquatic resources, supporting biodiversity, and building active, equitable social communities. Moreover, as learning landscapes, such campuses actively promote sustainable design by engaging faculty, staff, and students in the design and implementation process as a part of the pedagogy of place. This progressive focus positions universities as leaders educationally and environmentally. (Way et al., 2012, p. 25)

Geographers Finlay and Massey argue that "[u]niversities and colleges are important sites of transformation as centres of discourse and vehicles of social change" (Finlay and Massey, 2012, p. 150). With this in mind, one of the core goals of the Kendeda building, which is situated on the campus of the Georgia Institute of Technology in Atlanta and is an experimental and educational structure, was to "demonstrate that a regenerative building can be built in a hot and humid climate" (Georgia Tech, "Key Living Building Details", Fig. 2). Most greening initiatives on US-American campuses focus on green buildings by implementing LEED certification or waste/facility management procedures (Way et al., 2012, p. 25). The number of "greening campuses" initiatives has steadily increased over the last decade (Steele 2019).

The Kendeda Living Building was built by means of the Kendeda Fund, which is a private family foundation based in Atlanta, Georgia. It combines many innovative and state-of-the-art efforts in green building. The Kendeda building was formerly a parking lot and "acts like a forest with 90% of average rainfall managed on site" (Georgia Tech, "Key Living Building Details"). In regard to renewable energy, the photovoltaic system supplied 225% of the building's energy needs. This far exceeds the 105% LBC requirement. The project further "kept economic benefits close to home by sourcing at least 50% of products and services from within 621 miles (i.e.,

Fig. 2 Georgia Tech "Kendeda Building," Photographer: Jonathan Hillyer, CC BY-NC-ND 2.0

1000 km)" and "construction of the building diverted more waste from the landfill than sent to the landfill through the use of salvaged materials" (Georgia Tech, "Key Living Building Details"). On behalf of the aforementioned features, the Georgia Institute of Technology frames the Kendeda building as follows: "Rather than being less bad for the environment, The Kendeda Building fosters regenerative and restorative relationships amongst humans and nature where people give back more to the environment than they take" (Georgia Tech 2023e, "The Kendeda Building"). The building accordingly strongly follows regenerative principles through restoring and replenishing resources. In this context, the current director of the Kendeda building Shan Arora articulated in an interview:

> Let's just talk about this mind shift of thinking about buildings as giving back more resources to nature—what we call regenerative buildings—and I add buildings that are purposefully seeking to have a restorative impact on human's relationship with each other. As humans we are creatures that live in buildings, and so I think buildings can actually make human beings be better people; happier and healthier people. (Arora 2022)

Arora argues that buildings should be conceptualized as healthy, human-friendly, regenerative places. The statement further evokes another green principle: circularity—the idea of closed loops, where materials and resources are reused and repurposed. In consequence, the Kendeda building has not only adopted one, but two, green building certification standards: LEED version 4 and LBC version 3.1. In this regard, Atlanta is a pioneer for green building in the United States, ranking third in

LEED-certified office space, and tenth in LEED-certified residential buildings in 2019 (CBRE 2019). The different standardization systems such as LEED or living buildings ensure standardization and accountability, and many of them pursue the goal of resilience. Within LEED certification, it is encouraged through pilot credits and rewarded through points to establish and conduct "climate related risk management planning" or "emergency preparedness planning." Especially in green buildings, there is an awareness of resilience on an environmental and economic level. Green and resilient buildings therefore provide future-oriented solutions to limit global warming and contribute to eco-conscious lifestyles. In this regard, the Kendeda building is also used for educational and showcasing purposes to propagate environmentally friendly building techniques and blueprints to students (Georgia Tech 2023e, "The Kendeda Building").

Especially LBC's place, energy, water, and materials petals incentivize ecological resilience by focusing on biodiversity, urban agriculture, net positive water, or non-toxic, low-emitting, and repurposed materials. The requirements for LBC require buildings to be built on greyfields or brownfields[8] and "projects may not be built in ecologically sensitive habitats, including wetlands, primary dunes, old-growth forest, virgin prairie, prime farmland, or within the 100-year floodplain" (Georgia Tech 2023d, "Place Petal"). This sensitivity to floodplains increases socio-ecological resilience as well as the imperative of net positive water. Net positive water "requires one hundred percent of occupant water to come from captured precipitation or closed loop water systems" (Georgia Tech 2023f, "Water Petal"). Rainwater is turned into drinking water on-site without the use of chemicals. Water is a "regionally critical issue" in Atlanta due to droughts and limited available groundwater (Georgia Tech 2023f, "Water Petal"). In the design process, the data suggested that a "50,000-gallon cistern would provide ample capacity to overcome drought and provide water resiliency after studying 31 years of drought and precipitation data and stress testing the model for varying occupancy and water use patterns (high, moderate, low)" (Georgia Tech 2023f, "Water Petal").

In the LEED criteria of "sustainable sites," the Kendeda buildings receive all points with regard to "heat island reduction," "open space," "site development (protect or restore habitat)," and "rainwater management" (USGBC 2023b, "Kendeda Building"). In the criteria of "energy & atmosphere," the building receives all points for optimizing energy performance and renewable energy production (USGBC 2023b, "Kendeda Building"). LEED currently is testing pilot credits that are directly related to natural disaster mitigation. The credits (IPpc98) include hazard assessments for flooding, hurricane, tornado/high wind, earthquake, tsunami, wildfire, drought, and landslide/unstable soils (USGBC 2023c, "Resilience Workbook IPpc98 and IPpc99"). The goal is to identify the top three hazards for a site and

[8] Greyfield and brownfield are terms used in urban planning and real estate development to describe specific types of land or properties. Greyfield refers to commercially developed land that is currently underutilized, often due to economic reasons. Brownfield sites are properties that have been previously used for industrial or commercial purposes and may be contaminated by hazardous substances or pollutants.

ways they can be addressed. In addition to this prerequisite, one of two options has to be chosen which is "climate resilience planning" or "emergency preparedness planning." Credit IPpc99—Design for Enhanced Resilience—then evaluates if the hazards detected in IPpc98 are tackled via appropriate mitigation strategies (Wilson 2015).

Respecting the environmental heritage is a relevant aspect of the resilient city—especially preserving natural soils for rainwater run-off and plants. The return to native plants and local climates further adds to the achievement of resilience. The Place Petal planning description for the Kendeda building therefore describes the land it is built on as "rich and storied:"

> The natural and cultural history of the land on which The Kendeda Building is built is rich and storied. While the Beauty Petal and its Inspiration + Education Imperative asks teams to look at the site and its surrounding region's cultural heritage, the Place Petal focuses first and foremost on the environmental heritage. The Piedmont region, of which our site is a part, naturally features a variety of plants and soils—not an expanse of concrete. (Georgia Tech 2023d, "Place Petal")

The Water Petal focuses on achieving net positive water. This means that the water used on-site must be collected on-site. This also resonates with aspects of resilience that consider local rainwater supply as well as future predictions based on changes through global warming processes (Georgia Tech 2023f, "Water Petal"). Atlanta has very little available groundwater and relies on streams, rivers, and reservoirs for the city's supply (Georgia Tech 2023f, "Water Petal"). The Kendeda building can convert rainwater into drinking water. The graywater is used to "recharge" the landscape (Georgia Tech 2023f, "Water Petal").

With regard to the energy petal, the production of energy relies on several renewable resources and spreads risks through photovoltaics, passive design, shading, and efficient energy conservation.

The stair treads in Fig. 3 consist of heart pine joists dating back to the 1880s that were repurposed for the Kendeda building during the 2016 renovation of the iconic Tech Tower (Georgia Tech 2023c, "Materials Petal"). In the construction process, emphasis was laid on reusing and repurposing materials and sourcing them locally—contributing to the idea of circularity and regeneration.

Inside the Kendeda building, spaces for trading clothes (indirectly) point to the necessities and opportunities of "degrowth" in the sense of reducing consumption. Circularity is strongly promoted through the focus on recycling and related educational efforts promoting the reuse of materials. In the living building challenge rating's "Materials Petal," one goal is to pursue "local economic growth" through sourcing materials within a certain distance in order to grow a local ecosystem with short transportation routes (Georgia Tech 2023c, "Materials Petal"). With regard to social sustainability, six Georgia Works! graduates helped build the Kendeda nail laminated deck. Georgia Works! trains formerly homeless individuals for construction work. In this regard, the building is a prime example of how green buildings can go hand in hand with social sustainability.

In an interview with Shan Arora, current director of the Kendeda building, Arora spoke about how the building could be used as a space of refuge in case of disaster

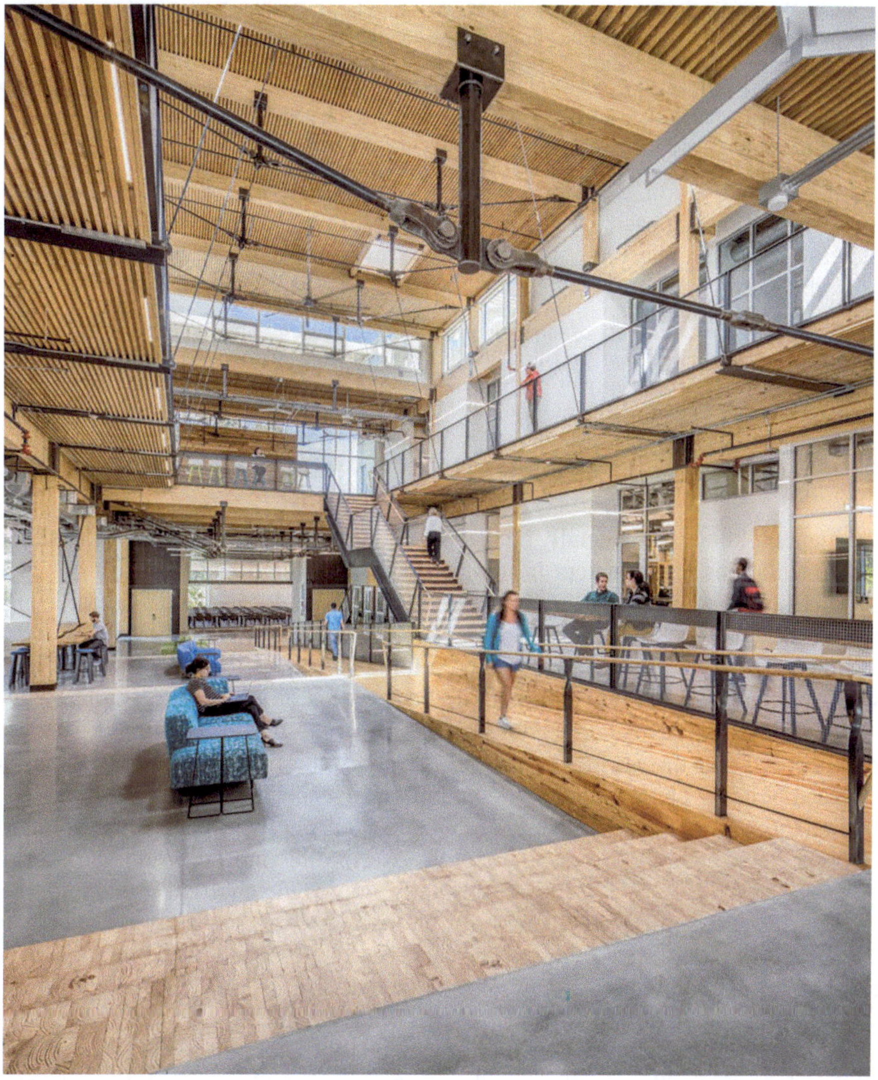

Fig. 3 "Kendeda" building interior, Photographer: Jonathan Hillyer, CC BY-NC-ND 2.0

due to its relative energy and water independence. The foam-flush toilets consume very little water, rainwater can be treated on-site and turned into drinking water, and the building is net energy- and water-positive. In 2021, the building provided 201% of its energy demand on-site (Georgia Tech 2023b, "2021 Annual Report"). In the interview, Shan Arora articulated:

> We have to think about resiliency for our communities. We have to think about creating places of refuge when the infrastructure fails. And there are certain types of buildings that naturally lend themselves to be those places of refuge: schools and houses of worship, com-

> munities of worship. And the reason is, they tend to have the things needed for people to congregate. And they tend to be known by the community in which they are. People know where their school is regardless of whether they actually go to the school; they probably know where the elementary school is, the middle school, the high school. They probably know where churches, synagogues, mosque, temples, etc. are. And so, we need to think about creating space that is multi-use infrastructure. (Arora 2022)

In Shan Arora's vision, the Kendeda building could serve as a site of refuge in times of crisis. Arora emphasizes that the resilience of the building would have to be increased for it to become such a place. Environmental engineers Dargin and Mostafavi articulate:

> Considering the complexity of modern critical infrastructure systems in terms of its interdependencies and external pressures, including increasing demand, aging, and climate change, the risk and severity of disruptions are becoming more likely. Community resilience planning is tasked with ensuring the equitable access and delivery of critical infrastructure system services in cities by reducing the disproportionate risks of service disruptions to the most vulnerable members in a community. (Dargin and Mostafavi 2020, p. 2)

Resilient buildings can crucially contribute to this infrastructure. Especially vulnerable populations must be kept in mind when dealing with natural catastrophes or extreme heat. In this regard, the Kendeda building is not located in a residential area but on the Georgia Tech campus, where it will most likely mostly serve Georgia Tech staff and students. The Kendeda building director's claims therefore seem slightly overstated. In my interpretation, the appraisal of the Kendeda building culturally relates to US-American individual heroism—the idea that one "heroic" building will save the world. The building certainly contributes to the dissemination of green building cultures, practices, and innovative technologies, but the way it is framed through the "core goal" of "amplify[ing] impact by inspiring change on the Georgia Tech campus, across the Southeast building industry, and the world" (Georgia Tech, "Key Living Building Details") raises the question if this impact is really achieved. Similar statements give the building more of a status as an "eco-cathedral," that serves as a flagship project. But the lessons learned are not transferred enough to more low-cost projects or beyond the Georgia Tech campus in my viewpoint. Flagship projects are absolutely necessary, but more emphasis should be given to the transferability of these inspiring results. Otherwise, builders with less financial means will only deduce that "greenness" and sustainability are only achievable with extensive financial and academic capacities.

Beyond Materiality: Narratives About the "Kendeda" Building

Despite the aforementioned criticism of the Kendeda building, the experimental structure and character of the building nevertheless serve as a living lab as well as a "space of possibility." Narratives and stories told about the Kendeda building can contribute to larger discourses on the achievement of climate neutrality or its inhibition and showcase this "space of possibility." The way green buildings are perceived will unequivocally contribute to their success. During my stay in Atlanta and tour of the Kendeda building, I was able to collect student's reactions to the building:

> First, I like the outdoor space next to it, and I like how open it is. Having all the open windows is really, really nice. And even having the patios where the door is open and so you sort of have like an indoor–outdoor patio. And it's just nice to be out there. Even when you're inside, you're still kind of outside and with nature and then again, having the rooftop is super wonderful. It's also a great way to build community. I guess that'll be my second reason I really like it. And that a lot of the involvement in the building is a collaboration and so students are always working with each other. (Georgia Tech Student 2022)

Anecdotally, staff and students recite stories about the building when speaking to each other but also through commentaries on platforms such as reddit.com. Some responses to the Kendeda building on the web forum "reddit.com" under the subreddit "r/gatech" explain that "the wood environment feels like a huge treehouse" (NotCreative11 2020) but "that the compost toilets were weird to get used to but pretty interesting." Another user states:

> The design of the building is really cool and it has that freshly cut wood smell, which is awesome. Nice place to get some work done or hangout. Some of the "Living Building" requirements make it inconvenient to work in. There are like 3 outlets in an entire lab, so we had to bring laptops fully charged and hope they lasted the whole lab. There are no printers in the building. I understand saving electricity… but you kinda need it in most lab settings. (jaketuber 2020)

The main complaints were about the lack of electric outlets but the main positive story mentioned by the reddit users was about the comfort of the fresh-cut wooden smell and the profitable learning environment. And undoubtedly, there are trade-offs between saving electricity and practicability in a laboratory setting. Some commentaries also valued the efforts in sustainability as "very cool" (sereca 2020). From the perspectives voiced on reddit, the stories told about the building are very mixed. They could be read like this account by one reddit user:

> The lecture hall is horribly distracting. My instructor was often distracted by the giant windowed garage door at the back of the classroom and would stop teaching. The students are often distracted by the giant picture windows that are behind the projector screens. Outlets are few and far between as well. All in all it's a pretty building and it's new, but there's definitely a lot of oversights to making it a practical learning environment for a classroom. (geoffbezo 2020)

The adjectives used range from "horrible" to "pretty"; the situations explicated range from distraction to fruitful learning environments. The building seems to generate quite mixed narrative responses. For one user, the Kendeda building is the "favorite building on campus" (sereca 2020). Generally, the building evokes strong reactions due to its experimental character. Many practical features are unusual like the compost toilets and therefore elicit mixed responses. The Kendeda building contributes to the imaginations of green future cities through its architecture, materials, and narrativity. Students and staff working on-site experience one materialization of the possibilities of the resilient city firsthand—its advantages and disadvantages in comparison with conventional buildings.

Interpretation of Results and Conclusion

The Kendeda building is a relevant flagship project that has the ability to inspire beyond the realms of Atlanta, GA, and promote LEED and the living building challenge as an ambitious and holistic endeavor. Especially, the educational efforts, through the manifold tours of the building, its use as a classroom, and its student ambassadors, showcase this. Similar spaces are much needed in times of the climate crisis. Nevertheless, Georgia Tech remains a relatively privileged space due to tuition costs that are unaffordable for many US-Americans with less economic means. Phrased in a critical manner, the appraisal of the Kendeda building culturally relates to US-American individual heroism—the idea that one "heroic" building will save the world. But individuals rarely have the power to evoke systemic change on their own. Certainly, some individual figures have proven to be very influential and inspirational such as Martin Luther King for desegregation, Mahatma Gandhi for non-violence and Indian independence, or Greta Thunberg for the climate movement. But scattered "eco-cathedrals" will not suffice to deal with the climate crisis in a scalable manner. Buildings like Kendeda are representational structures for their owners, but they inspire nevertheless through their incredible ecological achievements.

The Kendeda building demonstrates what is possible through private donations with university know-how and planning. Unfortunately, in the United States, green building solutions for the masses, especially for lower-income people, are not prioritized. In the end, green building certifications tackle the problem of increasing natural disasters and extreme weather from two sides: cause and effect—at least to a certain degree. These standards must be further developed from the perspectives of diverse disciplines including the humanities in order to prevent a narrow focus on resilience from an engineering point of view. Including concepts such as adaptation urbanism or negotiated resilience will prove to be crucial for achieving more social sustainability. What can be done in order to reach climate goals and build greater climate resilience in Atlanta and the United States overall? One aspect is information (for example through the media) and direct experience. Despite disasters becoming more frequent, climate change remains distant and abstract to many Americans (Smith and Leiserowitz 2012, p. 1028). Yet, enforcing stricter and more resilient and sustainable building codes, standards, and certification systems could help with climate adaptation. This can only be successful, however, when building codes are framed in a way that is appealing not only to political liberals but conservative audiences as well, for example, by arguing through the lens of individual profits and protection. Homeowners from any political spectrum quite directly understand the benefits of securing their homes against possible disasters and weather extremes such as heat. In the United States, insurance companies are increasing their involvement in risk reduction, offering premiums and discounts, for example, for participation in Insurance Institute for Business and Home Safety (IBHS's) Fortified Home program (Czaijkowski 2019, p. 315). Nonetheless, the underlying cultural issues must be dealt with—refraining from the partisan narrative and emphasizing the benefits of living in clean and healthy environments.

In 2018, 600 million tons of construction and demolition waste were generated in the United States (US EPA 2016, "Sustainable Management of Construction and Demolition Materials") and 27 percent of carbon emissions were generated by the building industry (Logan 2023). The building sector has great potential to adopt principles of circularity, repurpose resources, and implement urban mining.[9] The Kendeda building sets an important example in this regard. Resilient, green, and healthy buildings are necessary prerequisites for the mitigation of climate change and natural catastrophes. The Kendeda building demonstrates how green building standards can help mitigate the effects of natural disasters due to climate change from the perspectives of resilience, social and ecological sustainability, and cultural responsibility. Put in a very generalized way, building certification systems say "what needs to be done" and building code regulates "how it is done." LEED could be implemented nationally for corporate buildings of a certain size, for government buildings, and in a simplified form for use in homes. Incentivizing the use of the living building challenge (LBC) for government buildings, to set a visionary example, could move engagement even further. Building green and resiliently also has the potential to reduce energy consumption and incentivize on-site energy production and heating for instance through passive-house or plus-energy house structures with solar energy or heat pumps, which might become more attractive for Americans in a world where fossil fuels continue to be a crucial issue in geopolitical competition.

References

Arora, Shan. Interview by Katharina Wood. *Resilience and Sustainability: The Kendeda Building as a Case Study.* 1 Oct. 2022.

Baines, Donna, and Ian Cunningham. 2013. Using Comparative Perspective Rapid Ethnography in International Case Studies: Strengths and Challenges. *Qualitative Social Work* 12 (1): 73–88. https://doi.org/10.1177/1473325011419053.

Béné, Christophe, et al. 2018. Resilience as a Policy Narrative: Potentials and Limits in the Context of Urban Planning. *Climate and Development* 10 (2): 116–133. https://doi.org/10.1080/17565529.2017.1301868.

[9] Urban mining is the process of extracting valuable resources, such as metals and rare earth elements, from waste and discarded materials found in urban areas, including electronic waste, construction debris, and old appliances. This practice aims to reduce the reliance on traditional mining methods, conserve natural resources, and minimize environmental impact by repurposing existing materials.

Benfield, Kaid. *As Good and Important as It Is, LEED Can Be so Embarrassing*. 25 Oct. 2016., https://kaidbenfieldarchive.com/20130118-as-good-and-important-as-it-is-leed-can-be-so-embarrass.html. Accessed 9 Sep. 2023.

Callicott, J. Baird. 2011. The Temporal and Spatial Scales of Global Climate Change and the Limits of Individualistic and Rationalistic Ethics. *Royal Institute of Philosophy Supplement* 69:101–116.

Cassidy, Robert. White Paper on Sustainability: A Report on the Green Building Movement. Green Building Design & Construction, 2003.

CBRE. *Green Building Adoption Index for Office Buildings*. 2019.

Czaijkowski, Jeffrey. 2019. Strong and Well-Enforced Building Codes as an Effective Disaster Risk Reduction Tool: An Evaluation. In *The Future of Risk Management*, ed. Howard C. Kunreuther et al., 303–324. University of Pennsylvania Press. Critical Studies in Risk and Disaster.

Dargin, Jennifer S., and Ali Mostafavi. 2020. Human-Centric Infrastructure Resilience: Uncovering Well-Being Risk Disparity Due to Infrastructure Disruptions in Disasters. *PLoS One* 15 (6): e0234381. https://doi.org/10.1371/journal.pone.0234381.

Evans, Brad, and Julian Reid. *Resilient Life: The Art of Living Dangerously*. Repr. Polity, 2015.

Faltermann, Pascal. Was Bremen Von Gent Lernen Kann: Filip Watteeuw Im Interview. *WESER-KURIER*, 20 Feb. 2021., https://www.weser-kurier.de/bremen/was-bremen-von-gent-lernen-kann-filip-watteeuw-im-interview-doc7e3d5m1hmb98aexwrd. Accessed 23 Mar. 2023.

Finlay, Jessica, and Jennifer Massey. 2012. Eco-campus: Applying the Ecocity Model to Develop Green University and College Campuses. *International Journal of Sustainability in Higher Education* 13 (2): 150–165. https://doi.org/10.1108/14676371211211836.

Folke, Carl, et al. 2002. Resilience and Sustainable Development: Building Adaptive Capacity in a World of Transformations. *Ambio* 31 (5): 437–440.

Geerlin, Liliane, et al. 2021. Adaptation Urbanism: An Urban Resilience Policy Framework. In *Adaptation Urbanism and Resilient Communities: Transforming Streets to Address Climate Change*, ed. Billy Fields and John Renne, 62–71. Routledge. Advances in Urban Sustainability.

geoffbezo. 2020. *Kendeda/Living Building*. https://www.reddit.com/r/gatech/comments/fnp6u8/kendedaliving_building/.

Georgia Tech. *Equity Petal: The Kendeda Building for Innovative Sustainable Design*. 1 Jan 2023a, https://livingbuilding.gatech.edu/living-building-challenge/equity-petal. Accessed 22 June 2023.

Georgia Tech. *2021 Annual Report: The Kendeda Building for Innovative Sustainable Design*. 16 June 2023b, https://livingbuilding.gatech.edu/annual-report-2024. Accessed 23 June 2023.

Georgia Tech. *Materials Petal: The Kendeda Building for Innovative Sustainable Design*. 16 June 2023c, https://livingbuilding.gatech.edu/living-building-challenge/materials-petal. Accessed 23 June 2023.

Georgia Tech. *Place Petal: The Kendeda Building for Innovative Sustainable Design*. 16 June 2023d, https://livingbuilding.gatech.edu/living-building-challenge/place-petal. Accessed 22 June 2023.

Georgia Tech. *The Kendeda Building: The Greenest Classroom Building in the U.S.* 16 June 2023e., livingbuilding.gatech.edu/. Accessed 23 June 2023.

Georgia Tech. *Water Petal: The Kendeda Building for Innovative Sustainable Design*. 16 June 2023f, https://livingbuilding.gatech.edu/living-building-challenge/water-petal. Accessed 22 June 2023.

Georgia Tech. *Key Living Building Details*. https://livingbuilding.gatech.edu/about.

Georgia Tech Student. Interview by Katharina Wood. *Kendeda Building Impact*. 1 Oct. 2022.

Günter, Roland. *Im Tal der Könige: Ein Handbuch für das Ruhrgebiet*. 5, fortgesetzte und erw. Aufl. Grupello-Verl., 2010.

Harris, Leila M., et al. Negotiated Resilience. *Resilience*, 2017, 1–19. https://doi.org/10.1080/21693293.2017.1353196.

Hassler, Uta, and Niklaus Kohler. 2014. Resilience in the Built Environment. *Building Research & Information* 42 (2): 119–129. https://doi.org/10.1080/09613218.2014.873593.

Holling, C. 1996. Engineering Resilience. In *Engineering Within Ecological Constraints*, ed. Peter C. Schulze, 31–44. National Academy Press.

International Living Future Institute. *What Is the Living Building Challenge? - International Living Future Institute.* 6 June 2023., living-future.org/lbc/. Accessed 22 June 2023.

jaketuber. 2020. *Kendeda/Living Building.* https://www.reddit.com/r/gatech/comments/fnp6u8/kendedaliving_building/.

Logan, Andrew. Predicting Building Emissions Across the US. *MIT News,* 22 June 2023., news.mit.edu/2021/predicting-building-emissions-across-us-0921. Accessed 22 June 2023.

McBean, Gordon. 2004. Climate Change and Extreme Weather: A Basis for Action. *Natural Hazards* 31 (1): 177–190. https://doi.org/10.1023/B:NHAZ.0000020259.58716.0d.

Meerow, Sara, and Joshua P. Newell. 2019. Urban Resilience for Whom, What, When, Where, and Why? *Urban Geography* 40 (3): 309–329. https://doi.org/10.1080/02723638.2016.1206395.

Miller, Toby, and Richard Maxwell. Is American Individualism Bad for the Environment? *Psychology Today,* 8 Feb 2017., https://www.psychologytoday.com/us/blog/greening-the-media/201702/is-american-individualism-bad-the-environment. Accessed 22 June 2023.

Milligan, Richard, et al. 2021. Improvising Against the Racial State in Atlanta: Reimagining Agency in Environmental Justice. *Environment and Planning C: Politics and Space* 39 (7): 1586–1605. https://doi.org/10.1177/23996544211038944.

Nachtigall, Werner. 2003. Statements Zur Architekturbionik Und Bauökologie. In *Bau-Bionik: Natur – Analogien – Technik*, ed. Werner Nachtigall, 169–178. Berlin, Heidelberg: Imprint; Springer.

NOAA. 2023*: A Historic Year of U.S. Billion-Dollar Weather and Climate Disasters.* 17 Mar. 2024, https://www.climate.gov/news-features/blogs/beyond-data/2023-historic-year-us-billion-dollar-weather-and-climate-disasters. Accessed 17 Mar. 2024.

NotCreative11. 2020. *Kendeda/Living Building.* https://www.reddit.com/r/gatech/comments/fnp6u8/kendedaliving_building/

Ratcliffe, Susan, ed. 2017. *Oxford Essential Quotations.* 5th ed. Oxford University Press. Oxford Reference, https://www.oxfordreference.com/display/10.1093/acref/9780191843730.001.0001/acref-9780191843730.

Rega, Carlo, and Alessandro Bonifazi. 2020. The Rise of Resilience in Spatial Planning: A Journey Through Disciplinary Boundaries and Contested Practices. *Sustainability* 12 (18): 1–18. https://doi.org/10.3390/su12187277.

sereca. 2020. *Kendeda/Living Building.* https://www.reddit.com/r/gatech/comments/fnp6u8/kendedaliving_building/.

Shapiro, Shari. 2016. The Realpolitik of Building Codes: Overcoming Practical Limitations to Climate Resilience. *Building Research & Information* 44 (5–6): 490–506. https://doi.org/10.1080/09613218.2016.1156957.

Smith, Nicholas, and Anthony Leiserowitz. 2012. The Rise of Global Warming Skepticism: Exploring Affective Image Associations in the United States over Time. *Risk Analysis* 32 (6): 1021–1032. https://doi.org/10.1111/j.1539-6924.2012.01801.x.

Stad Gent. *Urban Green.* 23 Mar. 2023, stad.gent/en/city-governance-organisation/city-policy/ghents-climate-actions/urban-green. Accessed 23 Mar. 2023.

Steele, Jeffrey. University Campuses Studying up on Sustainable Building. *Forbes,* 14 Feb. 2019., https://www.forbes.com/sites/jeffsteele/2019/02/14/university-campuses-studying-up-on-sustainable-building/?sh=62d1d5881bed. Accessed 27 June 2023.

Stott, Peter. 2016. CLIMATE CHANGE. How Climate Change Affects Extreme Weather Events. *Science (New York, NY)* 352 (6293): 1517–1518. https://doi.org/10.1126/science.aaf7271.

Thomas, Mary. 2016. *The Living Building Challenge: Roots and Rise of the World's Greenest Standard.* Ecotone Publishing.

U.S. Green Building Council. *LEED Works: Myth-Busting | U.S. Green Building Council.* 7 Sep. 2023., https://www.usgbc.org/articles/leed-works-myth-busting. Accessed 7 Sep. 2023.

US EPA. 2014. *Urban Waters and the Proctor Creek Watershed/Atlanta (Georgia).* 1 Jan 2014., https://www.epa.gov/urbanwaterspartners/urban-waters-and-proctor-creek-watershedatlanta-georgia. Accessed 21 Mar. 2023.

US EPA. 2016. *Sustainable Management of Construction and Demolition Materials.* 1 Jan. 2016., https://www.epa.gov/smm/sustainable-management-construction-and-demolition-materials. Accessed 22 June 2023.

USA Today. In U.S. Building Industry, Is It Too Easy to Be Green? *USA TODAY,* 24 Oct. 2012., https://eu.usatoday.com/story/news/nation/2012/10/24/green-building-leed-certification/1650517. Accessed 9 Sep. 2023.

USGBC. *LEED Rating System.* 21 Mar. 2023a., https://www.usgbc.org/leed. Accessed 21 Mar. 2023.

USGBC. *Kendeda Building: Scorecard.* 23 June 2023b., https://www.usgbc.org/projects/kendeda-building. Accessed 23 June 2023.

USGBC. *Resilience Workbook IPpc98 and IPpc99.* 23 June 2023c., https://www.usgbc.org/resources/resilience-workbook-ippc98-and-ippc99. Accessed 23 June 2023.

USGBC. *Building Decarbonization: Can We Build Zero Emissions Buildings?* 17 Mar. 2024a., https://www.usgbc.org/education/sessions/building-decarbonization-can-we-build-zero-emissions-buildings-12848369. Accessed 17 Mar. 2024.

USGBC. *What's New in LEED V5.* 17 Mar. 2024b., https://www.usgbc.org/articles/whats-new-leed-v5. Accessed 17 Mar. 2024.

Way, Thaisa, et al. "Greening the American Campus: Lessons from Campus Projects." Planning for Higher Education, 40, 2, 2012, pp. 25–47.

Webber, Sophie, et al. 2021. Wheeling Out Urban Resilience: Philanthrocapitalism, Marketization, and Local Practice. *Annals of the American Association of Geographers* 111 (2): 343–363. https://doi.org/10.1080/24694452.2020.1774349.

Wilson, Alex. *LEED Pilot Credits on Resilient Design Adopted!* 1 Jan. 2015., https://www.resilientdesign.org/leed-pilot-credits-on-resilient-design-adopted/. Accessed 22 June 2023.

Wright, Daniel B., et al. 2012. Hydroclimatology of Flash Flooding in Atlanta. *Water Resources Research* 48 (4): https://doi.org/10.1029/2011WR011371.

Zaretsky, Michael. 2010. LEED After Ten Years. In *New Directions in Sustainable Design*, 191–201. Routledge. https://www.taylorfrancis.com/chapters/edit/10.4324/9780203843031-28/leed-ten-years-michael-zaretsky.

MIX
Papier aus verantwortungsvollen Quellen
Paper from responsible sources
FSC® C105338
FSC
www.fsc.org

If you have any concerns about our products,
you can contact us on
ProductSafety@springernature.com

In case Publisher is established outside the EU,
the EU authorized representative is:
Springer Nature Customer Service Center GmbH
Europaplatz 3, 69115 Heidelberg, Germany

Printed by Libri Plureos GmbH
in Hamburg, Germany